AF517037

SOILGUIDE

A Handbook for Understanding and Managing Agricultural Soils

Related Book

Title	Author	ISBN	Price
• A Manual on Conservation of Soil and Water	*USDA*	9789388043175	450
• Agrochemicals in Plant Disease Management	*N.G. Ravichandran*	9789388043564	3250
• Agrotechnology for Dryland Farmaing *2nd Revised Ed*	*A.M. Dhopte*	9789386102737	3950
• Biofumigation and Solarization for Management of Soil-Borne Plant Pathogens	*P.P. Reddy*	9789388043205	650
• Comprehensive Assessment of Soil Health *3rd Ed*	*SIPS*	9788193698525	950
• Crop Production in Salt Affected Soils	*I.C. Gupta*	9789387893832	1260
• Crop Production in Waterlogged Saline Soils	*S.K. Gupta*	9788172331597	1450
• Crop Protection Strategies under Climate Change Scenarios	*P.P. Reddy*	9789387307087	3550
• Drainage Engineering: Principles and Practices	*S.K. Gupta*	9789388172028	3500
• Dryland Resources and Technology (Vol. 8)	*L.L. Somani*	9788172330828	1195
• Fertilizer Manual	*IFDC*	9789383692637	7500
• Fundamentals of Organic Farming and Gardening: An Instructors Guide	*A. McGregor*	9789383692996	1950
• Fundamentals of Soil Science	*A. Rathinasamy*	9789386652171	1650
• Genesis and Management of Sodic (Alkali) Soils	*S.K. Gupta*	9789386102775	2495
• Glimpse on General Agriculture	*P. Laxman Rao*	9789386652454	225
• Handbook of Saline and Alkali Soils Diagnosis Reclamation and Management	*S.K. Gupta*	9789388812276	1450
• Indiras Objective Agriculture	*R.L. Arya*	9788172339081	650
• Manual on Fundamentals of Agronomy	*L.K. Jain*	9789388172011	1475
• Nanotechnology	*J.C. Tarafdar*	9788172337582	1850
• Organic Farming Theory & Practice	*S.P. Palaniappan*	9788172335380	1150
• Principles in the Quantitative Analysis of Water	*U.S.S. Ramulu*	9789386652485	250
• Principles of Seed Technology	*Dhirendra Khare*	9789388043588	475
• Rainwater Management Theory and Practice	*M.L. Jat*	9789386347237	3250
• Salinity Tolerance in Plants: Methods, Mechanisms and Management *2nd Ed*	*B.K. Garg*	9789388449342	2950
• Salt Affected Soils: Reclamation and Management	*S.K. Gupta*	9789388043489	525
• Sanrakshit Kheti Ke Antargat Satat Fasal Prabhandhan	*K.V.R. Rao*	9789386347220	650
• Soil and Plant Analysis	*C.S. Piper*	9788172336202	1850
• Soil and Water Conservation in Semi-Arid Areas	*N.W. Hudson*	9789383692859	1250
• Soil Chemical Analysis	*M.L. Jackson*	9789383692354	950
• Soil Conservation Technical Handbook	*D.H. Hicks*	9789388043496	575
• Studies in Arid Land Management	*T.S. Chouhan*	9788172330644	3750
• Sustainable Agriculture	*R.S. Meena*	9789388043625	2750
• Sustainable Development of Dryland Agriculture in India	*R.P. Singh*	9788172339616	550
• The Soil-Plant System in Relation to Inorganic Nutrition	*M. Fried*	9789388043526	575
• The Use of Saline Waters for Crop Production	*FAO*	9788172330934	950

SOILGUIDE

A Handbook for Understanding and Managing Agricultural Soils

Compiled and Edited by

Geoff Moore
Natural Resource Management Services
Agriculture Western Australia

With assistance from the following editors
Jo McFarlane
Brian Purdie
Georgina Wilson

Reprinted in India by:

UNITED BOOK PRINTS
5-B, Bhagat ki Kothhi, Near Police Chowki
New Pali Road, Jodhpur 342 001 (INDIA)
E-mail: info@scientificpub.com

Print : 2019

ISBN: 978-93-83692-88-0

Reprinted July 2001
First published March 1998

Moore, G. (2001)

Soilguide. A handbook for understanding
and managing agricultural soils. Agriculture Western Australia
Bulletin No. 4343.

Printed in India

CONTENTS

ACKNOWLEDGMENTS

This handbook is one of the outputs from two National Landcare Program funded projects: 'Farmscale land capability' and 'Productive life of regional soils'. Geoff Moore worked on the projects for six years and is responsible for the concept and overall compilation of *Soilguide*. Paula Needham worked for three years and was responsible for the sections on soil structure decline, hard layers in soils and subsurface compaction.

A publication of this type is a major task and would not have been possible without the assistance of many people who have contributed as authors, or by contributing technical information, reviewing chapters, or in some cases all three.

I would particularly like to thank all the authors for their contributions. In alphabetical order they are: W.K. Anderson, J. Bartle, E.G. Barrett-Lennard, W.M. Blacklow, P.S. Blackwell, K.J. Bligh, M.D.A. Bolland, R.F. Brennan, P. Carmody, D.J. Carter, N.A. Coles, L. Cransberg, H. Daniel, P.J. Dolling, M. Dracup, N.K. Edwards, P.M. Evans, M.A. Ewing, P.A. Findlater, D.J. Gillespie, E.A. Griffin, D. Hall, R.J. Harper, L. Leonard, S.P. Loss, D.J. McFarlane, M.G. Mason, P. Nelson, P.J. Needham, P. Nichols, M. Norton, B.J. Nutt, T. Piper, W.M. Porter, I. Pritchard, B.R. Purdie, M.M. Riley, J.J. Russell, G. Scholz, K.H.M. Siddique, K. Smetten, R. Snowball, R.N. Summers, P.J. Tille, G.H. Walton, D.M. Weaver, T.J. Wiley, K.J. Young, Officers of the Natural Resources Assessment Group and Officers of the Viticulture Branch.

In addition, there were contributions to specific sections from many people either in the form of providing technical information, refereeing or giving permission to include their unpublished data. The assistance of the following people (in alphabetical order) is gratefully acknowledged: E.L. Armstrong (field peas), R.K. Belford (waterlogging), R. Bell, D. Bicknell (windbreaks), P.S. Blackwell (subsurface compaction, soil structure decline), K.J. Bligh (water erosion), M.D.A. Bolland (soil acidity), J.W. Bowden (pH buffering capacity), R.F. Brennan (molybdenum), D.J. Carter (water repellence), B. Cartwright (boron), H. Cochrane (soil structure decline), W.J. Collins (pastures), D. Deeley (eutrophication), C.M.M. Franco (water repellence), J.W. Gartrell (boron), C. Gazey (soil acidity), J.S. Gladstones (lupins, viticulture), G. Hamilton (wind erosion), E.J. Hauck (runoff and water erosion), C.W.L. Henderson (plant available water capacity), R. Kelly (water erosion control), P.M. King (water repellence), I. Laing (runoff and water erosion), J.F. Leys (wind erosion), B. Lloyd (water erosion control), J.F. Loneragan (plant nutrition), D.A. McGhie (water repellence), N.J. McKenzie (root growth in clayey subsoils), R.A. Nulsen, C. Raphael (pH buffering capacity), P. Rengasamy (soil salinity, soil alkalinity), A.D. Robson (plant nutrition), M.E. Rogers (white clover), N. Schoknecht, G. Scholz (interpreting laboratory data), W. Scott (wind erosion), M. Strawbridge (salt tolerant pasture species), R.N. Summers (water repellence), W. Verboom (soil structure decline), B.A. Wren (interpreting laboratory data) and officers of the Chemistry Centre WA (interpreting laboratory data).

Special mention must go to Miss Karin Patten for editing the first draft and Mr Trevor Stoneman who reviewed the entire first draft. Mr Brian Purdie meticulously edited the second draft and made many excellent recommendations. Many thanks to Mrs Georgina Wilson for editing the third draft and Mrs Jo McFarlane for editing the final draft. Simon Eyres and Peter Maloney of the Photographic Unit, Agriculture Western Australia supplied the photographs unless otherwise noted. The technical illustrations were prepared by Suzie Thompson and Dot Hewgill (Dot to Dot Publishing Projects), apart from four maps which were prepared by Phil Goulding. Thank you to Jacqui Mallard for coordinating the graphic design and printing.

I am grateful to the editors of various technical journals and publishing companies for permission to reproduce their figures.

It is with sadness that we note the passing of Brian Purdie in April 1999 and acknowledge his great contribution to both this publication and soil science in Western Australia. He was a great mentor to many scientists and is greatly missed.

Geoff Moore
2001

CHAPTER 1

ABOUT SOILGUIDE

1.1 INTRODUCTION

Soil is the basis of almost all agriculture but given the very low rates of soil formation under the present climate in Western Australia, our soils are essentially a non-renewable resource. It is therefore essential that our stewardship of the land resources of the State is based on the best possible understanding of our soil and land resources.

Soil is our basic resource. This is obvious, but do we know enough about it to maximise production and minimise land degradation?

This handbook integrates the current knowledge of soils in south-western Australia in a user-friendly form. It describes how to assess which soil properties influence production and land degradation in the agricultural area and summarises management options to remedy or minimise soil limitations. The potential for growing a large range of crops and pastures can be assessed. In particular, the links between soil morphology, soil properties, management and agronomy are emphasised; links that are often understood poorly.

The manual is designed for use at the paddock scale or for 'site assessment', but it can also be used at a catchment or regional scale. It is specifically designed for rainfed agriculture in south-western Australia, however many sections will be relevant elsewhere as the manual works from first principles and does not require prior knowledge of the nature and distribution of soils.

Traditionally, soils in Western Australia have been classified according to the native vegetation growing on them, or placed into broad categories such as light and heavy land. When cropping large areas with an expectation of a low average yield, there is little need for a detailed understanding of the soils. However, there has recently been an important shift in emphasis in production away from quantity alone towards quality and quantity. To achieve these goals, understanding soil behaviour is crucial, especially if the farming systems are to be sustainable.

The shortage of quality information at a scale relevant to decision-making on farms indicated the need for a system that could be used on-site when assessing soil properties that affect management, production and land degradation. This is what we have endeavoured to deliver in *Soilguide: A handbook for understanding and managing agricultural soils.*

CONCEPTUAL FRAMEWORK

"The two important offices in vegetable life which the root is designed to fulfil are obvious to every one, viz., to support the plant in its position, and to imbibe from the soil the food and moisture requisite for its growth. How well God has adapted its structure and instincts to this twofold purpose observation is continually showing" (Wood 1864).

The framework for this manual is that soil is a medium for plant root growth. The growing plant relies on the soil for support, to exchange gases with the atmosphere and for a supply of water and nutrients. The plant of course needs to grow a root system to carry out these functions, thus properties which affect root growth also need to be considered. The role of the soil in providing anchorage for the plant is comparatively minor for agricultural crops which only need a soil depth of 10 to 15 cm to fulfil this requirement, but for forestry and perennial horticulture it is an important consideration.

If a soil can supply all of these requirements without degrading, then it is non-limiting to plant growth and potential yield is determined solely by the climate. Most soils (>80%) in south-western Australia are not in this category because they have physical and/or chemical limitations that require management to minimise their effect on crop yields.

This manual considers soil properties that directly affect root growth or the supply of water and nutrients to the plant, or that affect gaseous exchange with the atmosphere. In addition, the issue of the sustainability of the soil resource is covered in Chapter 7. Crop and pasture requirements are similarly discussed in terms of their tolerance of a limited supply of water and nutrients, poor gaseous exchange with the atmosphere and the ability of the root system to withstand unfavourable physical and/or chemical conditions.

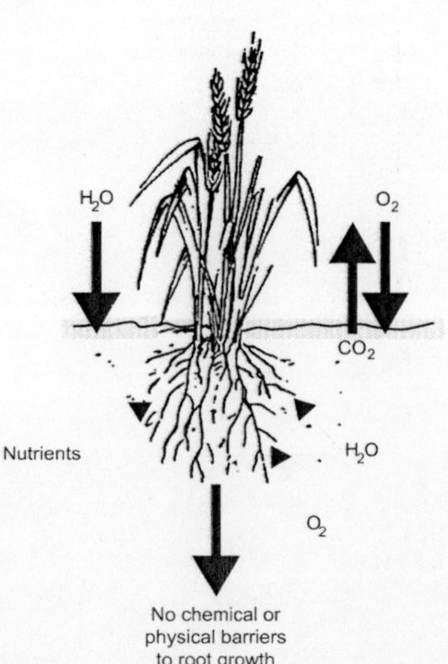

Figure 1.1.1 *The plant's requirements from the soil.*

HOW TO USE SOILGUIDE

This handbook is designed to identify the soil properties affecting management, production and sustainable land use at the paddock scale. It provides links between different soils, their properties and management options (Figure 1.1.2). The information can be expanded to a catchment or regional scale using knowledge of the spatial distribution of soils. Users of this handbook do not require a detailed technical knowledge of soils, however, simple skills such as field texturing and a basic knowledge of soil science are required.

Site assessments can be viewed at various levels depending on the type of information required. Interpretations can be made from three viewpoints:

(i) Specific soil properties

Soil properties (or behavioural properties) are described in five chapters:

Physical factors affecting water infiltration and redistribution, such as water repellence (Chapter 3),

Physical restrictions to root growth, such as subsurface compaction (Chapter 4).

Chemical factors affecting plant growth, such as soil acidity (Chapter 5).

Plant nutrition, which covers the essential elements for plant growth (Chapter 6).

Sustainable soil management, which covers the assessment and management of land degradation issues such as wind erosion and eutrophication (Chapter 7).

If you wish to consider a particular soil property (e.g. susceptibility to waterlogging), check the contents page to find the section.

There is a succinct summary of the principles in each section to give the reader an understanding of the processes, rather than simply presenting a series of methods to follow. Technical references are provided to allow the reader to follow up important points. In many sections, more than one method of assessment is described, with different methods often having varying degrees of precision and/or technical input. The preferred method may vary with the situation and the objective of the investigation.

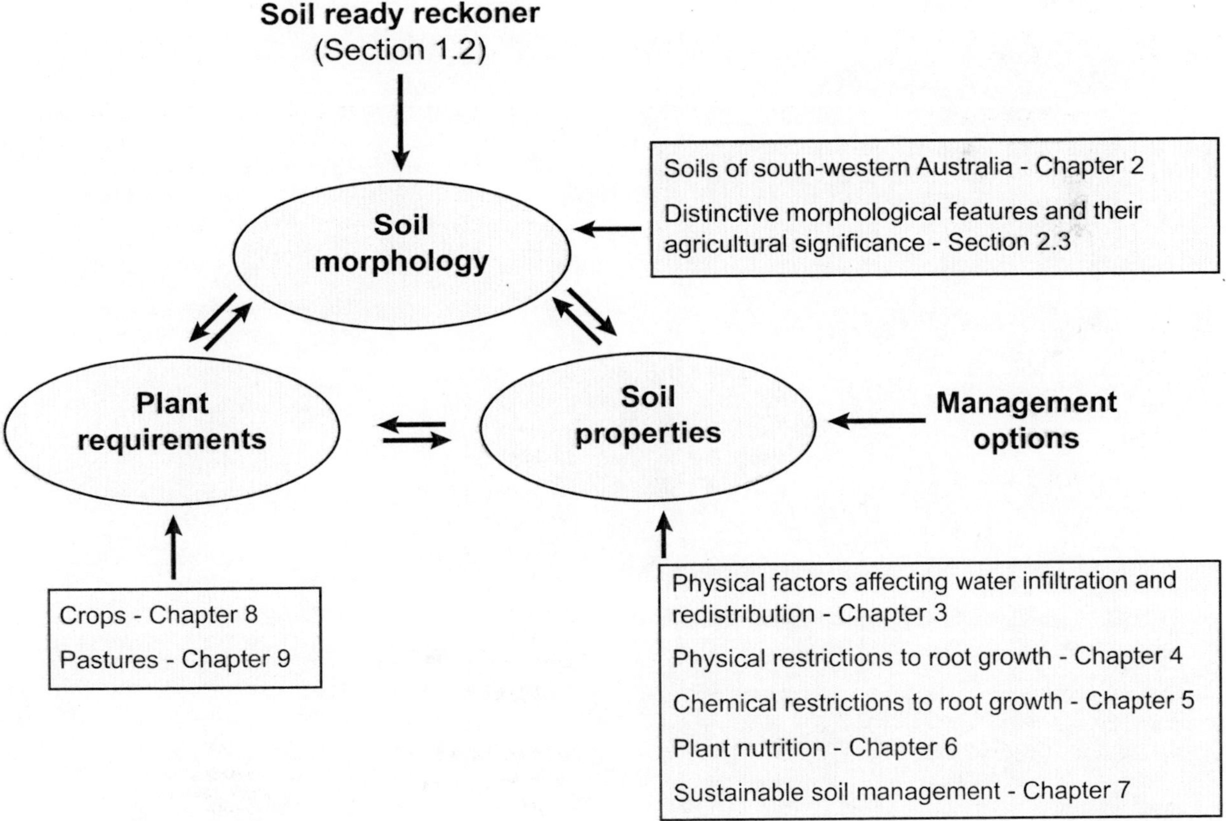

Figure 1.1.2 *The basic structure of Soilguide, showing the links between soil morphology, behavioural properties of the soil, plant requirements and management options.*

(ii) Individual plant requirements

The soil and climatic requirements are discussed in two chapters:

Crops - in which the most widely grown crops are described (Chapter 8).

Pastures - in which the most widely grown pasture species are described (Chapter 9).

For the various crop and pasture species, only the soil and climatic requirements are described.

(iii) Site assessment

When assessing a site, experimental trial or a farmer's paddock, the *Soil ready reckoner* (see next section) provides a straightforward system for quickly identifying which soil properties need to be managed.

The *Soil ready reckoner*, uses four broad groups based on profile hydrology which can be identified readily from morphological properties (i.e. texture down the profile). A diagram for each group summarises the soil properties. The relevant sections describing soil properties and/or land uses can then be consulted.

EXTENSION OF RESULTS

When the results of an investigation or trial are extended to a wider audience, some form of soil classification is required. The appropriate type of soil classification will vary according to the audience (i.e. local farmer group, technical, or national). For south-western Australia, a system for classifying the major soils into 37 groups based on their surface texture and texture profile is described in Section 2.1.

INTERPRETING LABORATORY DATA

Understanding and interpreting soil chemical and physical data from a laboratory can be complex. Chapter 10 summarises what the samples represent, the methods that may be used for the analyses and the meaning of the results in terms of characterising the soil and indicating factors that need to be considered in land management.

INFORMATION SOURCE

This handbook is a compendium of information on soils and their management and is designed to complement other sources:

- **Management information.** Considerable resources are available from Agriculture Western Australia and other organisations to extend information on factors affecting land management. This handbook does not attempt to be a comprehensive land management manual, but contains a summary of management options at the end of each section. This is complemented by technical references and a list for further reading.
- **Crop and pasture agronomy.** Crop and pasture agronomy is described in many texts and Farmnotes, to which the reader is referred. For instance, the agronomy and management of wheat is described in *The wheat book* (Perry and Hillman 1991) and there are similar publications for barley, lupins and field peas.
- **Earthworks design.** The technical aspects of earthworks design and runoff estimation are described in the *Soil conservation earthworks design manual* (Bligh 1989); *Australian rainfall and runoff* (Institute of Engineers of Australia 1987); and *Flood estimation procedures for Western Australia* (Flavell *et al.* 1987).
- **Soil description.** The standard methods for describing soil morphology, landform and native vegetation are in the *Australian soil and land survey field handbook* (McDonald *et al.* 1990).
- **Soil physical measurements.** The methods for many soil physical measurements are described in *Soil physical measurement and interpretation for land evaluation* (Coughlan *et al.* 1997) or soil physics texts. Simple tests for use on the farm are described in the *Farm monitoring handbook* (Hunt and Gilkes 1992).
- **Land resources.** The major soil groups, their extent and distribution are briefly described in Chapter 2. *Reference soils of south-western Australia* (McArthur 1991) provides morphological descriptions plus chemical data for 136 soil profiles under native vegetation. Agriculture Western Australia has a soil profile database containing approximately 15,000 soil/site descriptions.
- **Fundamentals of soil science.** Basic soil science is not covered in this manual. Readers are referred to texts such as: *The nature and properties of soils* (Brady 1996); *An introduction to the principles and practice of soil science* (White 1987); *Soil science, An introduction* (Leeper and Uren 1993); and *Soils, Their properties and management* (Charman and Murphy 1991).

1.2 SITE ASSESSMENT: SOIL READY RECKONER

When trying to identify which soil properties affect production and land degradation, it is not necessary to assess every soil property at each site. Instead, the soils can be divided into four broad groups based on their hydrological properties (water storage and transmission). The groups are called *profile hydrology groups* and are described on the following pages. For each group, there is an outline of the soil properties and a diagram that provides a visual summary of factors to consider. Further information can be found in the relevant sections of this manual.

The properties which need to be examined can be quickly identified and followed up in the relevant sections.

The four profile hydrology groups listed below account for most soils, although there will be a few which do not fit neatly into any group (e.g. organic soils). Some soils are best described as intergrades between some of the groups. With these soils, determine the optimum management by considering all relevant properties.

- ### UNIFORM COARSE-TEXTURED SOILS

  Coarse-textured soils with a uniform or gradational texture profile (e.g. deep sands, loamy sands, clayey sands and sandy earths). They are usually deep, but if shallow often overlie a substrate which is semi-permeable to water. These soils tend to be excessively drained and water continues to drain through the profile at the suction used to measure the upper storage limit.

- ### PERMEABILITY CONTRAST SOILS

  Soils with an abrupt change in texture between the surface and a clayey subsoil, e.g. sand over clay. The clayey subsoil has a much lower permeability than the surface layer(s), so perched watertables above the subsoil are a characteristic feature. Permeability contrast soils are defined in terms of texture and hydrology, i.e. an abrupt increase in texture plus a marked decrease in permeability in the B horizon.

- ### CRACKING CLAYS

  Uniform clay soils which swell and shrink in three dimensions to produce cracks at least 5 mm wide. They have varying degrees and depths of surface self-mulching.

- ### MEDIUM TO FINE-TEXTURED SOILS

  Soils with a medium to fine texture throughout the profile, e.g. uniform and gradational loams and clays. This group also includes loamy texture contrast soils which do not have a marked difference in permeability between the contrasting texture layers, and therefore do not develop perched watertables.

If a site has any of the following characteristics, they will almost certainly be major constraints to production (it is then not necessary to assess other properties):

Shallow soil over bedrock - refer directly to soil water (Section 3.3)

Poor drainage - refer directly to waterlogging (Section 3.4)

Salinity - refer directly to the assessment of soil salinity (Section 5.3)

Steep slopes (>15%) - refer directly to water erosion (Section 7.2).

UNIFORM COARSE-TEXTURED SOILS

The uniform coarse-textured soils (e.g. deep sands, loamy sands, clayey sands and sandy earths) are freely draining soils with a low capacity to hold water and nutrients. They are usually deep, with the soil profile often extending to more than 2 m.

Hydrology - Following rain the wetting front gradually moves through the soil profile. These soils do not reach a relatively steady moisture content 24 to 48 hours after a significant rainfall event, because water continues to drain through the profile at the suction used to measure the upper storage limit. Therefore the textbook model of water movement in rigid soils is less useful on these soils.

Deep drainage below the root zone can be frequent and extensive. Soil water can drain rapidly below the root zone unless there are actively growing plants. High rates of recharge are expected on highly leached coarse-textured soils with a very low water storage and where physical or chemical properties limit root growth.

Infiltration - Water entry is rapid unless the surface soil is water repellent. A water repellent surface will result in redistribution of water within the profile and can also cause significant runoff and erosion during summer storms.

Water storage - Low to very low water storage capacity (20 to 80 mm/m), which means crops or pastures are subject to moisture stress when rainfall is irregular. Water storage is not directly comparable with many other soils as mentioned under hydrology above.

Waterlogging - Generally freely draining, but may be waterlogged in seepage areas or where there is a high watertable.

Rooting conditions - Conditions for root elongation are favourable unless the soil is compacted or has a highly acidic subsurface e.g. aluminium toxicity. Traffic pans commonly develop at a depth of 10 to 30 cm. Soils with a cemented organic or ironstone pan can limit root growth.

Plant nutrient supply - Inherently infertile soils, multiple nutrient deficiencies are common, e.g. N, P, K (in the high rainfall zones), S, Cu, Zn, Mn (in lupins) and Mo (in highly acid soils).

Small areas of alkaline sands and loamy sands occur along the western and southern coasts.

Leaching - Loss of N, K, S and to a lesser extent P, by leaching, from some soils.

Toxicity - P toxicity can occur on highly leached sands if fertiliser is drilled with the seed.

Environmental factors - Highly susceptible to wind erosion as most uniform coarse-textured soils have a loose, single grain surface, so maintaining an adequate groundcover is a priority at all times. Water erosion is generally minor because there is minimal runoff unless the soils are water repellent. Nutrient pollution of groundwater, especially with P, is a major concern as the nutrients may contribute to eutrophication of rivers, estuaries or wetlands.

Land use considerations - Suitable crops or pasture species should have a deep root system to take advantage of the large volume of soil and to minimise moisture stress and deep drainage. These soils are not suitable for crops with stubbles that are easily detached and removed by wind, leaving the soil surface bare.

UNIFORM COARSE-TEXTURED SOILS

A) Wind erosion
Assess susceptibility and monitor paddock status [7.1].

B) Infiltration
Assess susceptibility to **water repellence.** If susceptible assess paddock status [3.1].

C) Runoff /water erosion
Assess if the soil is water repellent or the surface is firm to hardset over summer and the slope is >3% [7.2]

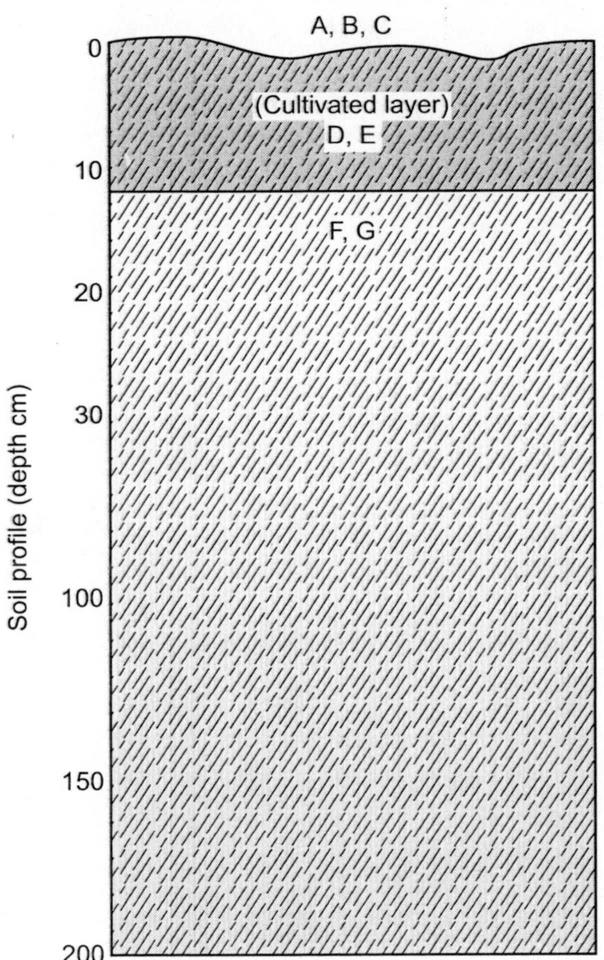

D) Nutrients
Possible deficiencies of **N, P, K, S, Cu, Zn, Mo** and **Mn** [6]

E) Soil acidity
Measure pH [5.1]

F) Subsurface compaction
Assess susceptibility to **traffic pans** and paddock status if susceptible [4.2]

G) Subsurface acidity
Measure the pH (at 10-20 and 30-40 cm). If above the critical level, assess the susceptibility [5.1].

Soil water storage (whole profile)
Measure or infer the water storage and relate this to the estimated depth of root growth for each crop or pasture species [3.3].

Note: *The numbers in parentheses '[ ]' refer to the relevant sections or chapters.*

PERMEABILITY CONTRAST SOILS

The permeability contrast soils are moderately well to poorly drained soils usually formed on highly weathered parent materials. The group includes many, but not all of the *duplex** soils as defined by Northcote (1979), and *texture contrast** soils as defined by Isbell (1996). Surface texture can vary from sand to sandy clay loam, with a sharp boundary to a clayey subsoil with limited permeability. The depth to the clayey subsoil can vary from 0.1 to 0.8 m.

Hydrology - The slow permeability of the clayey subsoil (B horizon) dominates the hydrology, by limiting water entry into the lower profile. Matrix flow is generally very restricted, so deep drainage is thought to be predominantly by saturated flow through preferred pathways from water perching on the top of the clay B horizon.

Infiltration - Water entry depends on the surface condition, surface texture and antecedent moisture content. Soils with a sandy surface (<5% clay) are susceptible to water repellence, while soils with more than about 8 to 10% clay can be susceptible to surface crusting and hardsetting.

Water storage - This is often low, but varies considerably. Water may be stored in the A horizon as perched water. The major factors affecting plant available water are: depth to the clayey subsoil, whether the subsoil is favourable for root growth, and crop tolerance of waterlogging.

Waterlogging - Waterlogging is a major constraint to crop production especially in wet years. The major factors affecting waterlogging intensity are: climate (amount and distribution of rainfall, temperature), permeability of the clayey subsoil, depth to clayey subsoil and position in the landscape.

Rooting conditions - Conditions for root elongation in the coarser textured surface and subsurface layers (A horizon) vary from poor to excellent. The main limitations include high strength (i.e. hardset surface horizon, traffic pan), highly acidic subsurface (often Al toxicity) or a perched watertable which is common during winter (June, July and August).

The clayey subsoil is often a major limitation to root growth. The root growth of annual species is often restricted to the top 0.3 to 0.8 m because of poorly structured, sodic subsoils, even when the depth of weathering extends many metres.

Plant nutrient supply - Inherently infertile soils, multiple nutrient deficiencies are common, e.g. N, P, S, Cu and Zn and occasionally Mo and/or Mn deficiency in cereals. Potassium deficiency is increasingly common, but varies with the ability of the clayey subsoil to supply K.

Leaching - Loss of N, K, S and P (if PRI <2) by leaching, from some soils. The nutrients are not necessarily lost from the profile as they may only be leached to the top of the clayey subsoil.

Environmental factors - Moderate to high susceptibility to wind erosion, depending on surface texture and condition. Water erosion can be a problem with saturation excess runoff during winter or runoff from summer thunderstorms.

Nutrient loss, especially of P attached to soil particles, is a major concern where the runoff enters rivers, estuaries or wetlands, because it can lead to eutrophication (e.g. Kalgan River, Oyster Harbour at Albany).

Land use considerations - Suitable crops or pasture species should have some tolerance of waterlogging. A deep root system is not essential, especially where the clayey subsoil is within 50 cm of the surface. Measures to reduce waterlogging will reduce recharge from these soils.

* *The term duplex as defined by Northcote (1979) refers to soils which have an increase in texture of one and a half texture groups or greater between the A and B horizons over a vertical distance of <10 cm. Texture contrast soils (Isbell 1996) have a clear, abrupt or sharp boundary between the A and B horizon together with an increase in clay in the B horizon of at least 20%.*

PERMEABILITY CONTRAST SOILS

A) Wind erosion

Assess susceptibility and monitor paddock status [7.1].

B) Infiltration

If surface is loose in summer then assess the susceptibility to **water repellence** [3.1].
Otherwise assess the susceptibility to **soil structure decline** [3.2].

C) Runoff /water erosion

Assess if the slope is >1% or longer than 400 m [7.2].

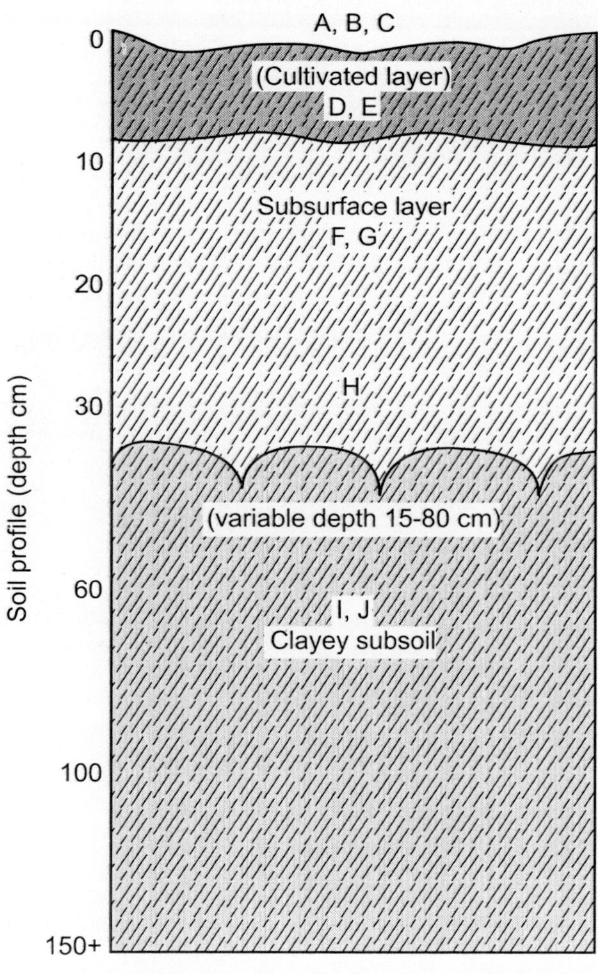

D) Nutrients

Possible deficiencies of **N, P, K, S, Cu, Zn** and **Mn** [6].

E) Soil acidity

Measure pH [5.1].

If A2 horizon is present

F) Subsurface compaction

Assess susceptibility and paddock status for **traffic pans** [4.2].

G) Subsurface soil acidity

Measure the pH (at 10-20 cm). If above the critical level, assess the susceptibility to acidity [5.1].

H) Waterlogging

Assess the site's drainage [3.4].

I) Subsoil structure

Assess the clayey subsoil for degree of restriction to root penetration [2.3].

J) Alkalinity

Assess if relevant [5.2].

Soil water storage (whole profile)

Measure or infer the water storage and relate this to the estimated depth of root growth for each land use [3.3]

Note: *The numbers in parentheses '[ ]' refer to the relevant sections or chapters.*

CRACKING CLAYS

Cracking clays (or vertosols) refer to all uniform fine-textured soils which exhibit three dimensional swelling and shrinking to produce seasonal cracking at the surface. Textures are usually light clay or finer throughout the profile, apart from possibly a thin layer (<5 cm) of coarser material on the surface. Land surfaces often show gilgai microrelief and the surface soil may be self-mulching. The shrinking and swelling is due to the presence of montmorillonite.

Cracking clays are limited in south-western Australia, but have been separated as a group because they have distinct hydrological properties. They are more common in the Ord River Irrigation Scheme and in central NSW and Queensland.

Hydrology - The key characteristic is that water storage is limited by the infiltration rate rather than being determined by the moisture characteristic. Water storage of a given soil can be increased by improving the infiltration properties (e.g. by the addition of gypsum). In addition, initial wetting of a dry soil may be from the 'bottom up' with surface water flowing down the cracks to the subsoil rather than wetting from the surface down. As a result, cracking clays do not conform to the model of water movement in rigid soils. Deep drainage is likely to be minimal on these soils.

Infiltration - The infiltration rate will vary enormously depending on whether the surface cracks are open and on the condition of the soil surface. Surface structure, presence of a sodic layer or low electrolyte conditions are the main factors affecting the surface condition.

Water storage - High to very high, but can vary considerably under dryland agriculture. The principal determinant is infiltration (or water entry) as explained above, but rooting depth and subsequent cracking patterns are also important.

Waterlogging - Surface waterlogging can occur depending on position in the landscape.

Rooting conditions - Root growth will depend on the soil structure as the plant roots will grow preferentially around peds and down cracks rather than through the soil matrix. The depth of root growth may be limited by unfavourable chemical properties, compaction or by poor subsoil structure.

Root growth may be restricted by an accumulation of salt in the subsoil corresponding to the depth of wetting. Assess by measuring the electrical conductivity down the profile to see if there is a 'salt bulge' in the subsoil.

Cracking clays are often alkaline to highly alkaline. If the pH_w is higher than 8.5 the soil may be highly sodic, or sodium carbonate may be present.

Acidity is likely to develop only slowly on cracking clays that have a neutral soil reaction, because they have a high pH buffering capacity.

Plant nutrient supply - Cracking clays may be deficient in N and P.

Environmental factors - Low susceptibility to wind erosion unless the surface is loose or self-mulching (fine aggregates <1 mm). The topography is usually flat to very gently sloping, therefore water erosion is generally a minor issue, but it does depend on the slope, runon and stability of the surface when wet. Nutrient loss is only of concern when significant amounts of soil are lost by water erosion.

Land use considerations - Suitable crop or pasture species should be able to tolerate high pH if the soil is alkaline and may need to tolerate poor aeration.

CRACKING CLAYS

A) Wind erosion
Assess susceptibility if surface soil is loose, also monitor paddock status [5.1].

B) Infiltration
Assess the susceptibility to ***soil structure decline*** and present condition [3.2].

C) Runoff / water erosion
Assess if the slope is >1% [7.2].

H) Waterlogging
Assess if a level site or low lying position and poor surface structure or sodic surface layer [3.4].

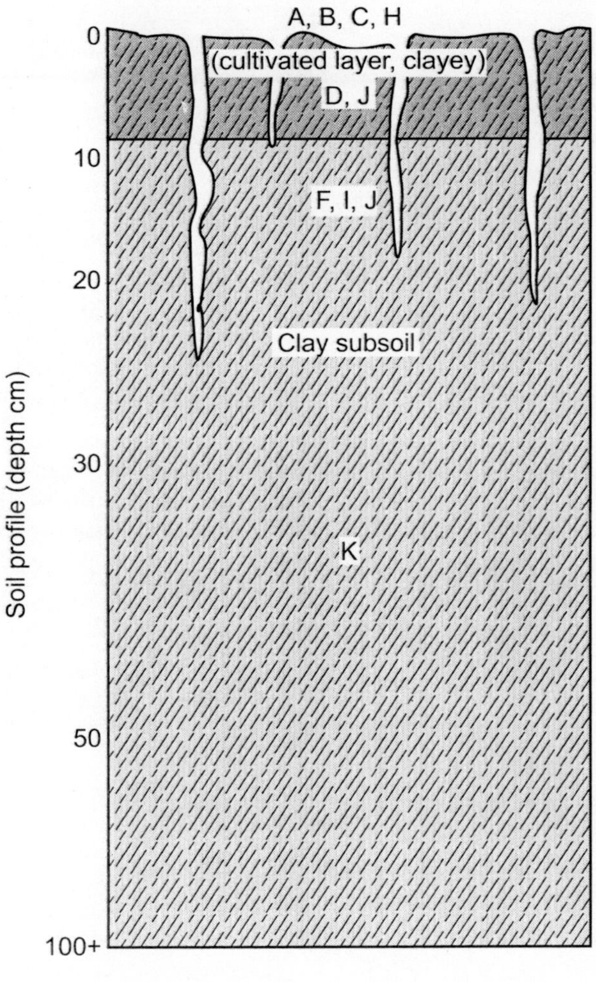

D) Nutrients
Possible deficiencies of **N** and **P** [6].

J) Alkalinity
Measure pH [5.2].

F) Compaction
Assess for presence of a plough pan [4.2].

I) Subsoil structure
Assess the clayey subsoil for degree of restriction to root penetration [2.3].

J) Alkalinity
Assess if relevant [5.2].

K) Salinity
Assess the salinity profile [5.3].

Soil water storage (whole profile)
Measure or infer the water storage and relate this to the estimated depth of root growth for each crop or pasture [3.3].

Note: *The numbers in parentheses '[ ]' refer to the relevant sections or chapters.*

MEDIUM TO FINE-TEXTURED SOILS

Includes a range of soils with medium to fine surface textures and usually a uniform or gradational profile. Soils vary from well structured loamy soils formed on dolerite, to calcareous loams with a 'fluffy' surface adjacent to salt lakes. There are also a few soils which technically qualify as texture contrast (Isbell 1996) or duplex (Northcote 1979) soils, but are included because there is not a marked decrease in the permeability of the subsoil. The subsoils are comparatively permeable therefore water does not perch on the clayey subsoil.

Hydrology - The hydrology generally follows the textbook model of water movement in rigid soils. The surface soil is often the layer with the lowest permeability. Deep drainage is infrequent on these soils, and is usually only associated with flooding on valley floors, heavy unseasonal rainfall or years with a very high rainfall during the growing season.

Infiltration - Susceptible to soil structure decline (e.g. surface crusting and hardsetting), consequently the infiltration rate will vary depending on the stability, present condition (degraded or good condition) and recent management (cultivation) of the surface soil.

Water storage - May be moderate to high and is influenced by soil texture and in particular the degree of structural development. Plant-available water varies widely and is mainly affected by the depth of root growth.

Waterlogging - Many of the soils in this group are well drained, although some will have problems with surface and possibly subsurface waterlogging, if the soil has a low macro-porosity.

Rooting conditions - Variable, depending on whether the soil is massive and porous (earthy fabric), pedal, or is massive and impermeable (e.g. some non-cracking clays). The effective rooting depth may also be limited by unfavourable chemical properties (e.g. high or low pH, sodicity, salinity, high boron) or a physical barrier such as a red-brown hardpan, calcrete or bedrock. Soil acidity is approaching critical levels on some soils.

In low rainfall districts (areas receiving <350 mm annually) root growth may be restricted by an accumulation of salt in the subsoil corresponding to the depth of wetting. Also, many soils have alkaline subsoils containing calcium carbonate, either finely divided in the soil matrix and/or as concretions.

Plant nutrient supply - This group includes some of the more fertile soils in south-western Australia. Nutrient deficiencies of N, P and S (if superphosphate is not applied) are common and Zn deficiency (can be checked with a soil test) has been recorded occasionally.

Environmental factors - These soils have a low to moderate susceptibility to wind erosion depending on their texture and surface condition. Water erosion can be a major problem on sloping sites, especially if the surface is unstable when wet. Nutrient loss is a concern when significant amounts of soil are lost by water erosion.

Land use considerations - A range of crops and pastures can be grown. Suitable species will usually need to tolerate a hard surface, high pH (if the subsoil is alkaline) and may need some tolerance of waterlogging.

MEDIUM TO FINE-TEXTURED SOILS

A) Wind erosion

Assess susceptibility and monitor paddock status [7.1].

B) Infiltration

Assess susceptibility to ***soil structure decline*** and present condition [3.2].

C) Runoff /water erosion

Assess if the slope is >1% or longer than 400 m [7.2].

H) Waterlogging

Assess if site is level or low lying and has poor surface structure [3.4].

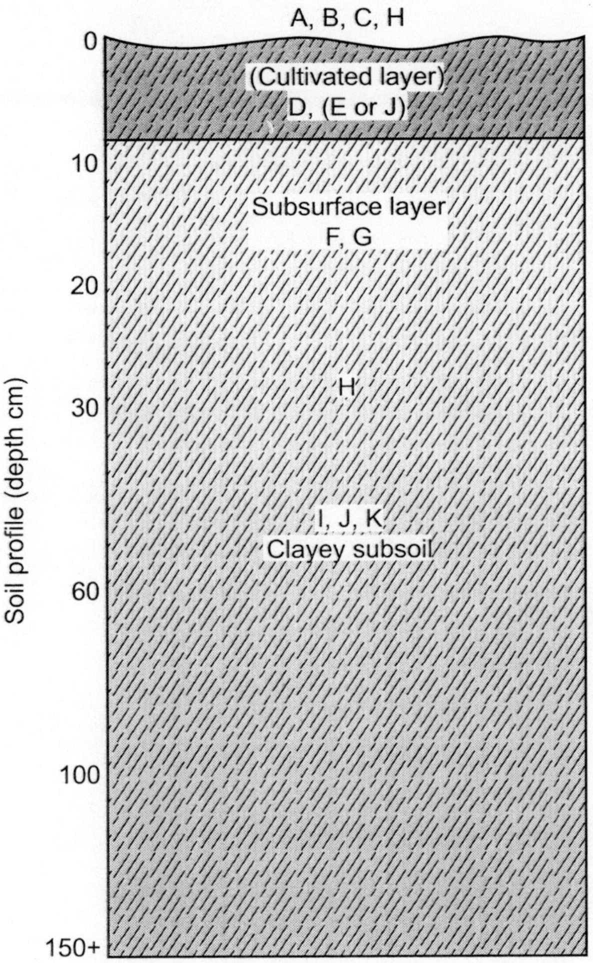

D) Nutrients

Possible deficiencies of **N, P, S** and **occasionally Zn** [6].

E) Acidity [5.1] or *J)* Alkalinity [5.2].

F) Subsurface compaction

Assess for presence of both traffic and plough pans [4.2].

G) Subsurface soil acidity

Measure the pH (at 10-20 cm). If above the critical level, assess the susceptibility [5.1].

H) Waterlogging

Assess the site drainage if there are signs of impeded drainage [3.4].

I) Subsoil structure

Assess the clayey subsoil for degree of restriction to root penetration [2.3].

J) Alkalinity

Assess if relevant [5.2].

K) Salinity

Assess the salinity profile if annual rainfall <350 mm [5.3].

Soil water storage (whole profile)

Measure or infer the storage and then relate this to the estimated depth of root growth for each land use [3.3].

Note: *The numbers in parentheses '[ ]' refer to the relevant sections or chapters.*

CHAPTER 2

SOILS OF SOUTH-WESTERN AUSTRALIA

2.1 SOIL GROUPS OF SOUTH-WESTERN AUSTRALIA

Natural Resources Assessment Group staff

A wide range of soils occur in south-western Australia. Many are morphologically and chemically similar to soils found in eastern Australia or overseas, but in WA there is a dominance of highly weathered soils and minimal soil renewal. As a result there are vast areas of infertile soils, and many are either deep sands or have coarse-textured topsoils. Shallow, sandy duplex soils and deep sands, which elsewhere would be considered non-arable, are major agricultural soils in WA. Agriculture is possible because of the relatively reliable winter rainfall under the region's Mediterranean climate.

This section describes the main soils of south-western Australia in terms of soil groups (Schoknecht 1997). The purpose of the soil groups is to present a system for communicating soil information on a functional basis that both technical and non-technical users can understand. In certain situations a more formal classification system may be required.

CHARACTERISTICS OF SOILS

In general, the characteristic features of soils in south-western Australia include:

- *A high degree of weathering.* The Archaean shield underlying most of south-western Australia is one of the oldest and most stable land surfaces on earth. Most of the landscape is highly weathered. This weathering dates back to the Cretaceous (65 million years ago) and is commonly 30 to 50 m deep. The general absence of tectonic, volcanic and glacial activity has resulted in minimal soil renewal. Many of the so-called 'younger soils' have formed from pre-weathered materials.

Native vegetation on a yellow deep sand showing the extensive root system.

- *Widespread laterisation.* Deep weathering has resulted in the formation of laterite profiles in many areas (see Section 2.2). Characteristics include a high degree of leaching, and accumulation of iron and aluminium near the surface. Many soils are formed on laterite, truncated laterite or erosional products from laterite.
- *Low fertility.* Many major and minor elements (e.g. N, P, K, S, Cu, Zn, Mn and Mo) are deficient for plant growth. Organic matter is also generally low (Robson and Gilkes 1981).
- *Coarse texture.* Coarse-textured soil materials (sands to sandy loams) dominate, especially in the surface horizons. In general, soils have a high sand and low silt content, with a bimodal particle size distribution (i.e. peaks in the sand and clay size fractions). The particle size distribution results from the intense weathering of laterite and granite (the dominant parent materials) leaving the highly resistant quartz.
- *Clay mineralogy.* Kaolinite is the dominant clay mineral (see Chapter 10), formed as a result of intense weathering of laterite and granite. It is present in almost all the soils and comprises more than three-quarters of the clay content in 80% of them (Singh 1991). Illite was found in 30% of samples, and was a major clay mineral only in calcareous soils of lacustrine origin. Smectite was only found in 15% of samples, mainly the poorly drained alluvial soils and soils formed from mafic rocks (Singh 1991).

SOIL CLASSIFICATION

There are numerous methods of classifying soils, from local schemes based on native vegetation and general terms like light and heavy land, to more formal systems. On an international scale the accepted standards for soil classification are *Soil Taxonomy* (Soil Survey Staff 1994) and the system developed by the FAO (FAO 1988). In Australia, the evolution of soil classification has included the Great Soil Groups (Stace *et al.* 1968), followed by *A factual key for the recognition of Australian soils* (Northcote 1979) and recently, the *Australian soil classification* (Isbell 1996).

SOIL GROUPS

The major soils of south-western Australia have been placed into 38 groups (Tables 2.1.1 to 2.1.8) as detailed in *Soil groups of Western Australia* (Schoknecht 1997). Surface texture and 'texture profile' (i.e. changes in texture down the profile) are used as initial criteria for identifying the groups, except for soils which are saturated for several months each year (called *wet soils*).

The 'soil group name' can be expanded by including a qualifier in brackets at the end of the soil group name, e.g. yellow deep sand (gravelly) is a gravelly phase of a yellow deep sand (gravelly is defined as >20% ironstone gravel within 15 to 80 cm of the surface), or red shallow loamy duplex (on granite) to indicate the soil overlies granite rocks.

The qualifiers are grouped into the following categories: texture, structure, subsurface, subsoil and substrate. Examples of each include:

Texture: - gritty, gravelly, stony

Structure: - massive, pedal (structured)

Subsurface: - bleached subsurface

Subsoil: - colour (e.g. yellow subsoil), mottled subsoil, saline subsoil, sodic subsoil

Substrate: - over weathered rock, over mottled zone, over clay at 1-2 m.

The use of only one qualifier is preferred, although it is possible to use multiple qualifiers if necessary (Schoknecht 1997).

A guide to the major soil groups of south-western Australia is provided in Figure 2.1.1. The groups and their properties are summarised in Tables 2.1.1 to 2.1.8 (indexed in Figure 2.1.1). Wet soils (profiles or horizons saturated for several months each year) are described in Table 2.1.8.

Grey deep sandy duplex.

Surface texture		Soil texture profile	Soil group
Ironstone gravelly soils >20% ironstone gravel within 15 cm	→	**Shallow gravels** Cemented gravels (ferricrete) at <80 cm and often <30 cm	**See Table 2.1.1**
		Sandy gravels Gravels with a sandy soil matrix	"
		Loamy gravels Gravels with a loamy matrix	"
Sandy surfaced soils Sand, loamy sand and clayey sand	→	**Shallow sands** Sands <80 cm and often <30 cm deep over rock, hardpan or other hard cemented layer	**Minor extent**
		Deep sands Sands >80 cm deep	**See Table 2.1.2**
		Sandy earths Sands grading to loams at 30-80 cm, may be clayey at depth	**See Table 2.1.3**
		Shallow sandy duplexes Sands over clays, or less commonly loams, at <30 cm	**See Table 2.1.4**
		Deep sandy duplexes Sands over clays, or less commonly loams, at 30-80 cm	"
Loamy surfaced soils Sandy loam to clay loam	→	**Shallow loams** Loams <80 cm and often <30 cm deep over rock, hardpan or other cemented layer	**See Table 2.1.5**
		Loamy earths Uniform loams and loams grading to clay loams or clays	"
		Shallow loamy duplexes Loams over clays at 5-30 cm	**See Table 2.1.6**
		Deep loamy duplexes Loams over clays at 30-80 cm deep	"
Clayey surfaced soils Sandy clay to heavy clay	→	**Cracking clays** Clays which crack strongly when dry	**See Table 2.1.7**
		Non-cracking clays Clays which do not crack strongly when dry	"

Figure 2.1.1 *A guide to the main soil groups of south-western Australia.*

Wet soils (profiles or horizons saturated for several months each year) are described in Table 2.1.8.

Table 2.1.1 Properties of ironstone gravelly soils.
Soils indicated by shading are of minor agricultural significance.

Soil texture profile	Soil group (plus common soil properties)	Profile hydrology group (Section 1.2)	Possible deficiencies* and toxicities	Main limitations to sustainable production	Distribution	Australian Soil Classification**
Shallow gravels Cemented gravels (ferricrete) at <80 cm and often <30 cm	*Shallow gravel* • Yellow/brown to grey (sometimes red) in top 30 cm • Neutral to acid pH • High topsoil gravel content (>20%)	Uniform coarse-textured soils, with some perching of water on the ferricrete	*Deficiencies* P, N, Cu, Zn, occasionally Mo *Toxicities* P if high P with seed, Al if pH_{Ca} <4.5	Very low soil water storage Wind erosion	West Midlands Darling Range Scattered throughout wheatbelt	Tenosols Rudosols
	For shallow gravels over clay at less than 30 cm, refer to Shallow sandy duplex, or Shallow loamy duplex					
Sandy gravels Gravels with a sandy matrix	*Moderately deep sandy gravel* • Yellow/brown to grey in top 30 cm • Neutral to acid pH • High topsoil gravel content (>20%) • Sandy clay loam to clay at 30-80 cm	Permeability contrast soils	*Deficiencies* P, N, (S), Cu, Zn, Mn *Toxicities* P if high P with seed, Al if pH_{Ca} <4.5	Low soil water storage Water repellence Waterlogging	Darling Range (jarrah gravel) South west Great Southern (wandoo gravel)	Chromosols Tenosols Kandosols
	Deep sandy gravel • Ironstone gravel dominant (>20%) • Yellow, brown and grey colours • Neutral to acid pH • Usually over sandy clay loam to clay, or cemented gravels at >80 cm	Uniform coarse-textured soils (if clayey subsoil absent)	*Deficiencies* P, N, (K, S), Cu, Zn, Mn *Toxicities* P if high P with seed, Al if pH_{Ca} <4.5	Very low soil water storage Wind erosion Water repellence (Water erosion)	Darling Range South west Great Southern (west)	Kandosols Tenosols Chromosols
Loamy gravels Gravels with a loamy matrix	*Loamy gravel* • Yellow, red or brown in top 30 cm • Neutral to acid pH • High gravel content (>20%) throughout • Usually grading to clay by 30-80 cm, occasionally deeper	Medium to fine-textured soils	*Deficiencies* P, N, (S)	Low soil water storage	South west Darling Range Great Southern (west) Northern wheatbelt	Kandosols Dermosols Chromosols Tenosols

* *Nutrient deficiencies which may occur, but are strongly influenced by land use and/or management, are shown in parentheses (e.g. S is related to superphosphate history, K to product removal).*

** *Isbell (1996).*

Table 2.1.2 Properties of deep sands.
Soils indicated by shading are of minor agricultural significance.

Soil texture profile	Soil group (plus common soil properties)	Profile hydrology group (Section 1.2)	Possible nutrient deficiencies* and toxicities	Main limitations to sustainable production	Distribution	Australian Soil Classification**
Deep sands Sands >80 cm deep	*Pale deep sand* • White, grey or pale yellow within top 30 cm • Neutral to acid pH • Gravel may be present • Coffee rock may be present at depth	Uniform coarse-textured soils	*Deficiencies* N, P, K, (S), Cu, Zn, Mn, occasionally Mo, Fe *Toxicities* P (if high P with seed), Al if pH_{Ca} <4.5	Very low soil water storage; Wind erosion; Water repellence; (Waterlogging)	West coast; West Midlands; South coast; Scattered in other areas	Rudosols; Tenosols; Podosols
	Yellow deep sand • Yellow within top 30 cm • Neutral to acid pH • Gravel may be present	Uniform coarse-textured soils	*Deficiencies* N, P, (S), Cu, Zn, occasionally Mo *Toxicities* Al if pH_{Ca} <4.5	Wind erosion; Subsurface compaction; Low soil water storage	West coast; West Midlands; Northern (e.g. Eradu sandplain), central and eastern wheatbelt; Scattered elsewhere	Tenosols; Rudosols
	Brown deep sand • Brown within top 30 cm • May be humic • Neutral to acid pH • Usually alluvial or lacustrine origin • Over rock, clay or other unconsolidated materials	Uniform coarse-textured soils	*Deficiencies* N, P, Cu, Zn	Wind erosion; Flooding; Subsurface compaction; Low soil water storage	Widespread, but of minor extent on valley floors	Rudosols; Tenosols
	Calcareous deep sand • White, grey or occasionally dark colours • Calcareous throughout • Often very deep in dune deposits or over limestone or calcareous hardpan	Uniform coarse-textured soils	Coastal dunes are non-agricultural soils, should not be cleared. Limited grazing of native vegetation may be possible. Soils further inland have nutrient deficiencies N, P, Cu, Zn Mo and Mn. Limitations include low soil water storage, alkaline and wind erosion.		Coastal areas (coastal dune sand, black wattle soil)	Rudosols; Tenosols

** Nutrient deficiencies which may occur, but are strongly influenced by land use and/or management, are shown in parentheses (e.g. S is related to superphosphate history, K to product removal).*
*** Isbell (1996).*

Table 2.1.3 Properties of sandy earths.
Soils indicated by shading are of minor agricultural significance.

Soil texture profile	Soil group (plus common soil properties)	Profile hydrology group (Section 1.2)	Possible deficiencies* and toxicities	Main limitations to sustainable production	Distribution	Australian Soil Classification**
Sandy earths Sands grading to loams at 30-80 cm, may be clayey at depth	*Acid yellow sandy earth* • Yellow throughout, may be grey at surface • Strongly acid pH • Gravel may be present • May be clayey below 80 cm	Uniform coarse textured soils (grading to medium to fine textured soils)	*Deficiencies* N, P, Mo, Cu, Zn *Toxicities* Al in subsoil if pH_{Ca} <4.5	Water stress due to limited root growth (Al toxicity) Wind erosion Subsurface compaction	Central wheatbelt Eastern wheatbelt (Wodjil soil)	Tenosols Kandosols
	Yellow sandy earth • Yellow within top 30 cm • Neutral to acid pH • May be clayey below 80 cm • Gravels may be present • Usually massive or poorly structured	Uniform coarse-textured soils (grading to medium to fine-textured soils)	*Deficiencies* N, P, (S), Cu, Zn, occasionally Mo *Toxicities* Al if pH_{Ca} <4.5	Subsurface compaction Wind erosion Low soil water storage	Wheatbelt	Kandosols
	Red sandy earth • Red throughout • Neutral to acid pH, occasionally calcareous at depth • Firm or loose surface • May have a red-brown hardpan at depth • Usually massive or poorly structured	Probably behave like medium to fine-textured soils, depending on the amount of coarse sand present	*Deficiencies* N, P, (S), Cu, occasionally Zn *Toxicities* Al if pH_{Ca} <4.5	Surface crusting Wind erosion Subsurface compaction	Northern wheatbelt Gingin district	Kandosols Tenosols
	Brown sandy earth • Brown within top 30 cm • Sand grading to loam by 80 cm • Neutral to acid pH • Usually alluvial origin	Probably behave like medium to fine-textured soils, depending on the amount of coarse sand present	*Deficiencies* N, P, occasionally Cu, Zn	Flooding Wind erosion	Widespread on alluvial flats, but of minor extent	Tenosols Kandosols

** Nutrient deficiencies which may occur, but are strongly influenced by land use and/or management, are shown in parentheses (e.g. S is related to superphosphate history, K to product removal).*
*** Isbell (1996).*

Table 2.1.4 Properties of sandy duplex soils.
Soils indicated by shading are of minor agricultural significance.

Soil texture profile	Soil group (plus common soil properties)	Profile hydrology group (Section 1.2)	Possible deficiencies* and toxicities	Main limitations to sustainable production	Distribution	Australian Soil Classification**
Shallow sandy duplexes Sands over clays or less commonly loams at <30 cm	*Alkaline grey shallow sandy duplex* • Grey surface layer(s), various colours in subsoil • Subsoil pH: alkaline, usually calcareous (usually sodic) • Usually not hardsetting	Permeability contrast soils	*Deficiencies* N, P, (S), Zn, occasionally Cu, Mn *Toxicities* B in subsoil if pH_w >9	Low soil water storage Waterlogging Wind erosion (Water erosion)	Great Southern South coast Scattered throughout wheatbelt	Sodosols
	Grey shallow sandy duplex • Grey surface layer(s), various colours in subsoil • Subsoil pH: neutral to acid • Gravel common	Permeability contrast soils	*Deficiencies* N, P, (K, S), Cu, Zn, occasionally Mn, Mo *Toxicities* Al if pH_{Ca} <4.5	Waterlogging Water repellence Wind erosion Low soil water storage (Water erosion)	South coast Great Southern Wheatbelt	Chromosols Sodosols
	Red shallow sandy duplex • Red within top 30 cm • Subsoil pH: neutral • Usually hardsetting surface • Clay may be underlain by rock or hardpan	Medium to fine-textured soils	*Deficiencies* N, P, (S) *Toxicities* Al if pH_{Ca} <4.5	Hardset surface Low soil water storage if shallow (Water erosion) (Wind erosion)	Northern wheatbelt	Chromosols Sodosols
Deep sandy duplexes Sands over clays or less commonly loams at 30-80 cm	*Alkaline grey deep sandy duplex* • Grey surface layer(s), bleached subsurface, various colours in subsoil • Subsoil pH: alkaline, usually calcareous	Permeability contrast soils	*Deficiencies* N, P, (K, S), occasionally Cu, Zn *Toxicities* Al if pH_{Ca} <4.5	Waterlogging Hardset surface Wind erosion (Water erosion)	Great Southern	Sodosols
	Grey deep sandy duplex • Grey surface layer(s) • Often bleached subsurface, various colours in subsoil • Subsoil pH: neutral-acid • Gravel often present, especially above clay	Permeability contrast soils	*Deficiencies* N, P, (K, S), Cu, Zn *Toxicities* Al if pH_{Ca} <4.5	Wind erosion Waterlogging Water repellence (Water erosion)	West Midlands South coast West coast Great Southern Scattered through wheatbelt	Chromosols Sodosols

** Nutrient deficiencies which may occur, but are strongly influenced by land use and/or management, are shown in parentheses (e.g. S is related to superphosphate history, K to product removal).*
*** Isbell (1996).*

Table 2.1.5 Properties of shallow loams and loamy earths.
Soils indicated by shading are of minor agricultural significance.

Soil texture profile	Soil group (plus common soil properties)	Profile hydrology group (Section 1.2)	Possible deficiencies* and toxicities	Main limitations to sustainable production	Distribution	Australian Soil Classification**
Shallow loams Loams <80 cm and often <30 cm deep over rock, hardpan or other cemented layer	*Red-brown hardpan shallow loam* • Red within top 30 cm • Red-brown hardpan by 50 cm • Often with stony surface mantle	Medium to fine-textured soils	*Deficiencies* P, N	Hardpan restricts root growth Hardset surface	Northern wheatbelt	Kandosols
Loamy earths Uniform loams and loams grading to clay loams or clays	*Calcareous loamy earth* • Usually calcareous throughout, but occasionally not in top 30 cm • Red or brown topsoil • May have limestone or calcrete at depth • Hardsetting or fluffy surface • Sometimes high salt content in subsoil	Medium to fine-textured soils	*Deficiencies* P, N, occasionally Zn	High alkalinity Wind erosion (Hardset surface)	Eastern wheatbelt Salmon Gums area (Kopi soil) Great Southern (east) (Fluffy surface i.e. Morrel soil in small areas near salt lakes)	Calcarosols Sodosols
	Friable red/brown loamy earth • Red to red-brown • Neutral to acid pH • Friable topsoil • Porous throughout • Gravel may be present	Medium to fine-textured soils	*Deficiencies* P, N, (S), Zn, Mo, occasionally Cu	Water erosion	South west (Karri loam)	Dermosols Kandosols
	Brown loamy earth • Brown or grey-brown topsoil • Neutral to acid pH • Grey-brown phases often mottled • Often formed in recent alluvium	Medium to fine-textured soils	*Deficiencies* P, N	Flooding (Waterlogging)	Widespread on alluvial flats, but of minor extent	Tenosols Kandosols

Table 2.1.5 continued.

Soil texture profile	Soil group (plus common soil properties)	Profile hydrology group (Section 1.2)	Possible deficiencies* and toxicities	Main limitations to sustainable production	Distribution	Australian Soil Classification**
Loamy earths (continued)	*Red loamy earth* • Red within top 30 cm • Usually massive or poorly structured • Neutral to acid pH, but sometimes calcareous at depth • Hardsetting or crusting • Sometimes with red-brown hardpan at >50 cm • Gravels may be present	Medium to fine-textured soils	*Deficiencies* P, N	Hardset or crusting surface	Northern wheatbelt Avon Valley	Kandosols
	Yellow loamy earth • Yellow within top 30 cm • Usually earthy fabric, but occasionally well structured • Neutral to acid pH • Gravels may be present in subsoil	Medium to fine-textured soils	*Deficiencies* P, N, occasionally Zn	Hardset surface	South west Eastern wheatbelt	Kandosols Dermosols

** Nutrient deficiencies which may occur, but are strongly influenced by land use and/or management, are shown in parentheses (e.g. S is related to superphosphate history, K to product removal).*

*** Isbell (1996).*

Table 2.1.6 Properties of loamy duplex soils.
Soils indicated by shading are of minor agricultural significance.

Soil texture profile	Soil group (plus common soil properties)	Profile hydrology group (Section 1.2)	Possible nutrient deficiencies* and toxicities	Main limitations to sustainable production	Distribution	Australian Soil Classification**
Shallow loamy duplexes Loams over clays at 5-30 cm	*Acid shallow duplex* • Shallow loam (or occasionally loamy sand) over pink, grey or brown clay • Subsoil pH: strongly acid • Often sodic • Commonly below breakaways	Permeability contrast soils	*Deficiencies* P, N, Cu, Zn	Water erosion Water repellence Hardset surface (Salinity)	Minor areas below breakaways in the south-west	Kurosol
	Red shallow loamy duplex • Red within top 30 cm • Subsoil pH: neutral • Firm to hardsetting surface	Medium to fine-textured soils (some permeability contrast soils)	*Deficiencies* P, N, (S)	Water erosion (Hardset surface)	Northern (Chapman Valley loam) and Central wheatbelt (Avon Valley loam) Great Southern	Chromosols Sodosols
	Alkaline red shallow loamy duplex • Red topsoil • Usually calcareous subsoil • Often hardsetting surface	Medium to fine-textured soils (some permeability contrast soils)	*Deficiencies* P, N	Hardset surface Wind erosion Waterlogging	Wheatbelt, Great Southern (e.g. Merredin soil, Salmon gum-gimlet soil, red-brown earth)	Sodosols Chromosols
	Alkaline grey shallow loamy duplex • Grey or brown topsoil • Subsoil pH: alkaline, usually calcareous • Hardsetting surface	Permeability contrast soils	*Deficiencies* P, N *Toxicities* B in subsoil if pH_w >9	Hardset surface Waterlogging Alkaline/sodic subsoil Low soil water storage Water erosion	Great Southern (Moort soil)	Sodosols
	Grey/brown shallow loamy duplex • Grey or brown topsoil • Subsoil pH: neutral • Firm to hardsetting surface	Medium to fine-textured soils (some permeability contrast soils)	*Deficiencies* P, N	Hardset surface	South west	Chromosols Kandosols
Deep loamy duplexes Loams over clays at 30-80 cm	*Alkaline red deep loamy duplex* • Red or red-brown topsoil • Calcareous subsoil • Often hardsetting surface	Medium to fine-textured soils	*Deficiencies* P, N	Hardset or crusted surface Calcareous subsoil, (Wind erosion) (Waterlogging)	Eastern wheatbelt	Sodosols
	Red deep loamy duplex • Red within top 30 cm • Subsoil pH: neutral • Firm to hardsetting surface	Medium to fine-textured soils	*Deficiencies* P, N, (S)	Hardset or crusted surface (Wind erosion (Waterlogging)	South west Chapman Valley	Chromosols Sodosols
	Grey/brown deep loamy duplex • Grey or brown topsoil • Subsoil pH: neutral • Firm to hardsetting surface	Medium to fine-textured soils (some permeability contrast soils)	*Deficiencies* P, N	Hardset surface Waterlogging	South west Yate loams (South coast)	Chromosols Sodosols

** Nutrient deficiencies which may occur, but are strongly influenced by land use and/or management, are shown in parentheses (e.g. S is related to superphosphate history, K to product removal).*
*** Isbell (1996).*

Table 2.1.7 Properties of clayey surfaced soils.
Soils indicated by shading are of minor agricultural significance.

Soil texture profile	Soil group (plus common soil properties)	Profile hydrology group (Section 1.2)	Possible nutrient deficiencies and toxicities	Main limitations to sustainable production	Distribution	Australian Soil Classification*
Cracking clays Clays which crack strongly when dry	*Self-mulching cracking clay* • Often grey, but also yellow, brown and red in top 30 cm • Deep cracks when dry • Self-mulching surface	Cracking clays	*Deficiencies* P, N	Waterlogging	Kimberley Scattered small areas throughout south-western agricultural area (e.g. Ravensthorpe)	Vertosols
	Hard cracking clay • Red, brown, yellow or grey within top 30 cm • Deep cracks when dry • Neutral to alkaline pH • Crusting or hardsetting surface	Cracking clays	*Deficiencies* P, N	Hardset surface	Northern wheatbelt South coast Great Southern	Vertosols
Non-cracking clays Clays which do not crack strongly when dry	*Grey non-cracking clay* • Often grey, sometimes yellow or brown within top 30 cm • Usually calcareous subsoil • Hardsetting surface	Medium to fine-textured soils	*Deficiencies* P, N *Toxicities* B in subsoil if pH_w >9	Hardset surface Poor subsoil structure Waterlogging	Great Southern	Dermosols Calcarosols Kandosols
	Red/brown non-cracking clay • Red or brown within top 30 cm • Hardsetting • Sometimes with calcareous subsoil	Medium to fine-textured soils	*Deficiencies* P, N *Toxicities* B in subsoil if pH_w >9	Hardset surface Limited root growth in subsoil due to poor structure, high salinity or high B Waterlogging	Mallee South coast	Dermosols Kandosols

* *Isbell (1996).*

Table 2.1.8 Properties of wet soils.
Soils indicated by shading are of minor agricultural significance.

Soil texture profile	Soil group (plus common soil properties)	Profile hydrology group (Section 1.2)	Possible nutrient deficiencies* and toxicities	Main limitations to sustainable production	Distribution	Australian Soil Classification**
Wet soils	*Non-saline wet soil* • Dark grey, brown or black topsoil • Acid to highly acid pH • High in organic matter • Usually sandy, but can be loams and clays • Variable subsoil, may contain bog iron ore	Sandy soils with a high organic matter content behave like uniform coarse-textured soils, while true 'peats' have a high soil water storage	*Deficiencies* N, P, K, (S), Cu, Zn, Fe, occasionally Mo, Mn *Toxicities* Low pH, but low [Al] compared with many other soils.	Waterlogging Wind erosion Water repellence	Soils of freshwater swamps, Peaty sands rare on south-west coast Freshwater seeps	Organosols Hydrosols
	Saline wet soil • Sands, loams and clays • Soils affected by secondary salinity	Either shallow watertable or seepage area	These soils should be fenced off and managed separately, with emphasis on maintaining ground cover (Section 5.3)		Saline valley floors and seepage areas	Hydrosols
	Salt lake soil • Highly saline • May be gypsiferous • Often calcareous		Non-agricultural land		Salt lake floors	Hydrosols

* *Nutrient deficiencies which may occur, but are strongly influenced by land use and/or management, are shown in parentheses (e.g. S is related to superphosphate history, K to product removal).*
** *Isbell (1996).*

2.2 SOILS AND LANDSCAPES OF SOUTH-WESTERN AUSTRALIA

Peter Tille[*], *Geoff Moore and Ted Griffin*[**]

The soils and landforms of south-western Australia have been described in many texts (e.g. Northcote *et al.* 1967; Mulcahy 1973; McArthur 1991). In general, the soils have an inherently low fertility which is related to a long history of weathering of the stable rock basement. While this is broadly true, there is a wide variation in not just the fertility but also the soil types, as illustrated in the previous section. The distribution of these soils is not random and different parts of the south-west can be characterised as having a particular suite of soils. This section describes how the soils have been influenced by parent material and some of the regional patterns of distribution.

Soil development is influenced by a complex interaction of many factors including rock or substrate type, landscape and landform, climate, hydrology and vegetation. The rock type influences many morphological, chemical and physical properties. For instance, red-brown loamy and clayey soils with relatively high inherent fertility often form on basic igneous rocks such as dolerite. Landscapes are greatly influenced by the manner in which parent materials have responded to erosion. On granitic materials, rounded hills with dendritic drainage patterns often form (e.g. between Kojonup and Katanning). Erosion is also influenced by rainfall, vegetation cover and landscape relief.

PARENT MATERIALS

Most of south-western Australia is underlain by the *Yilgarn Craton*[***] which is part of a vast Precambrian shield covering much of western and central Australia. The *Darling Fault* separates the Yilgarn Craton from the Perth Sedimentary Basin on which lies the Swan Coastal Plain (Johnstone *et al.* 1973). The ancient rocks of the Yilgarn Craton are the primary source of soil parent material over much of south-western Australia. Most of the sedimentary rocks and unconsolidated deposits have also been derived from weathering and stripping of these rocks. The major rock types are:

- Acid igneous and metamorphic rocks (e.g. granite, gneiss) which underlie most of south-western Australia.
- Basic rocks (e.g. dolerite dykes) which usually occur in small areas often associated with acid igneous rocks. Some broad areas, e.g. the Ravensthorpe Ranges, are formed on greenstone.
- Sedimentary rocks (e.g. sandstones, siltstones and shales) which cover significant areas such as the Perth Sedimentary Basin between Gingin and Geraldton. They are also found on the south coast together with spongeolite.

The unconsolidated sediments of more recent geological times are also important. These include:

- deposits of poorly sorted colluvium mainly on hill slopes,
- well-sorted alluvium in drainage lines, lake beds and coastal plains,
- aeolian dune deposits, mainly coastal.

During the long period of relative geological stability, there has been widespread, deep weathering and laterisation of the rocks (refer to *Laterite profile*, below). While the underlying bedrock has some influence on the nature of the soils formed on laterite, differences between parent rocks are reduced in highly weathered profiles.

A key to understanding the distribution of soil and landscapes in many parts of south-western Australia is that the deeply weathered profiles, often laterised, have been eroded to varying degrees and soils have formed on the resulting erosional and depositional surfaces. The profile properties change with depth, so that different degrees of stripping or erosion expose materials with different properties. Some areas are formed on largely intact laterite (e.g. Darling Plateau) and shallow stripping exposes the mottled zone (e.g. shallow duplex soils on slopes below breakaways), but in deep valleys the soils form on underlying basement rock (e.g. Avon Valley).

Colluvium and alluvium from eroded fresh rock or laterite are also the parent materials of many soils. Soils formed from colluvium, alluvium and aeolian deposits will reflect the properties of the source, although the transported material may come from a number of sources. Colluvium may overlie unrelated materials (e.g. colluvium from laterite overlying a soil formed on granite). Alluvial landscapes on valley floors and coastal plains are also highly variable, often consisting of several unrelated layers.

[*] *Agriculture Western Australia, Bunbury.*

[**] *Agriculture Western Australia, South Perth.*

[***] *Formerly called Yilgarn Block. This is not uniform and while granite is the dominant rock type, there are also significant areas of gneisses and greenstone. Dolerite dykes have intruded into the Yilgarn Craton in up to six different periods.*

Aeolian deposits such as dune landscapes in coastal districts have their own character. However, they can also contribute thin surface layers to other soils, e.g. parna on the valley floors in the eastern wheatbelt. The significance of desert dust from the arid interior for soils in low rainfall areas is largely unknown.

The main soil parent materials of south-western Australia and some of the corresponding soil groups and properties are summarised in Table 2.2 1.

BROAD SOIL PATTERN

The broad soil pattern is largely controlled by the parent material and the broad drainage pattern (outlined below). The drainage pattern influences the degree of erosion and stripping. There are three broad drainage divisions as illustrated in Figure 2.2.1 (Bettenay and Mulcahy 1972; Mulcahy 1973).

- In the eastern wheatbelt, ancient drainage lines contain extensive chains of salt lakes. These lakes only have connected flow in exceptionally wet years. In normal years, they act as a sump in which salts accumulate to form the playas (Bettenay 1962). In this ancient (or paleo-drainage) area there has been limited stripping and the weathering products tend to be retained on the broad valley floors.
- The Meckering Line (Mulcahy 1967) marks the change to areas with connected winter flow in rejuvenated drainage lines. These areas discharge to the west through valleys that have been deeply incised into the old plateau surface. In this area of so-called rejuvenated drainage there has been greater stripping and many soils have formed on truncated laterite or on exposed basement rock (e.g. Avon Valley). The eroded materials have been largely removed from the landscape.
- A divide (the Jarrahwood Axis) lies approximately parallel to the south coast, about 80 km inland. The divide separates the comparatively short rivers which flow into the Southern Ocean from the salt lake chains to the north. The rivers often form only shallow, narrow valleys.

In south-western Australia, the climate in earlier geological periods appears to have had a large influence on landforms and soil development. The main relationship between the present climate and soil properties is the degree of leaching of soluble salts (i.e. NaCl, $CaCO_3$) and the incidence of secondary salinity. There is limited leaching of soluble salts in the eastern wheatbelt due to the combination of low rainfall (<300 mm per annum) and low rainfall per wet day (Teakle and Burvill 1938). Secondary salinity is highest in the medium rainfall districts.

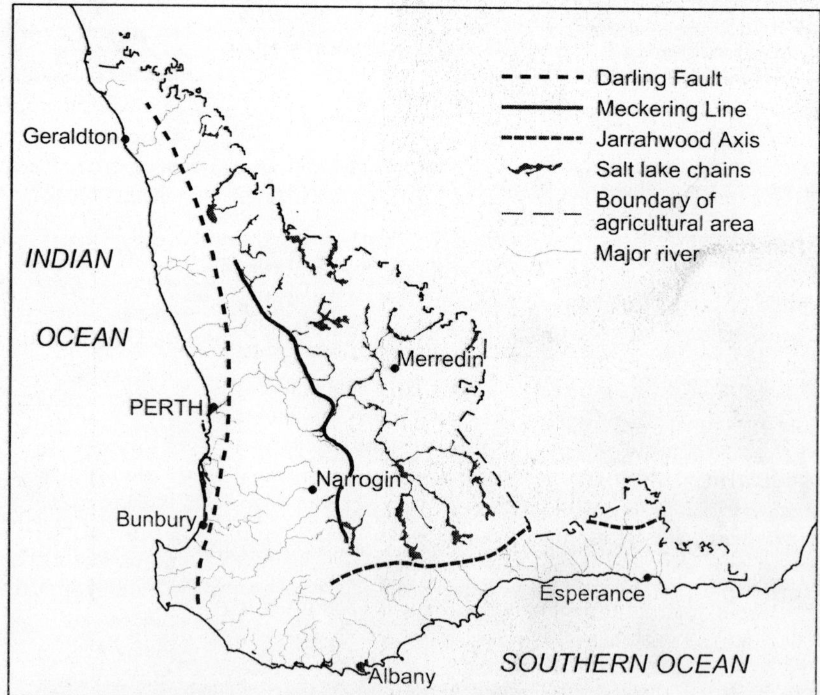

Figure 2.2.1 *The major geological features and drainage divisions of south-western Australia.*

Table 2.2.1 Main parent materials and intrinsic properties of soils of south-western Australia.

Parent material	Main soils[1]	Soil properties			Distribution
		General	Inherent nutrient deficiencies[2]	Clay mineralogy[3]	
Acid igneous and metamorphic rocks (e.g. granite, gneiss) weathered *in situ*	Red and brown, shallow and deep loamy duplex soils, red loamy earths Friable red/brown loamy earths Brown loamy earths Red and grey deep sandy duplex soils	Gritty (quartz, feldspars), acid to neutral pH	N, P, (S) occasionally Cu, Zn and Mo.	Kaolinite[4] (80-100%), illite (0-20%)	Widespread including Zone of Rejuvenated Drainage (e.g. Avon Valley), Chapman Valley, Pemberton, Zone of Ancient Drainage
Basic rocks (e.g. dolerite, greenstone, gabbro) weathered *in situ*	Alkaline red shallow and deep loamy duplex soils Red/brown cracking clays	High clay, low quartz (sand) content, red or brown colour, neutral to alkaline pH	N, P.	Kaolinite[4] (40-80%), illite (0-25%), smectite (0-40%), haematite/goethite (0-15%)	Single, or nests of dykes in many areas: especially Zone of Rejuvenated Drainage and Chapman Zone. Broad areas near Southern Cross and Ravensthorpe Ranges.
Laterite on granite (and gneiss etc)	Sandy gravels, loamy gravels Shallow gravels Yellow deep sands Acid yellow sandy earths Yellow sandy earths	High Fe and Al, deeply weathered profile	N, P, (S), Cu, Zn, Mo (if highly acidic), occasionally Mn.	Kaolinite[4] (60-90%), inhibited vermiculite (10-40%)	Widespread including sandplains of Zone of Ancient Drainage, Darling Range, remnants in many areas.
Laterite on sedimentary rocks	Sandy gravels, shallow gravels Yellow and red deep sands Pale deep sands	High iron, deeply weathered profile	N, P, (K, S), Cu, Zn, occasionally Mn, Mo (if highly acidic).	Kaolinite[4] (>80%), inhibited vermiculite (0-20%)	Dandaragan and Blackwood Plateaux, Collie Basin, south coast.
Shallow colluvium derived from laterite overlying truncated lateritic profile (i.e. mottled or pallid zone)	Grey shallow and deep sandy duplex soils Sandy gravels, loamy gravels	High spatial variability of soil properties, formed from pre-weathered material	N, P, (K), Cu, Zn, occasionally Mo, Mn.	Kaolinite[4] (90-100%), illite (0-10%)	Zone of Ancient Drainage, Zone of Rejuvenated Drainage.
Alluvium	Brown deep sands Brown sandy and loamy earths Sandy duplex soils Cracking clays	Highly variable due to differences in source materials and age	Variable depending on source materials but often N, P, possibly K, Cu and Zn.	Variable depending on source	Widespread including valley floors in Zone of Ancient Drainage, Zone of Rejuvenated Drainage, Swan Coastal Plain.
Aeolian	Pale and yellow deep sands Calcareous deep sands	Coarse-textured, low fertility	N, P, K, (S), Cu, Zn, Mn.	Kaolinite	Swan Coastal Plain, scattered areas in West Midlands.
Parna (Aeolian deposits from playas and arid interior).	Calcareous loamy earths (lunettes, gypsiferous dunes)	High $CaCO_3$, high silt and fine sand content	Low N, P, Zn, occasionally Mn, high K.	Kaolinite[4] (10-60%), illite (30-55%), smectite (0-20%)	Valley floors in Zone of Ancient Drainage, Salmon Gums Mallee Zone.

1. *Soil groups of Western Australia (Schoknecht 1997). Refer to Section 2.1.*
2. *Nutrient deficiencies which may occur, but are strongly influenced by land use and/or management, are shown in parentheses (e.g. S is related to superphosphate history, K to product removal).*
3. *Clay mineralogy from Singh (1991), McArthur (1991), L.G. Rogers (unpublished data), N. Lantzke (unpublished data). For more information on clay mineralogy refer to Chapter 10.*
4. *Kaolinite includes halloysite.*

LATERITE PROFILE

Many soils in south-western Australia are *lateritic** i.e. formed on laterite, truncated laterite (mottled, pallid zones), or colluvium and alluvium from laterite. Laterisation is consequently of widespread importance in soil development, soil properties and hydrogeology.

Remnants of a lateritic breakaway (G. Moore).

Laterisation results from intense weathering and fluctuating, high watertables and has commonly been associated with tropical areas. A number of theories concern laterisation and ferricrete formation (e.g. McFarlane 1976), but it is widely believed that the lateritic gravel in WA is an iron-enriched illuvial horizon which formed from weathering *in situ* when the climate was wetter and warmer (e.g. Mulcahy 1967). It is unlikely that all the laterite formed in the same period and it may still be forming in limited areas. The depth of weathering of the laterite profile is highly variable (2 to 50 m) and depends on a number of factors including local drainage conditions and properties (porosity, mineralogy) of the basement rock (Bettenay *et al.* 1980).

The classic horizons of a laterite profile are: *sandy and/ or gravelly topsoil, duricrust* and *mottled zone* over basement rock. There may also be a *pallid zone* and *saprolite*. The horizons are progressively more weathered towards the surface and their properties are described in Figure 2.2.2. Large areas of the intact laterite profile are only preserved in areas such as the Darling Range and the sandplain of the eastern wheatbelt. In many other areas, breakaways or small flat-topped mesas are the only remnants.

Granite is the predominant rock on which laterite has formed. There are also significant occurrences on sedimentary rocks and some on dolerite and metamorphic rocks. Laterite can also form on mixed parent materials, such as sediments overlying granite. The type of parent rock influences many soil properties and the hydrogeology. For instance, laterised dolerite can act as a barrier to groundwater movement because the mottled and pallid zones have low permeability (Johntson *et al.* 1983). Also, tree roots grew deeper into the pallid zone of laterised granite than laterised dolerite (Johnston *et al.* 1983).

The laterites are predominantly relict features and in many areas have been eroded extensively. Soils formed on truncated laterite, although younger, are formed on preweathered materials and reflect this in properties such as nutrient status.

Laterites in WA have been studied extensively with respect to their distribution and origins (e.g. Stephens 1946; Playford 1954; Mulcahy 1960, 1967), mineralogy and weathering (Gilkes *et al.* 1973; Davy 1979; Gilkes and Suddhiprakaran 1981; Singh 1991), fertility (Robson and Gilkes 1981) and hydrology (Johnston *et al.* 1983). However, much remains to be understood.

* *Laterite is used to describe deeply weathered profiles with a surface ferruginous horizon, usually underlain successively by mottled zone, pallid zone, saprolite and parent rock. However, lateritic profiles do not always follow this pattern and in some laterites formed on rock weathered* in situ *the pallid zone is absent (e.g. Gilkes* et al. *1973). The term* laterite *is derived from* 'later' *which means brick. Laterite is used as a building material in India which is soft when quarried but sets hard on exposure (Prescott and Pendleton 1952).*

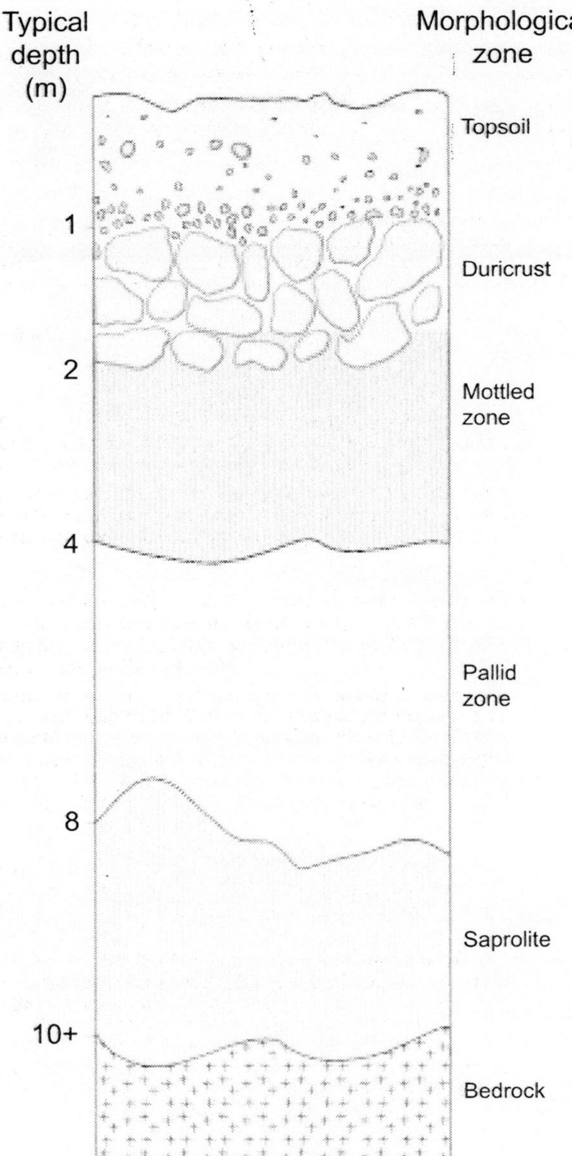

Description

Sand to sandy loam surface composed of quartz sand and containing varying amounts of loose ferruginous gravel.

Low clay content.

Accumulations of ferricrete which may be indurated (caprock, duricrust) or nodular (gravel). Often exposed as breakaways, because the duricrust is more resistant to erosion than the underlying clay. Loose gravel may exist above and/or below the cemented gravel.

High concentration of iron and aluminium oxides (goethite, haematite, gibbsite), low clay content.

Pale coloured (brown to grey) sandy clay loam to sandy clay, with red (haematite), or yellow to orange (predominantly goethite), mottles. These mottles may be slightly indurated and are precursors of concretions. Some loose gravel may be present, decreasing with depth.

Clay content is 30-40%, major clay minerals are kaolinite and halloysite, pH_w 4-5. The fabric of the parent rock is usually not evident, although quartz veins may appear.

Pallid zones and saprolite are only present when the laterite has formed from rock weathered in situ. *Distinctive pale (white to light grey) sandy clay loam to sandy clay often exposed during dam construction. Can be 30-100 m thick.*

Clay content is 30-40% (decreasing with depth), major clay minerals are kaolinite and halloysite. Often retains the fabric of the parent rock, bulk density 1.6-1.7 g/cm^3 (McCrea 1987). Water movement is mainly as saturated flow through preferred pathways, as demonstrated by Johnston *et al.* (1983) using fluorescent dyes. May have a low pH and high salt content (Dimmock *et al.* 1974).

Partially weathered rock in which the original fabric is still evident. A very thin layer in laterised dolerite. Developed by isovolumetric weathering in situ thus the clay (mainly kaolinite) retains the fabric of the underlying rock.

The clay content (10-20%) decreases with depth as unaltered primary minerals increase (mainly feldspar). Major clay minerals are kaolinite and halloysite (Gilkes *et al.* 1973).

Basement rock (or country rock), commonly granite (e.g. most of Yilgarn Craton), but can also be sandstone (e.g. Perth Sedimentary Basin), or occasionally dolerite and metamorphic rocks.*

Figure 2.2.2 *An idealised laterite profile with a description of the properties of each horizon.*

** The underlying rock can frequently be inferred from the colour and abundance of quartz in laterite:*

Granite *typically contains 25-30% quartz, which is considerably higher than the basic rocks, so soil textures are gritty with lower clay content (pallid zone 30-40%) compared with laterite formed on basic rocks.*

Dolerite *gravels and duricrust are redder with low quartz content (typically <10%), floaters of unweathered dolerite may be present. Textures grade from clay loam to clay down the profile, with 50-90% clay in the pallid zone. An abrupt transition between pallid zone and bedrock is common, i.e. very thin saprolite layer <10 cm (Davy 1979; Anand and Gilkes 1984; McCrea 1987).*

Sedimentary rock (e.g. sandstone) *gravel/duricrust can be high in quartz (especially evident when a fresh face is exposed). Colour is likely to vary with parent rock. In general, clay contents are lower and the dominant clay is kaolinite. Illite may be present in the less weathered zones (e.g. laterised sandstone developed on marine sediments containing glauconite at Dandaragan; Churchward 1970; Singh 1991).*

DISTRIBUTION OF PARENT MATERIALS AND SOILS

A system of provinces and zones can be used to describe the pattern of soil parent materials and therefore soils in south-western Australia. Bettenay (1983) identified five provinces which can be further subdivided into zones and systems (Purdie 1996). This section briefly describes the five provinces and the zones within each (Figure 2.2.3).

PROVINCES AND ZONES

Avon Province

The Avon Province lies largely on the Yilgarn Craton, (part of a large Archaean shield) and most of the area consists of lateritic plateau, or its remnants, overlying a granitic basement. There are seven zones including the Zone of Ancient Drainage (Figure 2.2.4) and the Zone of Rejuvenated Drainage (Figure 2.2.5) which cover a significant proportion of the land cleared for agriculture. In the south-west, significant areas of native forest remain.

Zone of Ancient Drainage

An extensive undulating plain characterised by a gently undulating plateau, with wide divides, long gentle sideslopes and broad valley floors (2-10 km wide), which have been in-filled by alluvium and colluvium. Drainage is into chains of salt lakes on valley floors which are a remnant of an ancient drainage system that flows only in very wet years. The valley floors have very low gradients, often 1:1,500 or less. The western boundary is the Meckering Line (Mulcahy 1967).

Soils are mainly formed on laterite, truncated laterite, parna (mainly from lake beds), rock weathered *in situ* (granite, greenstone, dolerite), colluvium and alluvium. There is a regular soil pattern on divides, sideslopes and valley floors (Figure 2.2.4). Soils are predominantly yellow deep sands and sandy gravels on the catchment divides, with grey shallow and deep sandy duplex soils on the valley slopes, and calcareous loamy earths and alkaline shallow duplex soils on the valley floors.

Land resource information is available for Merredin (Bettenay and Hingston 1961, 1964), Kellerberrin (McArthur 1992), Bencubbin (Grealish and Wagnon 1995), Northam (Lantzke 1992; Lantzke and Fulton 1993) and Lake Cairlocup (Harper 1994).

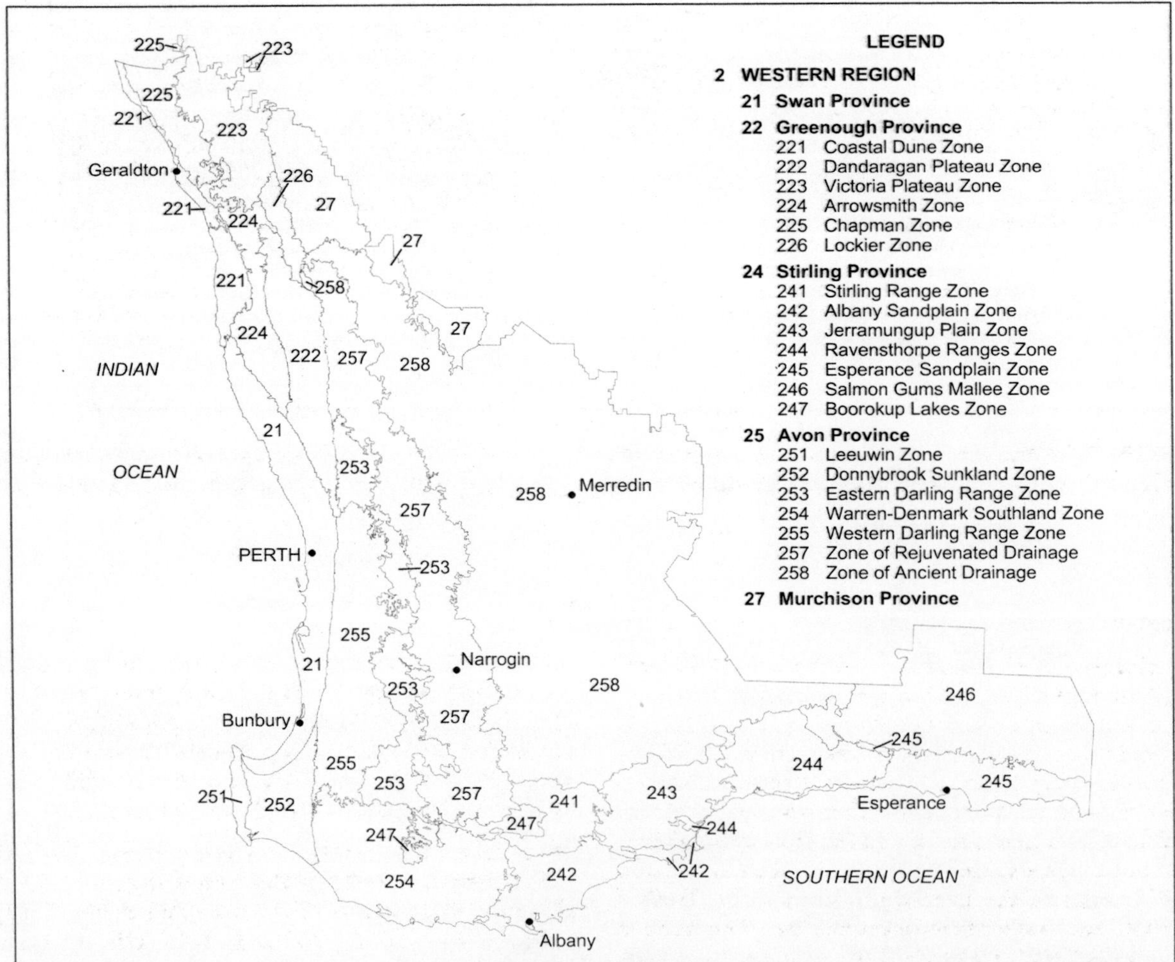

Figure 2.2.3 The provinces and zones of south-western Australia.

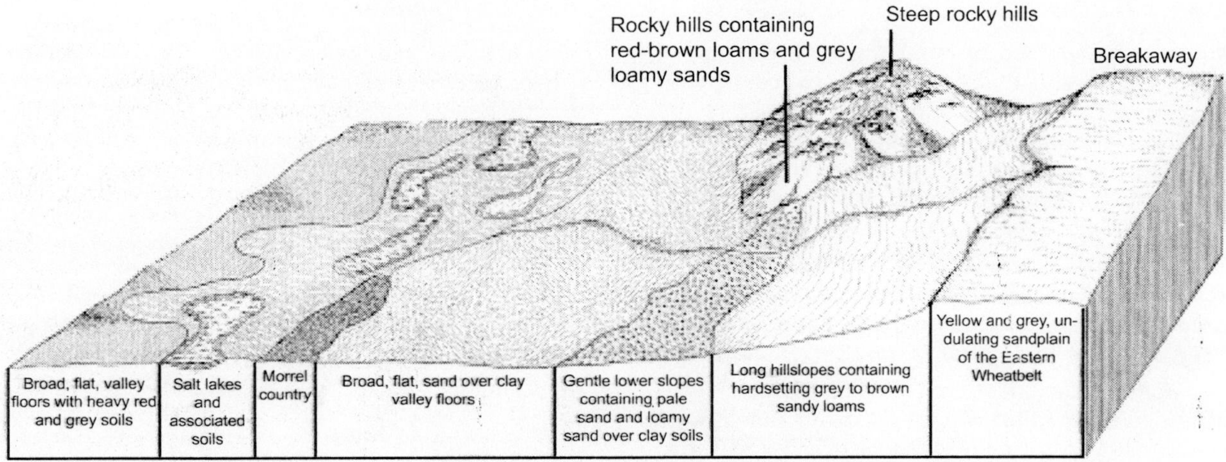

Figure 2.2.4 *Idealised block diagram showing the soils in the Zone of Ancient Drainage in relation to landscape position (from Lantzke and Fulton 1993). This diagram was developed for the Cunderdin-Tammin area. The proportion and types of soil may vary in other areas.*

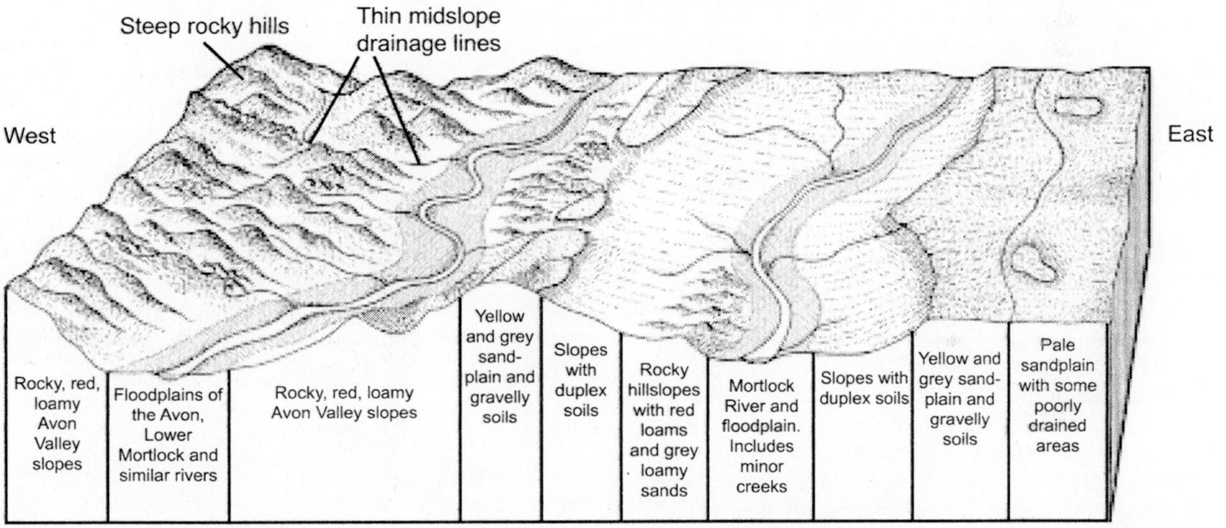

Figure 2.2.5 *Idealised block diagram showing the soils in the Zone of Rejuvenated Drainage in relation to landscape position (from Lantzke and Fulton 1993). This diagram was developed for the Northam area. The proportion and types of soil may vary in other areas.*

Zone of Rejuvenated Drainage

Gently inclined rises and low hills are mainly formed on dissected laterite, with narrow divides and only small scattered lateritic remnants. Some areas formed on acid igneous and metamorphic rocks weathered *in situ* (e.g. Avon Valley). It is more dissected than the Zone of Ancient Drainage, with narrower valley floors which contain creeks and rivers that flow every winter. Most of the landscape has formed on the mottled and pallid zones of the laterite profile, or from freshly exposed rock rather than intact laterite. This zone is bounded by the Meckering Line in the east and the Darling Range to the west.

At a superficial level there is a straightforward soil pattern from the uplands to the valley floors. But there is a high spatial variability of important soil properties affecting production (e.g. subsoil permeability), especially on the sideslopes, because of differential erosion and deposition. Refer to Figure 2.2.5.

Land resource information is available for York (Mulcahy 1959; Mulcahy and Hingston 1961), Northam (Lantzke 1993; Lantzke and Fulton 1993), Murray Valley (McArthur *et al.* 1977) and Tutanning Nature Reserve (Nyagba 1976).

Eastern Darling Range Zone

A gently undulating to undulating lateritic plateau, which is partly dissected by eastward flowing streams to form shallow valleys with broad (1 to 3 km) flat valley floors. Tributary streams flow into the major rivers which dissect the Darling Range. This zone forms an intergrade between the Western Darling Range Zone and the Zone of Rejuvenated Drainage.

Soils are mainly formed on laterite (over granite), truncated laterite, rock weathered *in situ* (granite), colluvium and alluvium. Soil pattern is closely related to topography and degree of erosion. On the uplands, sandy gravels and loamy gravels are dominant, while on the valley slopes there is a range of soils, depending on whether they are formed on truncated laterite (sandy gravels, grey deep sandy duplex soils, pale deep sands) or granite weathered *in situ* (brown loamy duplex soils, red deep sandy duplex soils).

Land resource information is available for west of Bolgart-Beverley (Lantzke and Fulton 1993), Murray Valley (McArthur *et al.* 1977) and the southern Darling Range (Tille 1996).

Western Darling Range Zone

An extensive undulating lateritic plateau (Darling Plateau) which is largely intact. The plateau has deeply incised valleys where it has been dissected by the major river systems of the inland zones. The western boundary is the Darling Scarp. To the east there is a gradual increase in dissection as this zone grades into the Eastern Darling Range Zone.

Soil pattern relates to the amount of erosion and valley forms (Mulcahy *et al.* 1972). Soils are mainly formed on laterite (mainly over granite, some dolerite) and colluvium (mainly from laterite) on the plateau and from rock weathered *in situ* (granite, gneiss, dolerite) in the incised valleys. On the plateau surface, gravels overlying duricrust are dominant (e.g. shallow gravels, moderately deep sandy gravels and loamy gravels). Pale deep sands and yellow deep sands are also found. Loamy earths and brown loamy duplex soils have formed on freshly exposed rock in the valleys.

Land resource information is available for the Darling Range (Mulcahy *et al.* 1972; Churchward and McArthur 1980; King and Wells 1990), Wungong area (Churchward and Batini 1975), Helena River catchment (Mulcahy *et al.* 1972), Collie (Bettenay *et al.* 1980), Cobiac Valley (Siradz 1985), Bridgetown-Greenbushes (Finkl 1971; Finkl and Churchward 1976) and the southern Darling Range (Tille 1996).

Donnybrook Sunkland Zone

Level to gently undulating lateritic plateau formed on sedimentary rocks of the Perth Basin. Shallow valleys occur where the plateau has been dissected by the Blackwood, Capel and Preston Rivers. Includes the poorly drained Scott River Plain in the south.

Soils are mainly formed on laterite (over sedimentary rocks), colluvium (mainly from laterite) and alluvium. Sandy gravels with some small areas of shallow gravel, pale deep sands and yellow deep sands on both the plateau and the dissected areas. On the Scott River Plain, pale deep sands, brown deep sands and wet soils are common.

Land resource information is available for Nannup (Churchward 1992), Wellington-Blackwood (Tille 1996) and Busselton-Margaret River-Augusta (Tille and Lantzke 1990).

Warren-Denmark Southland Zone

This zone falls gently from the southern edge of the Blackwood River catchment to the south coast. It includes lateritic plateau, dissected terrain formed on granite and gneiss, and swampy coastal plains. Some areas of the plateau are overlain by sedimentary deposits.

Soils are formed on laterite, colluvium (mainly from laterite), rock weathered *in situ* (gneiss, granulite) and alluvium. Dominant soils are loamy gravels and sandy gravels with pale deep sand and deep sandy duplex soils on the plateau. Red-brown and yellow-brown loamy gravels and friable red-brown loamy earths are found on rock weathered *in situ*. Pale deep sands, brown deep sands and wet soils are found on the poorly drained flats.

Land resource information is available for Manjimup (Churchward 1992), Northcliffe to Albany (Churchward *et al.* 1988) and Pemberton (McArthur and Clifton 1975).

Orchards and grazing country at Mullalyup, a dissected valley through the lateritic plateau in the Western Darling Range Zone (P. Tille).

Leeuwin Zone

This zone has formed on the Leeuwin Block, which is dominated by a lateritic plateau. The plateau has been dissected by a number of valleys, exposing the underlying granitic basement. There are coastal dunes and limestone ridges on the western margin. Extends between Cape Naturaliste and Cape Leeuwin.

Soils include gravels, friable red-brown loamy earths, yellow deep sands and deep calcareous sands (Tille and Lantzke 1990).

Greenough Province

The Greenough Province lies west of the Avon and Murchison Provinces, extending north from Muchea to the Murchison River. Most of the Province is a gently undulating landscape formed on dissected laterite overlying the Perth Sedimentary Basin. There are extensive areas of sandplain. The western margins are often dissected. Six zones have been identified in this province, which has been cleared extensively for agriculture.

Coastal Dune Zone

Coastal dunes and low hills parallel to the coast which have been breached and in-filled by alluvium from the Irwin and Greenough Rivers. A narrow strip along the west coast which extends about 200 km north from Greenhead to Horrocks Beach. Similar to the Swan Coastal Plain to the south, but Bassendean dunes are absent and the alluvial deposits are more recent.

It can be divided into three distinct areas. There is a regular soil pattern inland from the coast with coastal dunes (calcareous deep sands, yellow deep sands, shallow sands), alluvial plains (self-mulching grey clays, massive cracking clays, brown deep sands), and the Eneabba Plain of deep sandy duplex soils, pale deep sands, yellow deep sands and wet soils (e.g. at Geraldton; Rogers 1996).

Arrowsmith Zone

Dissected lateritic sandplain with spillway sands and gently undulating to undulating dissected lateritic plateau with broad valleys. The narrow valley floors are in-filled with wind blown sand and the area is mainly internally drained. In remnant plateau areas, outcrops of duricrust and breakaways are common, together with some flat-topped mesas. Where most of the laterite profile has been stripped away there are gently undulating rises to rolling low hills. The Dandaragan Scarp forms the eastern boundary, while in the south the Gingin Scarp forms the western boundary.

Soils are mainly formed on colluvium from laterite overlying sandstones, shales and siltstones with small areas on rock weathered *in situ*. Shallow gravels are found on the ridges while sandy gravels, pale deep sands, yellow deep sands and deep sandy duplex soils occur on the slopes. The depth of colluvial sands can be highly variable.

Dandaragan Plateau Zone

Gently undulating to undulating dissected lateritic plateau with broad, U-shaped valleys (overall relief 80-150 m) and sand-filled drainage lines. Extensive areas of sandplain with some breakaways. South of Dandaragan, the dissection of the western margin of the plateau has formed the Gingin Scarp and a number of smaller, more V-shaped valleys.

Soils are mainly formed on colluvial material from laterite (over sandstone), truncated laterite, rock weathered *in situ* (ferruginous sandstone) with minor areas formed on chalk (Gingin). The soil pattern is closely related to topography (Churchward 1970). Brown deep sands, yellow deep sands, pale deep sands, sandy gravels and shallow gravels are dominant, with red deep sandy duplex soils on the valley floors.

Land resource information is available for Dandaragan (Churchward 1970) and Gingin (Hosking and Greaves 1936).

Victoria Plateau Zone

A very gently to gently undulating lateritic sandplain, which is largely intact. The south is internally drained, while in the north the plateau has been weakly dissected by a type of relict drainage system resulting in long gentle slopes and alluvial surfaces. It is bound in the east by the Darling Fault and to the north-west by the Hardabut Fault. It is less stripped than the Dandaragan Plateau Zone, with fewer lateritic breakaways and sand-filled drainage lines.

Soils are mainly formed on laterised sedimentary deposits, with surface deposits of eluvial and colluvial sand 2.5-5.5 m thick. In the weakly dissected shallow valleys, alluvium and colluvium from Yilgarn Craton overlie red-brown hardpan. Dominant soils are yellow deep sands, yellow earthy sands, pale deep sands (more common in the south-west), also some pockets of red deep sands. There are red sandy earths and red-brown hardpan shallow loams on the broad shallow valleys (Rogers 1996).

Chapman Zone

Gently undulating to rolling low hills characterised by red loamy soils with some distinctive flat-topped mesas and ranges. Formed on the granulites, granites and migmatites of the Northampton Block which has been intruded by numerous dolerite dykes. Includes the Chapman Valley and the area around Northampton.

Soils are mainly formed on rock weathered *in situ* (granite, granulite, migmatite, dolerite), laterised sedimentary rocks (which have been eroded to form mesas), colluvium with minor alluvium. Dominant soils are red shallow loams, red loamy earths, stony soils and red loamy duplex soils, with smaller areas of red sandy earths and red shallow sandy duplex soils (Rogers 1996).

Harvesting wheat on rolling hills in the Chapman Zone.

Lockier Zone

Central portion of the Greenough Province consisting of alluvial valley plains with gently undulating country to the north where Permian glacial tillite occurs. This zone is drained by the Irwin, Lockier and Arrowsmith Rivers and is underlain by Proterozoic, Permian and Jurassic sediments and Precambrian granitic rocks. A few flat-topped mesas are outliers of the Victoria Plateau.

Dominant soils are hardsetting grey non-cracking clays and red-brown cracking clays. Includes the highly erodible soils in the Nangetty area (Rogers 1996).

Murchison Province

An extensive level to gently undulating plain, with residuals of the lateritic profile, overlying the Archaean granite and gneiss of the Yilgarn Craton. The widespread presence of red-brown hardpan (also called Wiluna hardpan) underlying shallow loamy soils is the main distinguishing feature between the Murchison Province and the Avon Province to the south. Only a small portion falls within the agricultural area.

Stirling Province

A strip along the south coast about 120 km wide from Torbay (just east of Albany) to Israelite Bay on the Great Australian Bight. Mostly formed over a basement of Archaean and Proterozoic granite, gneiss and migmatite which has been sporadically overlain by Tertiary marine and continental deposits. Much of the area has relatively low relief and has been extensively laterised. The northern boundary is the Jarrahwood Axis which separates drainage lines flowing north into the Swan-Avon catchment from the southerly flowing rivers. While much of this province has been cleared for agriculture, significant areas of native vegetation remain.

South coast sandplain

A level to gently undulating sandplain, which forms a narrow strip about 80 km wide along the south coast. Dissected by a number of short rivers flowing into the Southern Ocean. Some areas (e.g. sandplains east of Esperance) are internally drained into swamps. Can be divided into three distinct zones: Albany sandplain, Jerramungup plain and Esperance sandplain, the latter two zones being separated by the Ravensthorpe Ranges.

Albany sandplain. Dominant soils are grey deep sandy duplex soils and grey shallow sandy duplex soils, which are often alkaline and sodic. Sandy gravels and pale deep sands are also common.

Jerramungup plain. Dominant soils are alkaline grey shallow sandy duplex soils, alkaline grey deep sandy duplex soils, grey non-cracking clays and alkaline grey shallow loamy duplex soils. Sandy gravels and pale deep sands are also found.

Esperance sandplain. A sheet of fine sand of varying thickness overlying gravel or clay. Grey deep sandy duplex soils and grey shallow sandy duplex soils are dominant, often containing gravel. Pale deep sands are found on sand sheets and low dunes.

Land resource information is available for Esperance (Overheu *et al.* 1993) and Many Peaks (Teakle 1953).

Sheep grazing on the Esperance sandplain with coastal dunes in the distance (Part of the South Coast sandplain).

Ravensthorpe Ranges Zone

Rolling to undulating low hills formed on a greenstone belt of mafic and ultra-mafic rocks with distinctive red fine-textured soils. There is a rejuvenated drainage system with southerly flowing rivers. Centred on the town of Ravensthorpe and extends from West River to Jerdacuttup River.

Soils are mainly formed on rock weathered *in situ* (greenstone) and colluvium with minor areas of alluvium. Dominant soils are red cracking clays and alkaline red shallow loamy duplex soils. Stony soils are common on the Barren Ranges.

Salmon Gums Mallee Zone

A level to gently undulating plain with numerous salt lakes characterised by alkaline soils and mallee vegetation. Formed on ancient granitic rocks, uncomformably overlain by marine sediments. Defined drainage lines and rock outcrops are largely absent.

Soils are mainly formed on marine sediments, rock weathered *in situ* (granite) and parna (mainly carbonates from playas and the arid interior). Dominant soils are alkaline shallow sandy duplex soils and calcareous loamy earths (sometimes with a fluffy, powdery surface) and alkaline grey or red-brown finely textured soils. Calcium carbonate is present in most profiles. Some areas (e.g. 24% of Salmon Gum survey; Burvill 1988) have a very complex mosaic of soils, with 3 or 4 distinct types within a 50 m radius.

Land resource information is available for Salmon Gums (Burvill 1988) and Mount Beaumont (Scholz and Smolinski 1996).

Boorokup Lakes Zone

Poorly drained flats and gently undulating terrain with prominent lake systems, formed on Eocene sedimentary deposits. Extends from the North Stirlings to Lake Muir. Includes deep sandy duplex soils and wet soils.

Stirling Range Zone

Includes the Stirling Ranges, which are formed on Proterozoic metasediments, and gently undulating to undulating terrain formed over Archaean granitic rocks in the upper Pallinup Catchment. Soils range from stony soils to shallow sandy duplex soils and hardsetting grey clays.

Swan Province

Corresponds to the Swan Coastal Plain, a narrow (<40 km wide) plain covered by sedimentary material. The eastern boundary follows the Whicher Scarp in the south, then the Darling Scarp from Burekup to Muchea, then north-west along the Gingin Scarp. This province has been extensively cleared and supports most of Western Australia's population.

Three dune systems run parallel to the coast, with alluvial plains lying inland. Along the coast are beach ridges and parabolic dunes (the Quindalup dunes). Behind these lie dunes of siliceous sands overlying limestone (the Spearwood dunes). The third system (the Bassendean dunes) is a complex of low dunes, sandplains, poorly drained plains and swampy flats. On the eastern side of the coastal plain are flat and often poorly drained alluvial plains (Pinjarra Plain), which meet the gentle footslopes of adjoining scarps (Ridge Hill Shelf) along the eastern edge.

Sheep grazing on flats with the Stirling Ranges in the background (Stirling Range Zone).

Soils are formed on various materials (e.g. calcareous sands, limestone, alluvium, laterite). There is a regular pattern corresponding to each geomorphic element (e.g. McArthur and Bettenay 1960).

Quindalup dunes. Calcareous deep sands.

Spearwood dunes. Dominant soils are yellow deep sands, yellow to brown shallow sands.

Bassendean dunes. Dominant soils are pale deep sands with brown deep sands on the poorly drained plains. Areas of wet sand and bog iron soils are also found in swamps.

Pinjarra Plain. Dominant soils are deep sandy duplex soils, shallow sandy duplex soils, brown sandy earths, brown shallow loamy duplex soils and brown loamy earths. Minor areas of poorly drained cracking clays and red-brown non-cracking clays along the western margins. Brown sandy earths and brown loamy earths on recent alluvium.

Ridge Hill Shelf. Dominant soils are sandy gravels, yellow deep sands and pale deep sands.

Land resource information is available for the Swan Coastal Plain (Bettenay *et al.* 1960; McArthur and Bettenay 1960), Swan Valley (Pym 1955), Mandurah (Wells 1989), West Gingin (Smolinski and Scholz 1997) and Busselton (Tille and Lantzke 1990).

Further reading

Anon. (1988). 'Soils of south-western Australia'. Ministry of Education, Western Australia, 64 p.

McArthur, W.M. (1991). 'Reference soils of south-western Australia'. Soil Science Society of Australia Inc. (WA Branch), Perth, Western Australia.

2.3 DISTINCTIVE MORPHOLOGICAL FEATURES AND THEIR AGRICULTURAL SIGNIFICANCE

Geoff Moore

A description of soil morphology (soil profile description) provides information which can be used to formally classify soils, but its significance for soil management and crop production is often poorly understood. This section describes the agricultural significance of field texture, soil colour, mottles, ferruginous gravel and subsoil structure.

FIELD TEXTURE

Soil texture describes the proportions of sand, silt and clay (the particle size distribution). Sands are mineral particles with a size range of 2 to 0.02 mm, silt from 0.02 to 0.002 mm, and clay particles are smaller than 0.002 mm. Textures range from sand (<5% clay) through loam to heavy clay, which has more than 60% clay-sized particles.

The field (or hand) texture is a measure of the behaviour of a small handful of soil when moistened and kneaded into a ball (bolus) and then pressed out between thumb and forefinger to form a ribbon. The behaviour of the soil during bolus formation and the ribbon length determine the field texture grade (Table 2.3.1). Field texture is both an estimate of the particle size distribution as measured in the laboratory and a useful measure in its own right. Field textures should approximate the particle size analysis, but are influenced by organic matter, clay mineralogy, the amount of sodium on the exchange sites of the clay and the presence of calcium carbonate. In general they are a good guide to soil behaviour.

Texture is useful for predicting soil behaviour in terms of water availability (profile hydrology, available water capacity, waterlogging) and erodibility; important soil properties in the Australian environment. It is also significant in relation to other soil properties, including water repellence, nutrient deficiencies, nutrient leaching, subsurface compaction, soil structure decline and pH buffering capacity. There is not always a direct relationship between soil texture and plant growth. Rather, soil texture can be used to infer susceptibility to a particular process.

The *specific surface area* (the total area of all exposed particle surfaces) is closely related to the clay content, because clay-sized particles have a surface area which is two to four orders of magnitude greater than sand-sized particles. The close correlation between particle size and surface area results in soil texture providing a reasonable estimate of the specific surface area (not a routine laboratory measurement). Specific surface areas for various particle sizes and clay mineralogy are listed in Table 2.3.2. Specific surface area is an important property, affecting many chemical reactions in the soil, nutrient absorption, the amount of water held in the soil at lower storage limit (i.e. wilting point; Section 3.3) and susceptibility to water repellence (Section 3.1).

SOIL COLOUR

Colour is a very distinctive feature and may be diagnostic of other features. Colour is mainly due to the presence of iron oxides and organic matter. Organic matter consists of darkly coloured compounds, which if present in any quantity, tend to mask the colours of iron oxides. The presence of manganese oxides also darkens the soil. In a few soils the colour is derived directly from the parent material.

Iron is a component of almost all rocks and soils. Under aerobic conditions the ferric (Fe^{3+}) oxides, formed by oxidation and hydrolysis are relatively stable and immobile. Under anaerobic conditions mottles can form (refer to *Mottles* below).

Aerobic soils

Organic matter in the surface layer (A1 or Ap horizon) masks the colour of iron oxides, but in lower layers the iron oxide is dominant. In general, colour is related to the type(s) of oxides present, not the amount (Bigham *et al.* 1978). For example, the haematite content of the soil usually determines the redness.

The type of iron oxides present can be inferred from the soil colour.

In aerobic soils, the iron oxides goethite, haematite, maghemite, lepidocrocite and ferrihydrite have a distinctive range of colours from yellow to red. They each have a specific colour range, particularly with

Table 2.3.1 Determination of field texture grade, together with the approximate clay content as determined by particle analysis. Texture grades can be further subdivided with texture qualifiers* (adapted from McDonald *et al.* 1990).

Texture grade	Behaviour of moist bolus	Approximate clay content	Code
Sand	Nil to very slight coherence; cannot be moulded; single sand grains adhere to fingers	<5%	S
Loamy sand	Slight coherence; can be sheared between thumb and forefinger to give minimal ribbon of about 5 mm	about 5%	LS
Clayey sand	Slight coherence; sticky when wet; many sand grains stick to fingers; discolours fingers with stain; forms minimal ribbon of 5-15 mm	5-10%	CS
Sandy loam	Bolus coherent but very sandy to touch; dominant sand grains of medium size and readily visible; ribbon of 15-25 mm	10-20%	SL
Loam	Bolus coherent and rather spongy; no obvious sandiness or silkiness; forms ribbon of about 25 mm	about 25%	L
Sandy clay loam	Strongly coherent bolus; sandy to touch; ribbon of 25-40 mm	20-30%	SCL
Clay loam	Coherent plastic bolus, smooth to manipulate; ribbon of 40-50 mm	30-35%	CL
Clay loam, sandy	Coherent plastic bolus, sand grains visible in finer matrix; ribbon of 40-50 mm; sandy to touch	30-35%	CLS
Light clay	Plastic bolus, smooth to touch; slight resistance to shearing; ribbon of 50-75 mm	35-40%	LC
Light medium clay	Ribbon of about 75 mm; slight to moderate resistance to ribboning shear	40-45%	LMC
Medium clay	Smooth plastic bolus, handles like plasticine and can be moulded into rods without fracture; moderate resistance to ribboning shear; ribbon of 75 mm or longer	45-55%	MC
Medium heavy clay	Ribbon of 75 mm or longer, handles like plasticine; moderate to firm resistance to ribboning shear	≥50%	MHC
Heavy clay	Handles like stiff plasticine; firm resistance to ribboning shear; ribbon of 75 mm or longer	≥50%	HC

** Texture qualifiers (used as a prefix e.g.* coarse sandy *clay loam) can be used to refine a texture description:*

Coarse sandy — *Coarse to touch; sand grains are seen very readily with the naked eye.*
Fine sandy — *Can be felt and often heard when bolus is manipulated; sand grains are clearly evident under a hand lens of 10 times magnification.*
Gritty — *More than 35% very coarse sand and very fine (1-3 mm) gravel.*
Gravelly — *35-70% of gravel by volume.*
Stony — *35-70% of stones by volume.*

Table 2.3.2. Specific surface areas for different particle sizes and types of clay.

Particle size (clay mineralogy)	Specific surface area (m^2/g)
Coarse sand	0.01[a]
Fine sand	0.2[a]
Silt	2[a]
Clay	
- Kaolinite	11-36[b, c, d]
- Montmorillonite	47[b]
- illite	93-185[b, c]

a *L.A.G. Aylmore personal communication.*
b *Greenland and Quirk (1962).*
c *Aylmore (1974).*
d *Aylmore and Quirk (1966).*

regard to hue, but also with value and chroma. For a given mineral the chroma tends to increase with increasing concentration of iron oxide. The crystal size also influences colour. In general, lower values and chroma are associated with decreasing crystal size (Schwertmann 1993). Refer to box titled *Iron oxides and soil colour.*

Anaerobic soils

In waterlogged soils the anaerobic zones may lack the pigmentation of iron oxides, because they have been removed through bacterial reduction. Consequently, if the soil contains negligible amounts of organic matter, the soil is white, grey or green, reflecting the colour of the matrix material.

Soil colour and drainage status

The colour of clayey subsoils can be used to infer their drainage status. In general, permeability decreases as subsoil colour changes from red → brown → yellow → dark (black to very dark brown) → grey (light greys to bluish and greenish greys). Mottled soils are less permeable than whole coloured soils (Northcote 1983). It is worth noting that other factors such as sodicity and salinity also affect permeability.

IRON OXIDES AND SOIL COLOUR

Goethite
The most common iron oxide in soils and can form wherever weathering occurs. Its presence can be recognised by hues of 10YR to 7.5YR. Cementation into nodules darkens its appearance (Schwertmann 1993). In WA, goethite is widely distributed (Anand and Gilkes 1987; Singh and Gilkes 1992) giving the distinctive brownish yellow (10YR 6/8) colour to soils in the northern and eastern sandplains.

Haematite
Recognised and named because of its distinctive 'blood red' colour. The hue varies from 5YR to 5R, depending on crystal size. With a decrease in crystal size the colour tends towards purple (Schwertmann 1993). In WA, haematite is frequently associated with goethite in soils (Singh and Gilkes 1992). The redness of soil containing haematite increases with the content of haematite, or the ratio of haematite/goethite (Bigham *et al.* 1978; Singh and Gilkes 1992). The formation of haematite rather than goethite is favoured at high soil temperature, low water activity, high biomass turnover and high Fe-release from rocks (Schwertmann 1993).

Lepidocrocite
Lepidocrocite can be recognised by its orange colour. The hue is generally 7.5YR, with values more than 6 (Fitzpatrick *et al.* 1985). It may be redder (5YR) if poorly crystalline, or can be 10YR if present in very low concentrations. (Note: A hue of 7.5YR can also be given by ferrihydrite or a mixture of goethite with a very small amount of haematite. Haematite and lepidocrocite are unlikely to occur together). Lepidocrocite requires soluble Fe^{2+} in order to form, so the soil must have reducing conditions at some time during the year. It forms from the slow oxidation of Fe^{2+} and tends to be concentrated in orange mottles, bands, root-mats and pipe-stems (Fitzpatrick *et al.* 1985). It is rare in calcareous soils.

Ferrihydrite
Can be recognised by hues of 5YR to 7.5YR with values (6 (brown to dark reddish brown). It forms where redox reactions are active and the re-oxidation is fast. It is characteristic of gleyed soils and podsolised B horizons (Schwertmann 1993).

Other iron compounds
Jarosite can be observed as yellow (5Y) mottles in acid sulphate soils (Schwertmann 1993). These are thought to be very minor in WA (Section 5.1). *Maghemite* is usually a product of fires in a tropical environment, the hues are between 2.5YR and 5YR. Can be positively identified with a magnet as it is ferromagnetic (Schwertmann 1993).

For further information refer to *Soil Color* (Bigham and Ciolkosz 1993).

Gley colours are those on the Munsell Soil Gley Chart. The colour results from the almost complete reduction and subsequent removal of iron from the soil horizon. Gleyed colours are frequently found in swamps and other poorly drained areas.

MOTTLES

Mottles are a distinctive feature of many soils and are especially common in clayey subsoils. They are frequently used as indicators of waterlogging.

Mottles are concentrations of iron oxides (Fe^{3+}) which form as a result of the redistribution of iron oxides in waterlogged soils. The Fe^{3+} oxides (which give the soil its colour) are insoluble and remain in place unless they are reduced to Fe^{2+} under anaerobic conditions. The Fe^{2+} is soluble and moves readily in the soil solution, but is precipitated as Fe^{3+} oxides when oxidised in aerobic zones e.g. when there is a fluctuating watertable. The soil does not return to a uniform colour in the short-term, because the Fe^{3+} oxides are insoluble and therefore relatively stable. This has implications when assessing the drainage status at a site.

FERRUGINOUS GRAVEL

Ferruginous gravels are a common and distinctive feature of many WA soils, the result of widespread laterisation. The gravel was formed by fluctuating watertables and precipitation of iron. It is a relict feature formed when the climate was warmer and wetter (Mulcahy 1973). Refer to *Laterite profile*, Section 2.2.

Gravel can be a relatively minor component (<10%) of soil or it can be the dominant visual feature, as high as 60 to 80% by volume. It may be cemented, in which case it is called a ferricrete pan (Section 4.1) and if continuous can form a barrier to root growth.

Gravel affects the soil's physical and chemical properties as well as susceptibility to erosion (see also Chapter 10). Ferruginous gravels are chemically reactive because the gravel contains aluminium and iron oxides. A common property associated with gravelly soils is the ability to 'fix' phosphorus. On the other hand, gravel is considered to be physically inert because it has low porosity (high bulk density), and reduces the effective soil volume available for root growth, water and nutrient storage. Its effect on various soil properties is summarised in Table 2.3.3.

MOTTLES AND OTHER INDICATORS OF WATERLOGGING. ARE THEY A RELICT FEATURE?

Gottfried Scholz

Mottles and 'white colours' are frequent in Australian soils. The mottles range from dark red (5R 2.5/6) to yellow (2.5Y 8/8) to white and occasionally an olive colour (5Y). Most are relicts of a Tertiary weathering process that formed lateritic soils. They indicate paleo-, not current, waterlogging conditions.

The best indicators of current waterlogging are either a bleached layer just above a B horizon with low hydraulic conductivity, or rust-coloured precipitates near fine roots in the Ah, Ap and upper B horizons and/or rust coloured specks (less than 5 mm in diameter) in the 'white' horizon above the clay layer. The upper parts of clayey B horizons often show rust-coloured precipitates along fine roots and in fine pores. Under prolonged waterlogging, mottles of the Munsell Soil Gley Chart appear in the B horizon.

The rust coloured specks and precipitations along fine roots are ferrihydrite, with Munsell hues of 7.5YR to 2.5YR and values <6 (Schwertmann 1993). Severe waterlogging can be recognised through the smell of hydrogen sulphide or mercaptene gases which develop through anaerobic decay of organic matter.

The lateritic white clays with their pronounced red and yellow mottles may occur in low lying areas subject to waterlogging. The mottling is not a feature of current waterlogging. The white clays can be subject to further bleaching by high watertables, but it is difficult to decide whether waterlogging is contemporary. However, the white clays have often retained a good structure (prismatic to coarse blocky), developed during initial soil formation. The surfaces of these peds commonly have fine root-mats which in waterlogged conditions show either rust-coloured precipitations or have a distinct smell through anaerobic decay of organic matter. If there is no rust colouring or smell, further information should be considered before a definite decision is made about waterlogging, for example position in the landscape, observations of perched watertables and effects on crop growth (refer to Section 3.4 for further information).

Yellow and red sands may contain soft brown nodules and diffuse, whitish mottles when dry. These features do not indicate waterlogging.

A high gravel content reduces the effective soil volume, but only prevents root growth when the gravel is cemented (G. Moore).

Gravelly soils are often discussed as being either superior or inferior to similar soils where gravel is absent, but it is worth considering the following points:

- Gravel may be an incidental feature, e.g. a gravelly layer often indicates the presence of a clayey subsoil slightly deeper in the soil. This has implications for water availability and waterlogging depending on the depth to the clayey subsoil and seasonal conditions.
- In the A horizon of some soils, the texture of the matrix in a gravelly layer often contains a slightly higher clay content than in similar soils where gravel is absent. This improves nutrient retention and may increase water storage slightly (e.g. gravelly sandy duplex soils on the Esperance sandplain).
- In the Zone of Rejuvenated Drainage, the gravelly soils tend to be higher in the landscape which has favourable implications for drainage.

SUBSOIL STRUCTURE

This section concentrates on subsoil structure which is an intrinsic property. The surface structure in many agricultural soils reflects recent management practices, especially the amount and frequency of cultivation. The soil aggregates (or clods) are largely created by cultivation rather than being an inherent soil property.

Table 2.3.3 Effect of high gravel content on soil properties.

Soil property	Explanation of effect
Phosphorus adsorption	In many soils, gravels adsorb less phosphorus than the soil matrix (i.e. soil material in between the gravel) mainly because of the comparatively low surface area of the gravel (Weaver *et al.* 1992). However, the assumption that gravelly soils have a high P adsorption is usually correct, because the matrix material is often high in Fe and Al oxides.
Soil water storage	The main effect on the physical properties is the dilution of the soil mass (Babalola and Lal 1977). Therefore, when calculating the water storage on gravelly soils, the available water capacity is reduced in proportion to the amount of gravel present. Consequently, gravelly soils can be 'droughty' and/or susceptible to waterlogging depending on the depth of the gravelly layer and the permeability of the underlying soil. A gravel layer may disrupt pore continuity, which could improve water retention in the overlying soil layer (Miller 1969; Unger 1971). This has not been studied in WA.
Susceptibility to traffic pans	A high proportion of gravel in the A horizon is likely to reduce the susceptibility to traffic compaction in coarse-textured soils.
Soil evaporation	A surface layer of gravel acts as a mulch and reduces evaporation from the soil.
Root growth	Roots will elongate through dense gravel layers, unless the gravel is cemented, when they grow horizontally until they encounter a crack or fissure. Limited field observations suggest greater water movement and root growth into the clayey subsoil in some permeability contrast soils where gravel is present in the top of the clayey subsoil (B1 horizon).
Soil erosion	Surface gravel reduces susceptibility to erosion by wind and water in the same way as anchored plant material. Gravels and stones on the soil physically cover the surface and increase its surface roughness (Section 7.1).

Soil structure has a large influence on root growth, especially in the clayey subsoils of permeability contrast soils, medium to fine-textured soils and cracking clays. For instance, on permeability contrast soils at East Beverley the depth of root growth into the clayey subsoil was the main variable affecting the grain yields between the so-called 'good' and 'poor' growth areas (Belford *et al.* 1992).

Roots grow through the soil by either pushing soil particles aside, in the pore spaces (>0.1 mm), or in the cracks between peds (structural units). In medium and fine-textured soils the ability of the roots to push their way through the soil is limited, compared with a coarse-textured soil. Therefore, the roots of annual plants are generally restricted to *structural porosity* (pore space between soil structural units or peds; Nikiforoff 1941) and *biological porosity* (pore space produced by worms, ants, roots and other biological activity in the soil). Structural properties affecting root growth in clayey subsoils are summarised in Table 2.3.4.

In pedal soils, structure is described by the grade of pedality (degree of development and distinctness of the peds), ped size and type (shape of structural units, which are illustrated in Figure 2.3.1 and described in McDonald *et al.* 1990). Subsoil structure is usually described from a soil pit or large soil core. Descriptions from an auger hole are usually inadequate. When describing structure the emphasis is on the shape, size and arrangement of the solid particles, however it is the voids or structural porosity which affect root growth. The effect of various types of subsoil structure on root growth is summarised in Table 2.3.5.

Table 2.3.4 Structural properties affecting root growth in clayey subsoils.

Structural property	Effect on root growth
Type of structure	The type of structure (pedality) has a large effect on root growth (refer to Table 2.3.5).
Pedal soils	In structured or pedal subsoils there is preferential growth around, rather than through, the peds (Winters and Simonson 1951). The degree of preferential root growth varies with the soil strength and the angle of the cracks in relation to the preferred geotropic growth direction of roots (Dexter and Hewitt 1978; Whiteley and Dexter 1983). Roots can elongate more rapidly in cracks that are narrower than the root diameter than they can through undisturbed clods without cracks (Whiteley and Dexter 1983).
Ped size	The size of peds influences root growth. There is less root growth in a soil consisting of large aggregates than a soil with small aggregates, because small peds are more easily displaced. Root tips must exert large axial pressure to penetrate large aggregates (Whiteley and Dexter 1984; Misra *et al.* 1986, 1988).
Biopores	In WA, the native vegetation is generally adapted to the soil conditions and roots are able to penetrate the poorly structured clayey subsoils. After clearing, the old root channels (or biopores) provide preferential channels for growth of agricultural plants. However, the biopores are thought to gradually close with time due to the shrink-swell properties of the clay and eluviation of material from the A horizon. This has made the clayey subsoils in south-western Australia difficult for the roots of agricultural plants to penetrate.
Variability in root growth	Root growth is often highly variable in poorly structured subsoils, because root growth is greatly influenced by the presence of biopores, sand seams and the incidence of waterlogging (e.g. Cox 1988; Tennant *et al.* 1992). This is reflected in the saturated hydraulic conductivity, which frequently varies by 1-2 orders of magnitude in clayey subsoils (e.g. Seow *et al.* 1988; Cox 1988).
Structure of moist soil	Subsoil structure is usually assessed on dry to slightly moist soil, but this may not correspond with the void space encountered by roots growing into moist soil. The relationship between void space when the soil is dry and moist is related to the type of pedality, clay mineralogy (shrink-swell potential) and the stability (exchangeable cations, soluble salts) of the structural units.
Sodic subsoils	Soils with a high proportion of exchangeable sodium are considered *sodic* if the exchangeable sodium percentage (ESP) is >6% of the cation exchange capacity. Soils with ESP >15 are considered *highly sodic* (Northcote and Skene 1972). The effect of sodicity in the subsoil is different than for surface soils, because the subsoil is not normally disturbed by machinery (i.e. not subject to remoulding). Sodic soils which contain smectite (montmorillonite) clays have shrink-swell properties. High levels of sodium favour aggregate slaking and dispersion, and coupled with the shrink-swell properties, act to reduce pore space which then reduce permeability and root growth. Root growth is also affected by the development of a perched watertable.

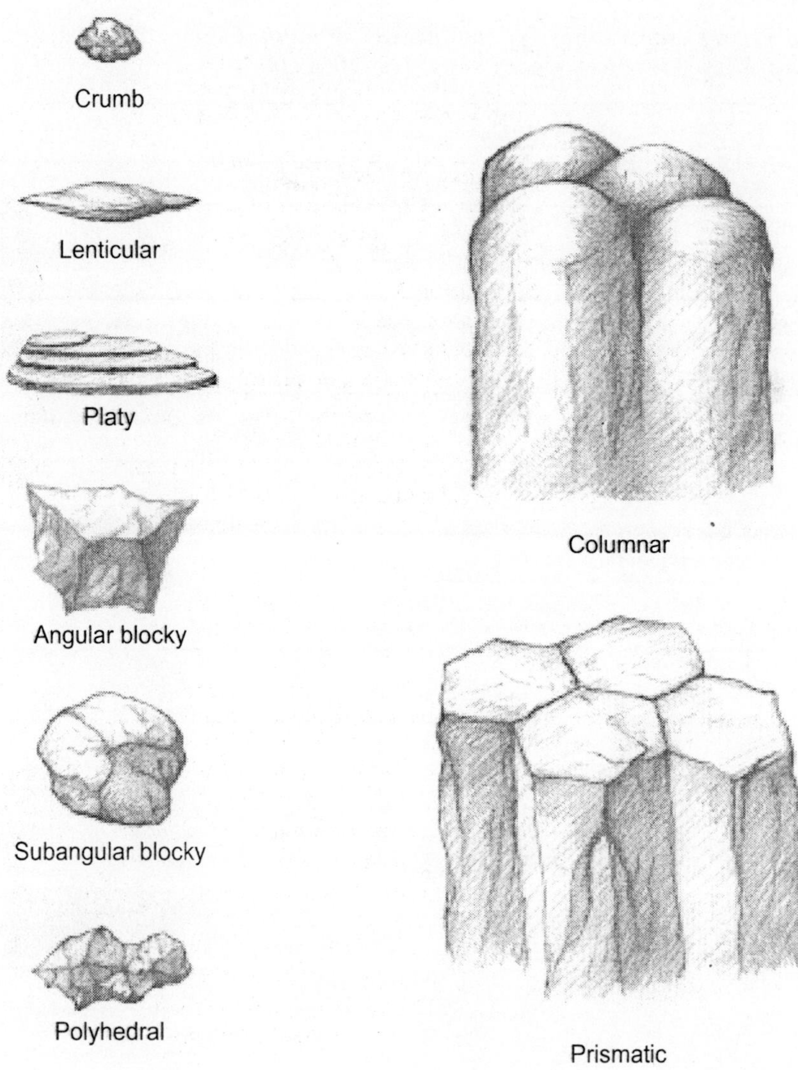

Figure 2.3.1 *The main types of soil structure (by N. Schoknecht).*

Ameliorating clayey subsoils

There are limited options for ameliorating unfavourable subsoil properties: mechanical and/or chemical treatments, or growing plants which create new pores.

Deep tillage and chemical amelioration. In high value, intensive horticultural industries (fruit and vegetables) it may be feasible to modify the subsoil porosity by deep ripping and incorporating gypsum (Jayawardane and Blackwell 1985; Jayawardane *et al.* 1995). In dryland agriculture, mechanical or chemical amelioration is unlikely to be economic. Subsoils with low to moderate porosity are likely to have problems with waterlogging. Drainage may increase production, but will not improve the porosity.

The potential for gypsum amelioration of clayey subsoils has not been widely tested on soils from south-western Australia.

Agronomic options. The concept of plant roots creating new pores (biological drilling) through weakly structured and dense subsoils and using the improved porosity to improve the root growth and yields of subsequent crops is an attractive alternative to deep tillage. Cresswell and Kirkegaard (1995) say there is little doubt that plant roots can penetrate and increase the pore size of dense subsoils. On the other hand, if the pores created are large and/or if the roots of a subsequent crop cannot penetrate the soil matrix, they think the benefits to a subsequent annual crop may be minimal.

There is limited evidence of successful biological drilling in WA. Anecdotal evidence from Jerramungup suggests lucerne (*Medicago sativa*) roots penetrated the columnar subsoil of a sodosol, resulting in improved yields of subsequent crops. In pioneering research in Victoria, lucerne is being included in crop-pasture rotations in the mid-central and western regions to improve root growth in the clay subsoils (Gardner *et al.* 1992). In WA, perched watertables in many of the permeability contrast soils or inadequate rainfall may restrict the usefulness of lucerne.

The ability of annual species with taproots to create biopores appears limited. In the Riverina region of NSW, the ability of canola to improve subsoil conditions was limited even though the roots penetrated 30 cm deeper than wheat. The increased yield of wheat following canola compared with continuous wheat was not a result of biological drilling (Cresswell and Kirkegaard 1995). Narrow-leafed lupins (*Lupinus angustifolius*) are another taprooted plant, although their ability to penetrate clayey subsoils is similar to or less than wheat (Dracup *et al.* 1992). The greater ability of lucerne and other perennial plants to penetrate clayey subsoils is probably related to the length of the growing period, rather than to differences in the rate of root penetration (Winters and Simonson 1951).

Table 2.3.5 Types of structure (pedality) and implications for root growth of annual agricultural plants. Applies to moderately to strongly structured clayey subsoils (>20% clay).

Type of structure*	Implications for root growth
Granular peds	No limitation (peds are non-accommodated**).
Subangular blocky, polyhedral	Growth will vary with the degree of structural development and size of the peds. If strongly developed structure (peds <20 mm), then root growth should be relatively unimpeded. Restricted if peds >20 mm.
Angular blocky	Root growth and aeration are likely to be impeded, even though the peds may be relatively small. This is due to the ped faces nearly all being accommodated***. Limited root growth if peds are >50 mm.
Prismatic	Will be impeded, with limited growth into horizons with a coarse (peds >100 mm) prismatic structure.
Columnar	Soil is most likely sodic. Root growth is likely to be limited to the sand seams between the domed caps of the columns. A hard layer containing some organic matter may be present on the top of the columns. Limited root growth may occur if there is some secondary structure within the columns.
Platy, lenticular	May indicate compaction when just below the cultivation layer. Vertical root growth will be impeded.

* *The types of structure are described in McDonald* et al. *(1990).*

** Non-accommodated peds. *These are not casts of the moulds formed by surrounding peds (i.e. the peds do not fit the available space closely). The shape of such peds is usually spherical, sometimes blocky. Soils with a granular structure have spherical, non-accommodating peds and high porosity (Griffiths 1986).*

*** Accommodated peds. *These fit tightly together when the soil is moistened and swells. They generally have plane or curved surfaces, which are moulds of the adjacent peds (i.e. the peds fit closely into the space created by the surrounding peds) and include peds which are platy, prismatic or blocky. Soils with a smooth ped fabric, especially those with a high lustre, are likely to fit together tightly resulting in a low void space between the peds. Soils with a rough ped fabric are likely to have higher structural porosity (Griffiths 1986).*

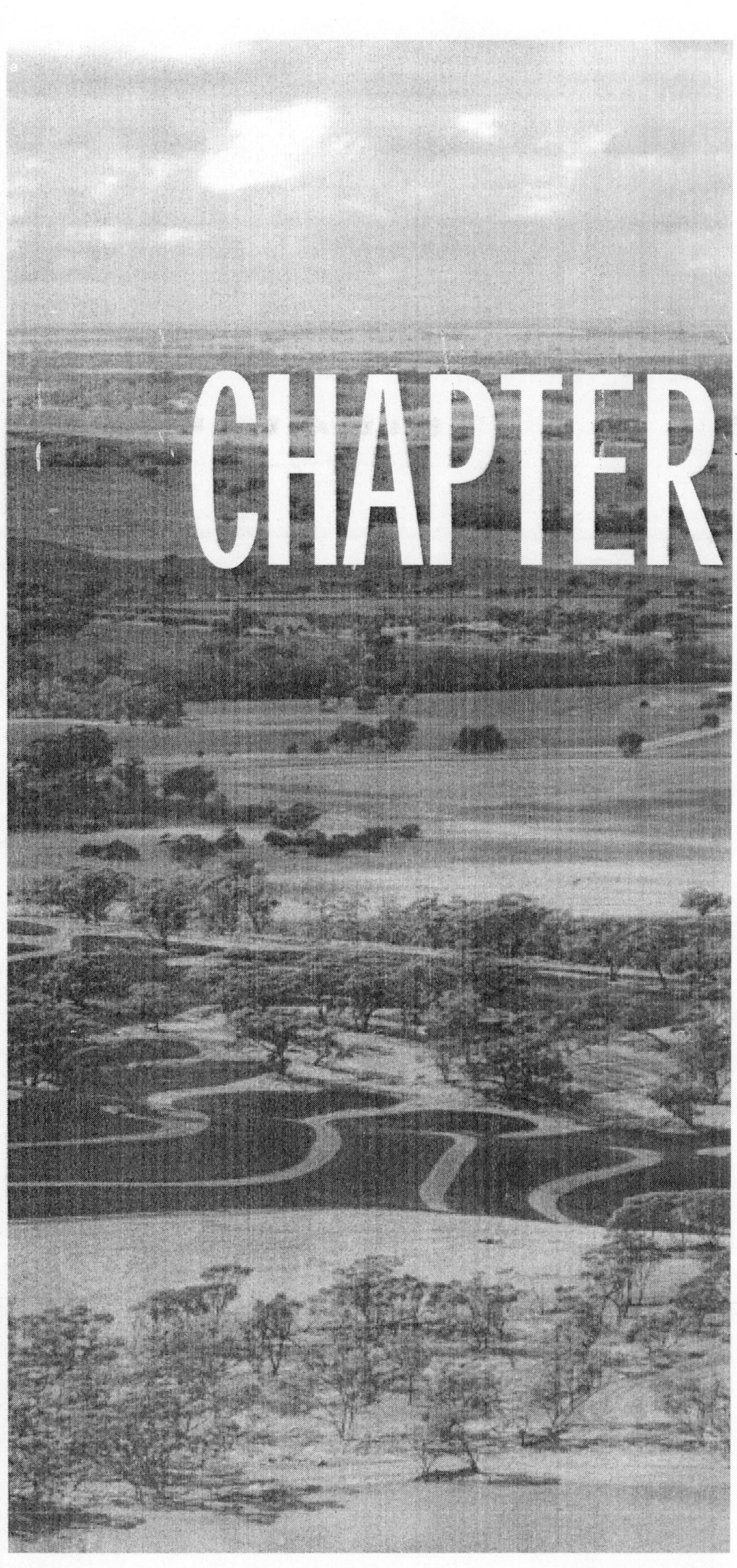

CHAPTER 3

PHYSICAL FACTORS AFFECTING WATER INFILTRATION AND REDISTRIBUTION

3.1 WATER REPELLENCE

Geoff Moore and Paul Blackwell*

Water repellence affects the wetting pattern of soils and results in an uneven wetting pattern in autumn. In the paddock, patches of wet soil alternate with patches of dry soil which results in the poor germination of crops and pasture. Water repellence can be demonstrated by placing a drop of water on the soil surface. (It is advisable to remove a few millimetres from the surface before doing this. See Assessing paddock status, below.) If a soil is repellent the droplet will form a bead and not penetrate quickly. Sandy soils are particularly susceptible.

A water droplet on a dry, water repellent sand.

In south-western Australia, about 5 million ha of land are either water repellent or susceptible to developing repellence. The main areas of susceptible soils are the 'sandplains' across the south coast from Albany to Esperance e.g. Albany sandplain, Jerramungup plain and Esperance sandplain, the Arrowsmith Zone in the Greenough Province (West Midlands) and the sandy soils on the Swan Coastal Plain (Figure 3.1.1).

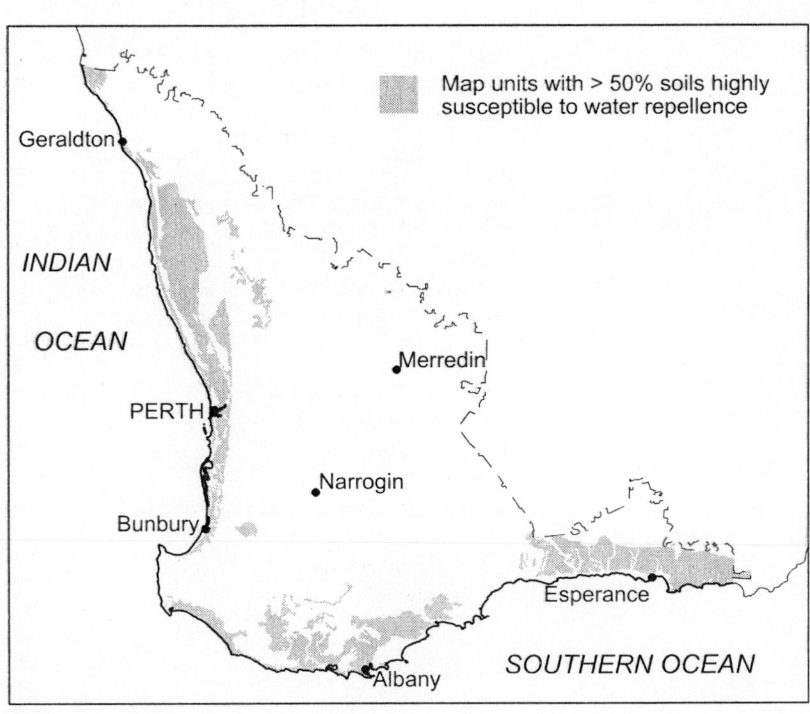

Figure 3.1.1 *Distribution of soils susceptible to water repellence in south-western Australia.*

PRINCIPLES AND CAUSES OF WATER REPELLENCE

For a soil to develop water repellence there are two requirements:

- the addition of sufficient hydrophobic organic matter; and
- a susceptible soil.

All soils develop repellence to some extent, but some textures, structures and associated vegetation are more susceptible (Wallis *et al.* 1991; Wallis and Horne 1992). This section concentrates on soils which have developed, or have the potential to develop, water repellence to the extent where it can affect production and/or management.

Water repellent behaviour is caused by dry coatings of hydrophobic material on soil particles or aggregates, as well as hydrophobic organic matter such as fungal hyphae and particles of decomposing plant material (Bond 1969; Roberts and Carbon 1972; Debano 1981).

Ideas about the type, origins and chemistry of the organic coatings are still being developed. Early theories suggested they originated

* Agriculture Western Australia, Geraldton.

from basidiomycete fungi associated with root systems (Bond and Harris 1964), humic acid skins (Roberts and Carbon 1971), *Eucalyptus* species (McGhie and Posner 1980) or certain plant materials (McGhie and Posner 1981). These theories are now seen as incomplete.

Recent research demonstrates that 'particulate organic matter' has a significant role in the development of water repellence. The particulate organic matter seems to act as both a highly hydrophobic soil component itself and as a 'carrier' and 'reservoir' of waxes. These waxes can diffuse out under heating/cooling and wetting/drying cycles (e.g. over summer) resulting in the redistribution of the hydrophobic materials. This imparts a low level of hydrophobicity to adjacent sand grains, but if they already have a coating of organic compounds then moderate to severe repellence can develop (Franco *et al.* 1995a, b). A conceptual diagram of how water repellence develops is illustrated in Figure 3.1.2.

A number of different organic compounds have been identified using different extractants. The waxy type substances consist of long chain hydrocarbons, fatty acids and alkanes (Roberts and Carbon 1971, 1972; Ma'shum and Farmer 1985; Ma'shum *et al.* 1988; Spadek *et al.* 1994).

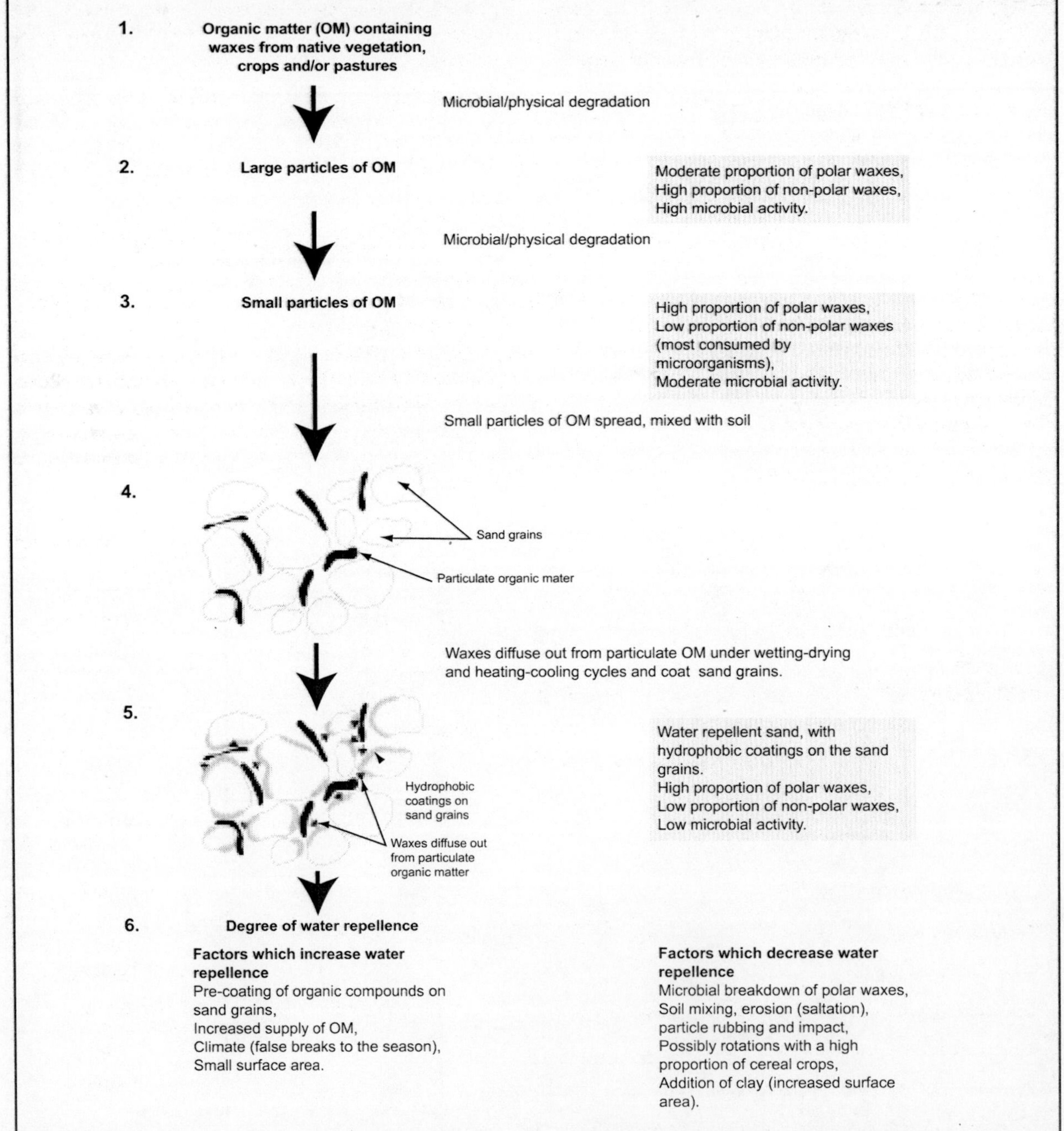

***Figure 3.1.2** A conceptual diagram of how water repellence develops in soils (adapted from Franco* et al. *1994).*

Microbial activity breaks down dead plant material in such a way that it contributes to the development of water repellence in susceptible soils. Franco *et al.* (1995a) suggest that non-polar waxes degrade selectively, resulting in a concentration of the more repellent polar waxes. Some micro-organisms are capable of degrading these stable polar waxes (Roper 1994). Such micro-organisms occur naturally, but in low numbers, because survival is difficult in harsh field conditions such as extremes of temperature, low soil moisture, nutrient supply and pH (Roper 1994; Michelson and Franco 1994).

Research by Capriel *et al.* (1995) suggests that coarse-textured soils are inherently more susceptible to water repellence, because they have a comparatively low microbial biomass and the organic matter contains a higher proportion of alkyl C (more hydrophobic) and less protein than it does in finer textured soils. Research continues for a biological solution to water repellence using similar principles to those of the biological clean-up of crude oil spills. Recent results with moisture conservation and the addition of small amounts of lime have been encouraging (M. Roper, personal communication).

Effect of land use

Hydrophobic organic matter is produced by the decomposition of plants, but some plants produce more dry matter than others. This is the main reason why some land uses/rotations, especially with legumes, induce greater degrees of water repellence. A common observation by farmers is an increase in non-wetting behaviour on sands with a low clay content, associated with an improvement in productivity. However, some farmers using no-till crop establishment methods have been observing improved soil wetting. This may be due to better conservation of moisture and also water movement along intact dead roots.

Field measurements from long-term rotation trials or permanent pasture sites (Table 3.1.1) are the only reliable method for assessing the effects of land use on water repellence. Some early studies demonstrated a close relationship between water repellence and certain land uses e.g. legumes (Roberts and Carbon 1972; McGhie and Posner 1981), but these experiments were not definitive, because the plant material did not always undergo natural decomposition and incorporation into the soil (Wallis and Horne 1992). There is some evidence that continuous cereal cropping may reduce repellence, providing the soils are not already severely repellent (McGhie and Posner 1981; Summers 1987; Carter *et al.* 1994).

Blue lupins (*Lupinus consentinii*) definitely increase the severity of water repellence. A sandplain paddock growing continuous blue lupins for about five years will develop moderate to severe repellence (Blackwell and Nicholson 1990).

Sheep camps tend to be more water repellent because of accumulation of organic matter. In addition, research indicates that the waxy substances are not broken down effectively when passing through the sheep.

Native vegetation can also induce water repellence (McGhie 1980a). Under native vegetation, repellence may be used by plants as a competitive advantage to improve water conservation (by channelling water deep into the soil profile), restricting the germination of

Table 3.1.1 The effect of land use on water repellence in sandy soils. Data are from long-term rotation trials and farmer paddock surveys (adapted from Blackwell 1993; Spadek *et al.* 1994).

Rotation	Site (years with same rotation)	Molarity of ethanol drop test* (M)	Water repellence rating
Continuous wheat	Esperance Downs Research Station - EDRS (26th year)	0.3	Low
	Badgingarra Research Station - BRS (3rd year)	-	Severe
	Newdegate Research Station NRS (16th year)	-	Nil
Clover/wheat rotation	EDRS	1.2-1.6	Moderate
Continuous clover	EDRS (20th year)	2.6	Severe
Wheat/lupin rotation	EDRS	1.1.-1.2	Moderate
Continuous lupins	EDRS (20th year)	1.1	Moderate
Permanent annual pasture	NRS (16th year)	-	Severe
	BRS (3rd year)	-	Severe
Continuous blue lupins	Geraldton farm survey (12th year)	~3	Very severe

* *Refer to* Laboratory tests, *below.*

Patchy pasture growth typical of water repellent sands (P. Blackwell).

competitive plants and reducing evaporation (through a partial dry layer at the surface). There is a strong link between particular native species e.g. *E. astringens* (brown mallet), *E. patens* (blackbutt), *Banksia speciosa* (showy banksia) and the induction of water repellence (Carter *et al.* 1994).

Consequences of water repellence (from King 1990)

Pasture

- Patchy emergence early in autumn typifies pastures on water repellent soils. Pasture is characterised by patches of plants on wetted areas with dry, bare ground inbetween. Light rains accumulate in hollows and hoof prints and wet the soil sufficiently for plants to survive for a few weeks.
- Weed seeds that lodge in small depressions can germinate early in the season. Seeds from aerial seeding plants (e.g. capeweed, geranium, silver grass) are more likely to accumulate in these surface depressions.
- Subterranean clover is at a competitive disadvantage compared with aerial seeding plants. The runners do not place the seeds preferentially in the hollows, so seeds placed just below the surface may or may not be wet by light rains. Even when they do germinate, the roots may find only dry soil, so the seedling dies.
- The pasture is characterised by having both young and established plants, which results in difficulties with the use of some herbicides.
- Insects (e.g. red-legged earth mite) are able to multiply quickly because of the early germination and survival of the pasture or weeds that are good hosts (e.g. capeweed).
- Heavy rain in summer or autumn can result in significant runoff and erosion on sloping sites. Rills and gullies have been observed on soils which would have negligible runoff if they were not repellent.

Crops

- Controlling the depth of cultivation and sowing equipment is difficult when moving from wet to dry soil.
- Fertiliser placed in the hollows could be subject to leaching, while in dry soil it is positionally unavailable.
- Water repellent soils require more rain before sowing to wet them than non-repellent soils. This means that seeding may be delayed, reducing the yield potential.

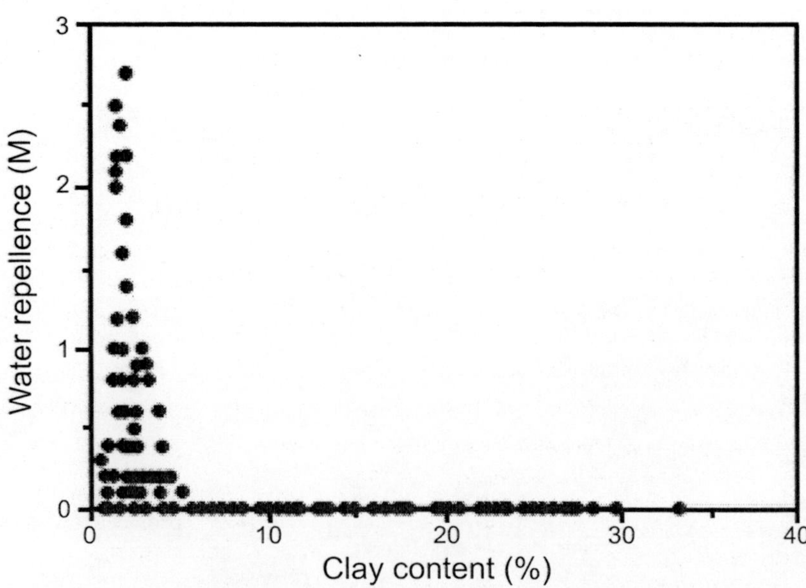

Figure 3.1.3 *The relationship between percentage clay (0-10 cm) and the degree of water repellence for soils in the Lake Cairlocup district (from Harper and Gilkes 1991).*

- Crops sown before the soil is wet evenly, result in patchy germination.
- The forced delay in sowing, dry soil patches and poor cover can increase the risk of wind erosion.
- Incorporated herbicides (e.g. trifluralin) are not always active in dry soils. Surface applied herbicides (e.g. simazine) can be washed into the furrows and damage crop plants, while weeds can germinate on the ridges.
- There is an increased risk of water erosion on sloping sites.

These factors tend to concentrate on the negative aspects of water repellence. Not all consequences are necessarily unfavourable and as described earlier some native plants may induce water repellence to give a competitive advantage in a harsh environment. Repellence can also reduce moisture lost by evaporation because of the 'dry mulch' effect of the repellent surface layers (Yang *et al.* 1996).

Soils and the development of water repellence

Water repellence in Western Australia has been reported mainly in sandy soils (Bond and Harris 1964; Roberts and Carbon 1971; Summers 1987). Two exceptions are the soils associated with the 'mallet hills' in the Great Southern (McGhie 1980b) and the highly calcareous 'fluffy', or kopi soils.

Within a soil profile, the severity of water repellence is related to the quantity of organic matter present (Wallis and Horne 1992; Harper and Gilkes 1994).

The soil property most closely linked to susceptibility to water repellence is the specific surface area (e.g. Roberts and Carbon 1971). McGhie and Posner (1980) found that coarse sands are more water repellent than fine sands, because the coarse sands have a smaller surface area. Experimental work in the Lake Cairlocup district demonstrated a close relationship between clay content (closely related to specific surface area; Section 2.3) and water repellence in the field (Figure 3.1.3; Harper and Gilkes 1991, 1994).

The depth of sand is not important in the *development* of water repellence, because 2 cm of sand over loam can become as repellent as a deep sand (Bond 1969). Likewise, Summers (1987) found that deep sands and shallow duplex soils on the Esperance sandplain were equally water repellent.

However, depth of sand can be important through the effects on re-wetting, especially in high rainfall districts. On shallow duplex soils, re-wetting of the surface from a perched watertable can reduce the effect of water repellence. The depth of sand over clay can also affect productivity, thus the amount of organic matter produced.

ASSESSING SUSCEPTIBILITY TO WATER REPELLENCE

The susceptibility of soil to water repellence is related to two main factors: surface area and supply of hydrophobic compounds. The supply of hydrophobic compounds varies with the productivity of the system; a complex interaction between climate, soil properties, land use and management. For example, a uniform coarse sand (clay <1%) which has a low surface area may not develop severe repellence if it is in an unproductive farming system.

Soil materials with low surface area are more susceptible. For instance, the amount of hydrophobic material needed to completely coat a sandy soil could only cover a small proportion of a clayey soil (surface area of sands 0.01 to 0.2 m^2/g, compared with clays 10 to 200 m^2/g). Therefore, most soils with a clay content of more than 5% (0-10 cm) have a low susceptibility. In general, the surface area of the soil is too large to be coated with hydrophobic organic compounds so the soils wet easily (exceptions are described below).

Surface soils with less than 5% clay

For sandy soils with a low clay content (<5% in the 0-10 cm layer), three categories of susceptibility to water repellence have been identified: low, moderate and high (Table 3.1.2). The *specific surface area* can usually be inferred from particle size analysis or field texture (Section 2.3).

Surface soils with more than 5% clay

Most soils with more than 5% clay in the surface are not susceptible to water repellence, however a few soils with ≈10-25% clay (0-10 cm) were water repellent before clearing.

Water repellence is not normally induced by agriculture on soils containing more than 10% clay (0-10 cm).

Coarse-textured soils with a loamy or clayey sand surface (e.g. yellow-brown clayey sands, sandy earths with 5-10% clay in the surface 10 cm) can be considered an intermediate category. In practice they rarely develop repellence to the extent that it affects management (Blackwell 1993), but moderate to severe repellence has been measured under long-term pastures of blue lupins (Blackwell and Nicholson 1990). Circumstantial evidence suggests that as farming practices change (increased production, less soil disturbance and mixing) a shallow (<5 cm) surface layer of 'fluffy' soil, which is difficult to wet, is developing on some soils. The clay in the surface could have been reduced to <5% if it was leached and/or organic matter increased as a result of conservation farming.

Two main groups of soils with >10% clay are water repellent before clearing:

- Soils associated with the mallet hills in the Zone of Rejuvenated Drainage. These contain 15 to 25% clay in the surface. Following clearing, water repellence may decline if the surface is eroded (McGhie 1980b).

  These soils develop water repellence under native vegetation because of two complementary factors. First, their finely aggregated structure makes them behave more like coarse sand, with the hydrophobic organic matter coating the aggregates (D. McGhie personal communication). This mechanism has also been found on humic clay soils in New Zealand (Wallis *et al.* 1991). Second, large quantities of extremely hydrophobic organic matter accumulate under *Eucalyptus astringens* (brown mallet) and wandoo (*E. wandoo*) which may be a competitive advantage for the species (McGhie 1980a).

- The highly calcareous, fluffy or morrel (kopi) soils in the Zone of Ancient Drainage (Teakle *et al.* 1940; Bettenay and Hingston 1964; Burvill 1988; Lantzke 1992). These soils often contain high concentrations of soluble salts. They are observed to be water repellent in the field, sometimes even in the subsurface soil. It is probably connected with the fungal mycelium frequently observed (G. Scholz personal communication). Water repellence in these soils has not been studied in detail.

ASSESSING PADDOCK STATUS

Water repellence is strongly influenced by past management and land use. A paddock with inherently susceptible soils may or may not be water repellent depending on the paddock history and the time of the year.

The only reliable method of diagnosing soil condition is to assess each paddock with susceptible soils.

There are two methods:

- Field observations are simple, but the degree of repellence may be strongly influenced by seasonal conditions. If field observations detect water repellence, a laboratory test is recommended.
- Laboratory tests are the most accurate method, and can be used to monitor change in the medium-term.

Field observations

A paddock with susceptible soils and a history of clover-based pasture, blue lupins, perennial pasture or regular legume crops is likely to have developed repellence to some extent. The timing of observations is important and in some years, water repellence is negated by the rainfall pattern early in the season. Water repellence also decreases with increasing temperature (King 1981).

Table 3.1.2 Susceptibility to water repellence: soils with <5% clay.

Typical texture	Nominal specific surface area[1] (m^2/g)	Example	Productivity[2]	Susceptibility[3]
Coarse sand (<1% clay)	<0.1	Grey coarse sands	High ⇑ Low	High Moderate-high
Coarse sand (1-2% clay) or fine sand (<2% clay)	0.1-0.5	Texture contrast soils with a fine sand surface (e.g. Esperance sandplain)	High ⇑ Low	High Low-moderate
Weak clayey sand (3-5% clay), or Loamy sand (5% clay)	0.5-4.0	Uniform yellow deep sands	High ⇑ Low	Moderate Low

1. *Specific surface area is surface area per unit mass.*
2. *Productivity, or the supply of hydrophobic compounds, can be inferred from the soil organic carbon (McKissock* et al. *1997), i.e. unless a change in land use will dramatically alter the productivity of the system. As a guide, 'high' refers to soils with OC levels >2%, while 'low' refers to soils with OC levels <1.2%.*
3. *Susceptibility to water repellence refers to the degree of hydrophobicity which develops rather than the expression of water repellence.*

Heavy rain early in the season when temperatures are high can wet even severely repellent soils. Wetherby (1984) defined a non-water repellent year as one in which soaking rains (>12 mm over 24 hours) totalling 25 to 50 mm fall during the first two months of the growing season.

On the other hand, early or late false breaks to the season, infrequent light showers and late breaks to the season which coincide with lower temperatures exacerbate water repellence (Blackwell *et al.* 1994a). If there is a late break to the season the consequences of water repellence can be more severe on the south coast than the northern wheatbelt and West Midlands.

The following symptoms indicate the soil is likely to be water repellent:
(adapted from Bond 1968, 1972)

- During opening rains, water ponds on dry sandy soils in patches. Runoff may occur from sandy soils on sloping sites.
- Dry patches of soil between depressions which are moist.
- Patches of dust in a 'wet' soil when cultivating.
- Staggered emergence (pasture, crops, weeds) with early growth in depressions.
- Patchy crop or pasture growth, with failure to germinate in areas of dry soil.

Following significant rainfall, water repellence can be confirmed by examining a cross-section of soil under these areas of uneven growth. The bare patches should correspond with zones of dry soil.

The degree of water repellence can be estimated from field observations (Table 3.1.3). In general, it frequently goes unnoticed in the field until moderate repellence (equivalent to a MED of 2, refer to *Laboratory tests*, below) has developed.

Laboratory tests

Soil sampling followed by a laboratory test is the most reliable method for determining the severity of water repellence. The Molarity of Ethanol Droplet (MED) test is recommended, but two different tests are compared in Table 3.1.4.

Time of sampling. Sampling can be done at any time, but the results are likely to be less variable if the samples are taken in summer when the soils are at their driest. This reduces the influence of drying temperatures on the results of the MED test (Carter *et al.* 1994).

Depth of sampling. Highly repellent soil is closely associated with the top 50 mm, depending on the amount of soil mixed by cultivation.

In many water repellent soils there is a thin (few mm) surface layer which is wettable even though the underlying soil is moderately to severely repellent (possibly due to volatilisation of the waxy substances from the surface). It is therefore normal practice to scrape away the top 3 to 5 mm before sampling.

In paddocks with a long history of continuous pasture or no-till, the sampling depth should be 5 to 50 mm. In cropped paddocks the same sampling depth can be used, but hydrophobic material is likely to be more evenly mixed through the cultivation layer, so sampling depth will be less critical.

The Molarity of ethanol droplet test (MED)

The MED test (King 1981) measures the molarity of drops of ethanol that infiltrate the soil within 10 seconds. The ethanol allows the droplet to infiltrate by lowering the surface tension of the liquid which in turn lowers the liquid-soil contact angle. The more water repellent the soil, the higher the molarity of ethanol needed before the drops can infiltrate. The test is straightforward and quick, so a number of samples can be tested to allow for the high spatial variability of water repellence in the field.

If ethanol is not available, diluted methylated spirits can be substituted (D. Maschmedt unpublished data). For example, to make a solution equivalent to 1M ethanol, add 12 mL of methylated spirits to 200 mL water; for 2M ethanol add 24 mL. The additives in methylated spirits are at a low concentration and do not adversely affect the results (R. Summers personal communication).

The MED test is sensitive to temperature and the initial moisture content of the soil, so laboratory analyses use oven-dried soils to give a standardised measure. Research has shown that the soil drying temperature can affect the results, with higher MED values associated with soils dried at high temperature. In general, 40°C for seven days (King 1981) gives similar results to 105°C for 48 hours (Carter *et al.* 1994).

Laboratory measurements should be made at 20°C on oven-dry (105°C) soils. Measurements made at varying temperatures between 0 and 42°C can be easily corrected using a simple relationship. For a description of the MED method refer to King (1981) or Carter and Hamilton (1997).

Table 3.1.3 Estimating the degree of water repellence from field observations.

Degree of water repellence	Field observations
Not significant to low	Either not observed or observed infrequently.
Moderate to severe	Observed in most years and a major problem 1 year in 3.
Very severe	A major problem in most years.

Table 3.1.4 Water repellence rating from laboratory tests, for soils tested at 20°C (adapted from King 1981).

Severity of water repellence	Test	
	MED (M)	Contact angle (degrees)
Not significant	-	<75
Very low	-	75-80
Low	<1.0	81-86
Moderate	1.0-2.0	87-92
Severe	2.0-3.0	93-97
Very severe	>3.0	>97

The MED test is not sensitive at the low end of the scale, so the water droplet penetration time (WDPT) method could be used as a complementary test in slightly repellent soils (described below under *Other laboratory tests*).

Angle of contact test

This measures the angle of contact between a water drop and the soil surface. It is reliable, but is unsuitable for rapid testing of a large number of soils (Emerson and Bond 1963).

Other laboratory tests

The water droplet penetration time test (Letey 1969) is only applicable to slightly water repellent soils. In the laboratory the MED test is preferred, except in soils with low repellence, which the MED test cannot distinguish. A water drop test in the field is unreliable, because of variations in temperature and antecedent moisture.

MANAGEMENT OPTIONS

Where feasible, manage water repellence in broadscale agriculture to take advantage of the positive effects, namely the increased subsoil moisture and reduced soil evaporation. This is only possible if the subsoil can store additional moisture and is a suitable environment for plant roots. The strategy is unlikely to be successful on a shallow sandy duplex soil with a sodic clay subsoil.

The decision pathway for managing soils susceptible to water repellence is outlined in Figure 3.1.4. When assessing the success of a given treatment or test strip, remember the 'status' of the paddock. If there is no difference between treated and untreated (or control) areas, it may simply reflect the rainfall pattern early in the season. For instance, a paddock with moderate water repellence would not have significant problems every year.

Water repellence can be managed in at least seven ways:

(i) **Using:-** water harvesting, furrow sowing (see below)

(ii) **Masking:-** adding clay to cover the hydrophobic layers (see below)

(iii) **Grow species adapted to water repellent soils:-** e.g. pines, tagasaste, blue lupins

(iv) **Avoiding dry or drying soil:-** furrow sowing onto moisture; cultivating in the rain (high risk of wind erosion)

(v) **Lowering surface tension:-** soil wetting agents

(vi) **Decomposition:-** microbial consumption of hydrophobic organic matter

(vii) **Dilution:-** physically mixing with non-repellent soil (high risk of wind erosion).

Lime has been tried as an ameliorant on water repellent soils, but is not recommended. It can improve wetting slightly, because of the fine particle size and corresponding surface area effect, but field trials have shown it is relatively ineffective, especially compared with clay amendment (Blackwell *et al.* 1994b; Carter and Hetherington 1994).

The basic principles of furrow sowing

Furrow sowing is being adopted extensively to manage repellence (e.g. Blackwell *et al.* 1994a; Blackwell 1997). No-till furrow sowing can use discs, blades or narrow points. Some furrow sowing has resulted in erosion and crop damage by herbicides, especially for lupins, but no-till techniques and new herbicides have helped reduce these problems. Furrow-sown lupins in the West Midlands have yielded on average 40% higher than level-sown lupins in seasons with false breaks. Responses on the south coast and with cereals have been less, but still provided encouraging benefits in seasons with unreliable rainfall.

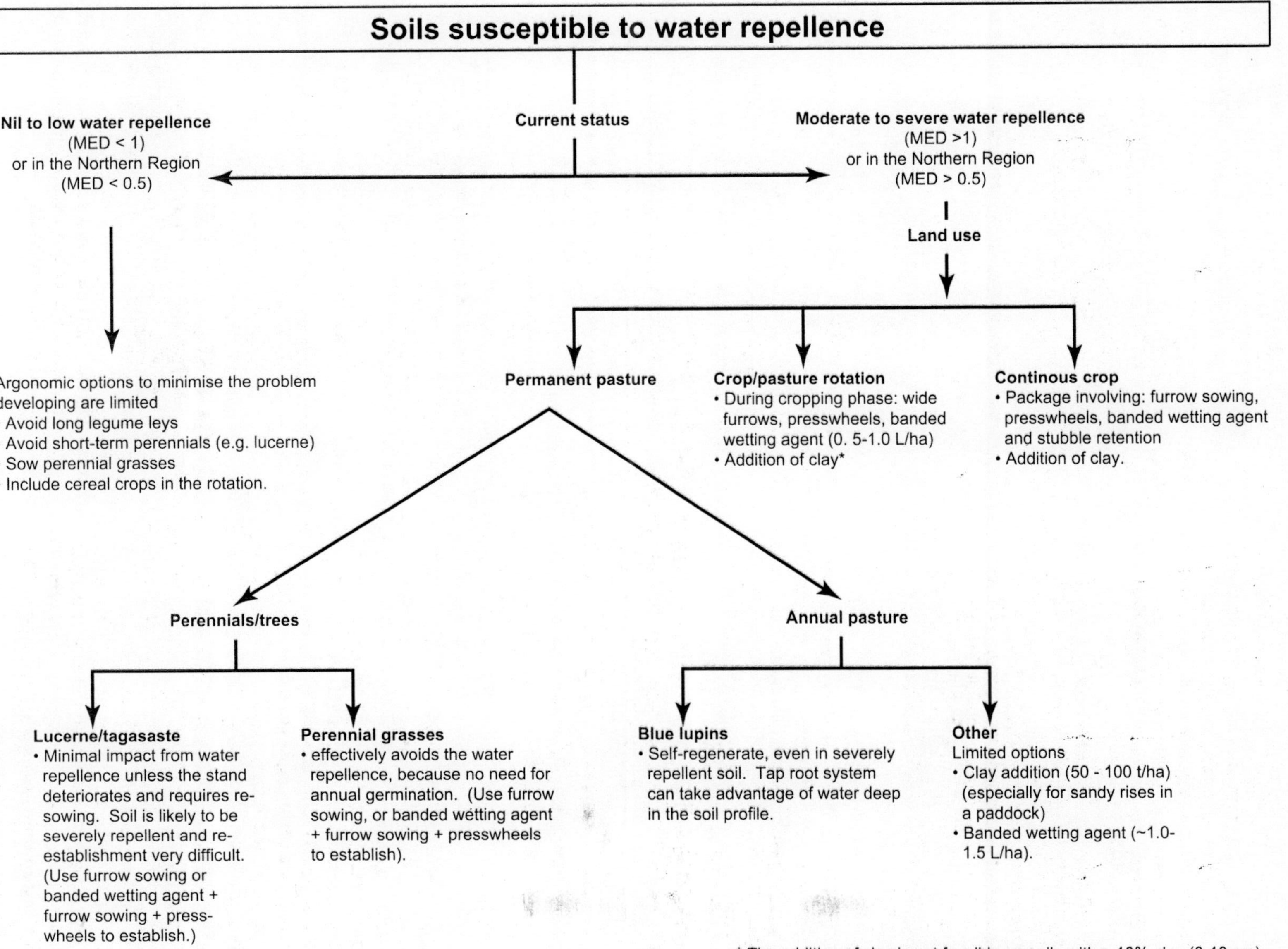

Figure 3.1.4 *Decision pathway for managing soils susceptible to water repellence.*

A furrow is a groove or corrugation in the soil surface. Furrows allow sowing onto moisture and water harvesting. Sand and clods can be thrown by points and blades or pushed by discs and presswheels to form the furrow (Figure 3.1.5a). Firming sand at the furrow base with a presswheel (Figure 3.1.5b) will improve water retention and nutrient uptake. A thin band of surfactant sprayed in the base of the furrow will further improve soil wetting (Figure 3.1.5c); especially in relatively dry and more repellent conditions.

The **benefits** are mainly from sowing onto moisture and water harvesting.

- *Sowing seed onto moisture.* Zones of dry sand are common in water repellent topsoil, but the lower topsoil or subsurface soil is often more consistently wet. Furrowing can place seed in or close to moist sand without the seed being too deep. (Level sowing at the same depth may plant the seed too deep). Deeper furrows will allow sowing onto deeper moisture, but extra care is needed because of erosion risks.
- *Water harvesting.* Water often sheds from upper parts of soil microrelief on water repellent soils. The repellence forms a 'hydrophobic seal' and sheds water into the lower parts of the microrelief such as furrows. Ponding of water can improve infiltration into water repellent sand by increasing the surface pressure head. Wider furrows harvest more water, but the sides of the furrow only need a low slope.

The **risks** are mainly erosion, pathogen transfer from bare soil, herbicide concentration, leaching of fertilisers and waterlogging.

- *Erosion; especially of large, loose ridges.* Wind shear at the ground surface and raindrop impact can erode material from ridges into furrows, especially before and during the first autumn rains. This can bury seed deeper than it was originally placed, which may make emergence more difficult, or impossible. Pathogens and herbicides can also be splashed onto plants from bare soil. Sufficient water movement along a furrow can also cause rill erosion and expose or remove seed. Plant residues, stubble and root-bound clods help to control ridge erosion when cover levels are more than 30 to 50%. Rill development will be reduced by sowing on the contour of a slope. Erosion risks, especially in pasture, are increased by large inter-row ridges.
- *Herbicide concentration (Simazine/diflufenican overdose of lupins).* Post-emergence application of simazine and diflufenican have resulted in severe lupin damage after furrow sowing (e.g. 40% plant morality from 1 L/ha simazine applied at the two-leaf stage of a dry furrow-sown lupin crop near Geraldton). The ridges were bare, the furrows wide and the herbicide was applied dry before a heavy shower. High stubble levels, narrower furrows, lower ridges and spraying onto wet soil reduce the risk considerably. Much safer methods use metribuzin/diflufenican mixtures onto a wet crop with 8 to 10 leaves. This treatment has worked well for radish, doublegee and capeweed in the northern wheatbelt, with no yield penalty. Metribuzin is more soluble and more easily metabolised by the crop than simazine, thus less likely to collect in the furrows and be concentrated at the roots.
- *Leaching.* Harvesting water may accentuate leaching of nutrients from around plants growing in a furrow, especially in soils with a poor capacity to retain nutrients, and in wet seasons. Wider furrows have a larger catchment and induce leaching more easily in higher rainfall zones.
- *Waterlogging* on shallow duplex soils, especially on the south coast.

More complex designs include firming by presswheels and banded surfactant:

- *Firming by presswheels.* Firming the sand will increase water storage and improve uptake of soil moisture by the seed, and nutrients by the plant. Trials in the West Midlands have shown that firming the sand improved lupin yields by 11% in a season with a false break.

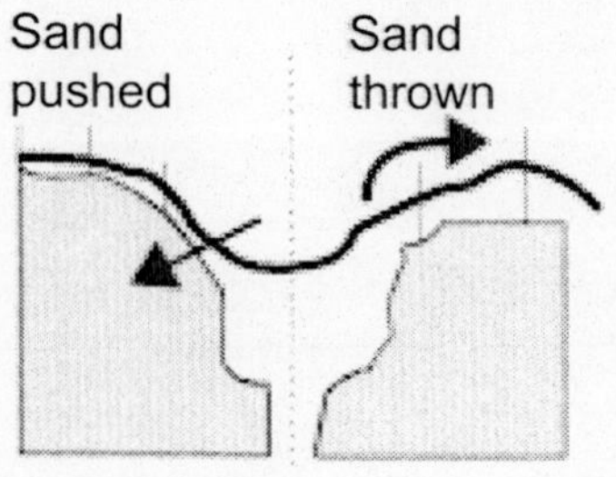

(a) Furrow forming (undisturbed soil shaded)

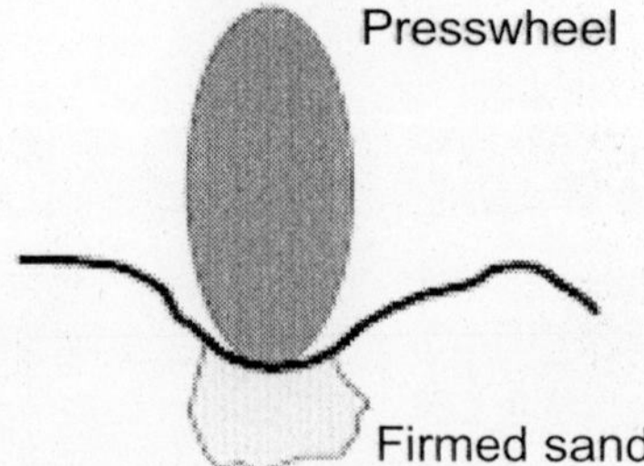

(b) Furrow pressing

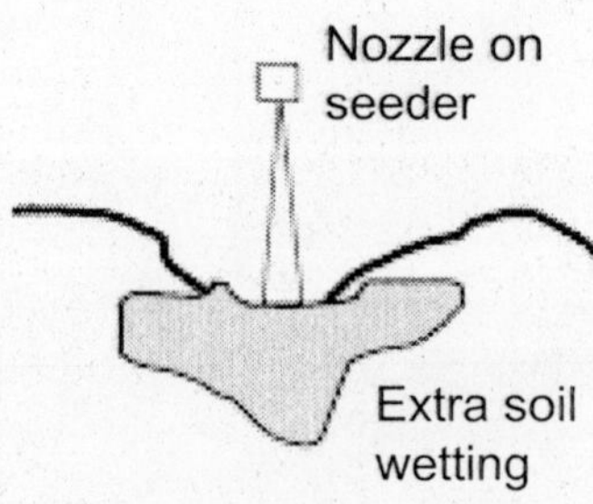

(c) Surfactant band, can be various shapes

Figure 3.1.5 *The principals of making furrows.*

Furrow sowing on a water repellent sand.

On sands containing moderate clay in the lower topsoil such as the more productive yellow sands, it is not advisable to have the pressure higher than about 2 kg/cm unless a deep working blade or point is used.

- *Banded surfactants.* Surfactants have many formulations. Early formulations emphasised the need for longevity to increase the cost benefits. They have been proven for establishment of pastures, crops and fodder shrubs. Statewide trials of the more persistent formulations have revealed very unpredictable crop yields with banded surfactant, despite improved crop establishment. Extreme cases have ranged from economical yield benefit to very large yield depression. Better responses have generally been on the south coast and poorer responses in the West Midlands and with wider furrows. Narrow furrows and low rates of persistent surfactants have sometimes given economic yield responses on the south coast. The persistence of the surfactant into the spring has reduced the water retention of sands, reduced nutrient uptake and diminished yields in most cases. New formulations are being tested which are less persistent, or counteract poor water retention.

Clay addition

Adding clay to the hydrophobic layers lowers susceptibility to water repellence. Assuming it can be evenly distributed through the soil, the net effect is to increase the specific surface area and lower water repellence. For example, adding 100 t/ha of clay to a sandy soil in South Australia increased the clay content in the top 5 cm from <1 to 5.6%. The clay has been effective in the long-term control of water repellence with MED values still very low (0.3), 20 years after application (Ward and Oades 1994). The effectiveness of clay addition has also been demonstrated in south-western Australia (Carter and Hetherington 1997).

The overall aim is to increase the clay content of the top 50 mm to about 5%. Important properties affecting efficacy include:

- *High clay content of additive.* An additive containing 60% clay will be twice as effective as one containing 30% clay, assuming other factors are the same.
- *High dispersion.* Dispersion results in quicker incorporation into the soil. Ideally the clay additive should be spontaneously dispersive in distilled water, i.e. classes 1-2 with the Emerson Aggregate Test (Emerson 1967) described in Section 3.2.
- *Clay mineralogy.* Ward and Oades (1993) found that a kaolinitic clay reduced water repellence, but a montmorillonitic clay was ineffective. Kaolinite did not strongly adsorb hydrophobic molecules, but scanning electron microscopy revealed that it coated the sand grains. On the other hand, montmorillonite particles formed micro-aggregates and only covered a small proportion of the sand grains due to edge-on attachment. Illitic clays have also been used successfully (Ma'shum *et al.* 1989). In theory, the higher the specific surface area (Section 2.3) the less clay required (Ma'shum *et al.* 1989). In practice, the requirement for a dispersive clay is of over-riding importance and kaolinitic clays with a comparatively low specific surface area have proved more effective than illitic and montmorillonitic clays.
- *Other benefits.* The clay may improve the nutritional status of the soil (e.g. K supply; Dellar *et al.* 1994); reduce susceptibility to wind erosion and increase herbicide efficacy because of a more uniform germination of weeds. Pasture can be improved by topdressing the clay, if it is applied in small amounts, about 10 to 20 t/ha. More clay can be applied after rain has incorporated the clay without crusting.

Further reading

Blackwell, P. (1993). Improving sustainable production from water repellent sands. *Journal of Agriculture Western Australia* **34:** 160-167.

Carter, D.J. and Howes, K.M.W. (eds.) (1994). Proceedings of the 2nd National water repellency workshop. 1-5 August 1994, Perth, Western Australia, 220 p.

Wallis, M.G. and Horne, D.J. (1992). Soil water repellency. *Advances in Soil Science* **20:** 91-146.

3.2 SOIL STRUCTURE DECLINE

Paula Needham, Geoff Moore and Gottfried Scholz***

Soil structure decline was first raised as an important agronomic issue in WA in the 1950s (Stoneman 1962). It is a phenomenon of surface soil, caused mainly by excessive tillage. Degraded soils typically have reduced infiltration, increased runoff, are more compact, require more tractor power and in many areas can only be cultivated within a narrow moisture range (Smith 1969; Sullivan *et al.* 1983).

The true extent of the problem is difficult to assess as structural decline is often gradual. However, one estimate suggests that 30% of soils (mainly medium and fine-textured soils) are susceptible (Carder and Grasby 1984). Surface soils with a clayey sand or coarser texture (<8-10% clay) have minimal secondary structure and are generally not susceptible. Various hard layers occur in soils, but they do not always indicate structural decline (refer to Section 4.1).

What is soil structure?

Soil structure is defined by the way soil particles are arranged and bound together. A simple geometric description is difficult, so it is usual to measure structure in terms of the following attributes:

(i) Porosity. Well structured soils contain a range of pore sizes. The larger pores provide pathways for roots and allow air and water movement; smaller pores store water. There also needs to be good continuity between pores.

(ii) Strength affects the ease of penetration of agricultural equipment, seedling emergence and root elongation through the soil.

(iii) Stability is a measure of the soil's resistance to break down under imposed stresses, either natural (like raindrop impact) or those associated with soil management (tillage, grazing).

(iv) Resilience is the ability of the soil to reform aggregates through natural processes (such as shrinking and swelling) as it goes through wetting and drying cycles (Kay 1990). The greater the resilience, the less sensitive a soil is to poor management. Self-mulching behaviour is a good example.

The ideal soil structure has a large proportion of aggregates from 0.5 to 2 mm which are not easily broken down. It has high porosity for water entry and gas exchange, low strength, is stable when wet or mechanically disturbed, and aggregates can reform if subjected to adverse management.

Soil structure can be viewed from the micro-scale (smaller than 10^{-4} mm) to the macro-scale (larger than 10^{-4} mm) (Dexter 1988). In profile descriptions, soil structure and fabric are observed with the naked eye at the macro-scale. A soil is described as massive or apedal when it has no visible peds, but it is not structureless at the micro-scale. On the contrary, it contains pores, micro-aggregates and bridges of very fine particles connecting coarser particles. These characteristics give apedal or massive soil strength to withstand compaction, deformation and collapse (Koolen 1982).

PRINCIPLES OF SOIL STRUCTURE DECLINE

Soil structure is a dynamic property. In general, it is much easier to destroy aggregation than improve it. The breakdown of aggregates into primary particles can cause structural decline, leading to a reduction in the volume and size of soil pores (especially transmission pores) and increased strength on drying.

Declining soil structure results in *crusting* and *hardsetting* of the surface:

Crusting is a thin layer (usually <10 mm) with higher density, finer pores and lower hydraulic conductivity than the underlying soil. When dry, it separates from and can often be lifted off the underlying soil, which is usually loose (Northcote 1979). Crusts are formed when unstable soil is exposed to direct raindrop impact or sudden wetting (e.g. sprinkler, flood irrigation) which cause surface compaction plus slaking and/or dispersion (Letey 1985; Shainberg 1985). Susceptibility to crusting is increased by the mechanical impact of raindrops and a low electrolyte concentration at the surface as a result of leaching (Agassi *et al.* 1981).

A crust often consists of two parts: a thin (<1 mm) compacted seal on the surface caused by raindrop impact; and a 'washed-in' layer immediately below (up to 3 mm thick), with fine material plugging the larger pores (McIntyre 1958a, b).

Crusts reduce infiltration, increase runoff and have high strength which can reduce crop emergence. The saturated hydraulic conductivity (K_S) of crusts is highly variable depending on the degree of structural decline, e.g. the K_S of cultivated loamy-textured soils with a surface crust in NSW ranged from 2 to >100 mm/h (Chartres 1992). Crusts act as a throttle at the surface reducing infiltration. As a result the soil below the crust is wet under tension, which reduces aggregate disruption and the strength of the underlying soil as it dries (Emerson and Bakker 1973; Kemper *et al.* 1975).

* Formerly Agriculture Western Australia, now Sussex, England.
** Formerly Agriculture Western Australia, now Scholz Environmental Consulting Perth.

Hardsetting soils are unstable when wet and slump after cultivation to a density similar to before cultivation. However on drying, the soil continues to shrink (mainly vertically) resulting in a massive layer, with few if any cracks. Hardset horizons are thicker than crusts and usually the whole surface layer (e.g. depth of cultivation) is affected. They have a low hydraulic conductivity because of dense packing and a low macroporosity (Mullins *et al.* 1990).

The stresses imposed on tilled soil by natural wetting and drying cycles are sufficient to cause hardsetting in many WA wheatbelt soils. Externally applied stresses, such as tillage and stock trampling, increase the severity of hardsetting (Cochrane *et al.* 1994).

Increased strength is one of the features of a hardset soil. Soil strength can be measured using a penetrometer. Hardsetting soils show resistance to a 6.35 mm diameter penetrometer of at least 5 kg/cm^2 when dry (Northcote 1979). However, it is worth noting there are degrees of hardsetting, because it is a continuous rather than a discrete attribute (Harper and Gilkes 1994). This is not considered above, but is important from the viewpoint of soil behaviour and crop growth.

Effect of tillage on seedbed condition directly after seeding: (a) Seedbed following multiple tillage under continuous wheat. (b) Seedbed following direct-drilling under continuous wheat (T. Proffitt).

Effects of soil structure decline

Soil structure influences seedling emergence, root growth, infiltration, water storage, aeration and soil workability, all of which can reduce crop yields, either directly or indirectly. The main effects are:

- *Reduced infiltration.* Reduced infiltration results in increased runoff and more evaporation from the soil surface. The net effect is less water in the root zone and this is directly related to the yield potential (French and Schulz 1984). There is also an increased risk of water erosion from the higher runoff.
- *Poor soil workability.* Soils with poor structure can cause difficulties in the preparation of an even seed-bed. High strength means more wear on points and machinery and higher fuel bills.
- *Delayed seeding.* Seeding is frequently delayed because cultivation is restricted to a narrow range of water contents (i.e. so-called 'Sunday' soils). Delayed seeding can reduce the yield potential (e.g. 20-30 kg/ha/day; Anderson and Smith 1990).
- *Reduced seedling emergence.* Surface crusts reduce emergence, especially of sensitive crops like narrow-leafed lupins. Seedlings that do manage to emerge can have damaged cotyledons (White and Robson 1989). Severely degraded soils can also result in patchy emergence of cereal crops.

TILLAGE TERMINOLOGY

No-till is direct seeding without rearranging the entire topsoil structure (Houghton and Charman 1986). Soil is only cultivated in the sown rows, leaving inter-row areas relatively undisturbed. The amount of soil disturbed varies between systems; 'disc' no-till machines disturb the soil less than narrow (<50 mm wide) points. (No-till replaces the term *zero tillage*, which is commonly used in North America.)

Direct-drill is tillage of the entire topsoil, usually in a single pass at the time of sowing. Typically, direct-drilling is carried out using wide points on a combine seed drill or air seeder, or using a culti-trash seeder or offset discs. (Direct-drill replaces the term *minimum tillage* which is imprecise.)

Reduced tillage is a single working before sowing.

Multiple tillage usually involves two weed control workings before sowing using wide points. (Multiple tillage replaces the terms *district practice* (or *conventional practice*), which are changing as more farmers shift to no-till or direct-drill systems.)

- *Reduced aeration.* Plant growth may be affected by a low oxygen supply in the root zone even when plant-available water is adequate and the bulk density is non-limiting.
- *Reduced trafficability.* The correct timing of operations, such as spraying for weeds, may not be possible without a high risk of bogging and/or damaging the soil structure. This can result in poor weed control or higher rates of herbicide to kill established weeds.

A lupin seedling pushing through a surface crust (H. Cochrane).

Processes of soil structure decline

There are two main processes which contribute to structural decline: *slaking,* which is mainly a physical process, and *dispersion* which is predominantly a chemical process.

Aggregate slaking occurs when aggregates in a loosened, dry soil are wetted and subsequently collapse. Rapid wetting to saturation and over-burden pressure promote slaking. During wetting, aggregates collapse to form micro-aggregates and primary particles (e.g. coarse sand), because they have insufficient strength to withstand the stresses induced by rapid water intake (e.g. Emerson 1967, 1977; Collis-George and Greene 1979; Mullins *et al.* 1990).

Slaking reduces infiltration because the aggregates block the macropores. Primary particles either form a new layer of smaller aggregates which have a lower porosity, or block (to varying degrees) the larger pores between partially slaked aggregates (Collis-George and Greene 1979). Propensity to slake is higher in soils with weak structural development, particularly those with low organic matter. It is also related to the clay mineralogy (slaking in kaolinite > illite > montmorillonite) and antecedent moisture content (Emerson 1977). Most air-dry soils will slake to some extent if wet rapidly, but for a given soil subjected to rapid wetting, slaking decreases as antecedent moisture content increases (Emerson 1977).

Clay dispersion is the complete breakdown of aggregates into primary particles of sand, silt and clay in saturated soils. It is slow, often taking hours to complete.

Soils disperse when the attractive forces between the particles are no longer strong enough to hold them together (Emerson 1977). Clay particles commonly carry a net negative charge and this is compensated by hydrated cations which maintain electrical neutrality. These cations tend to diffuse away from the surface to equalise the concentration in the soil solution. However, they are still electrostatically attracted to the clay particles, resulting in a diffuse cloud of cations on the exterior surfaces of clay particles (*diffuse electrical double layer*).

The thickness of the diffuse electrical double layer depends on the electrolyte concentration, valency of the cations and pH (variable-charge surfaces). It decreases as the electrolyte concentration increases and also when there is a high proportion of exchangeable divalent cations (e.g. Ca^{2+}). Conversely the presence of monovalent ions (e.g. Na^{+}) increases the thickness.

If the width of the diffuse layer exceeds a certain threshold, then the short-range forces are no longer sufficient to maintain a flocculated soil and the clay particles disperse (e.g. Sumner 1992). The dispersed clay can block pores, reducing infiltration and gas exchange. A characteristic sign of dispersion is muddy or cloudy water, the cloudiness being dispersed clay in suspension.

Kay (1990) identifies three groups of soils with respect to dispersion: *spontaneously dispersive soils*, i.e. those that disperse spontaneously on wetting; *potentially dispersive soils*, i.e. those which require some mechanical energy or remoulding before the clay disperses; and *flocculated soils* which are unlikely to disperse even when subject to considerable remoulding or shearing.

In many soils, both dispersion and slaking occur to some extent, but usually one process is dominant. Therefore, instability is caused predominantly by dispersion or predominantly by slaking. On severely degraded sites, dispersion is usually the dominant process.

Soil aggregates immersed in distilled water showing soils susceptible to slaking and dispersion, in comparison with a stable soil.

SUSCEPTIBILITY TO SOIL STRUCTURE DECLINE

Many hypotheses have attempted to define the most important properties contributing to structural decline. Some consider high amounts of very fine sand or silt as the critical factor whereas others favour increasing clay content. Clay mineralogy, organic matter, sodicity and electrical conductivity are also significant (e.g. Emerson 1983; Shainberg 1985; Mullins *et al.* 1987, 1990; Batey 1988). However, no universal relationships between these properties and their effect on structural decline have been established. In most soils, a combination of properties are responsible, with management having a pivotal role in all but the most unstable soils.

Susceptibility to structural decline is complex, however it is worthwhile highlighting three factors: soil texture, soil chemistry and the effect of management practices.

(i) Soil texture

Soil texture has a large influence on a soil's behaviour when it is cultivated. If a soil is predominantly composed of coarse particles (fine sand or larger), the number of aggregates will be small and the infiltration of water is likely to be high. In these coarse-textured soils there is usually minimal aggregation and that present is due to plant roots and organic matter binding the particles together, rather than clay aggregation. Clay dispersion may occur in soils with a low clay content (<10%), but in general the dispersed clay will influence the physical properties of the surface soil less than mechanical processes, such as packing density (Aylmore and Cochrane 1987).

Susceptibility to crusting is strongly related to texture. Crusts will form on most soils, but form more readily on sandy loams than on clay loams. Soils with a high proportion of silt or very fine sand are highly susceptible to crusting (Bradford and Huang 1992). With increasing clay and decreasing silt content the dominant effect of crusting is to reduce infiltration rather than cause a large increase in strength (Bradford and Huang 1992).

In general, soils with textures between sandy loam and clay loam (10 to 35% clay) are the most susceptible to structural decline (Figure 3.2.1, Mullins *et al.* 1990). As the clay increases, the positive effects on soil structure associated with shrinkage and expansion come into play.

(ii) Soil chemistry

Stability is strongly related to the chemical properties of the soil, especially the clay fraction. In general, soils disperse if they have a high exchangeable sodium percentage (ESP). The degree of dispersion is also affected by the electrical conductivity (EC) of the soil solution, the ratio of exchangeable calcium to magnesium, and the amount of organic matter (Table 3.2.1). A high EC can overcome the dispersive effects of high ESP and whether clay particles are flocculated or dispersed is often determined by the balance of these factors. Organic matter generally assists with the development of stable aggregates.

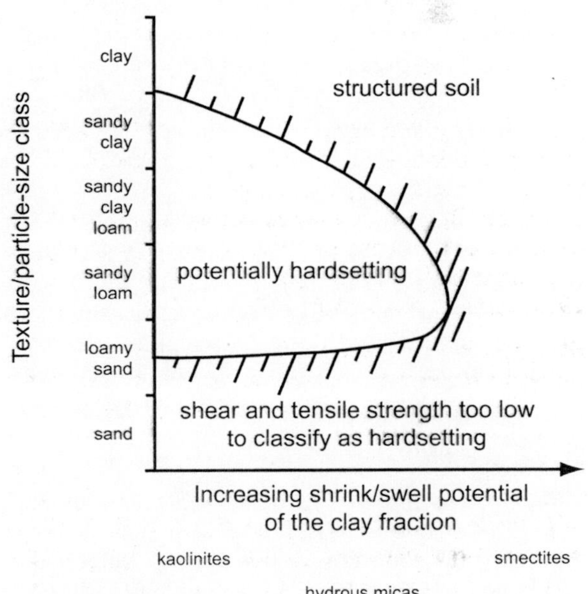

Figure 3.2.1 *Susceptibility to hardsetting as related to soil texture and clay mineralogy (from Figure 1, Mullins* et al. *1990).*

Table 3.2.1 Properties which increase or decrease the stability of soil.

Decreasing stability (negative effect)	Increasing stability (positive effect)
High exchangeable sodium (sodic)	Low exchangeable sodium (non-sodic)
Low electrolyte concentration	High electrolyte (salt) concentration, but not saline
Low ratio of exchangeable calcium to magnesium	High ratio of exchangeable calcium to magnesium
Low organic matter	High organic matter
	Finely divided calcium carbonate
	High concentration of aluminium and iron oxides and hydroxides
	Exchangeable aluminium (acid soils)

Calcareous soils, especially if the calcium carbonate is finely divided throughout the soil matrix, are usually stable because the concentration of Ca^{2+} in the soil solution is usually $>10^{-3}$M which is sufficient to flocculate the soil. Calcium carbonate also appears to have a direct cementing effect (Rimmer and Greenland 1976). Oxides and hydroxides of aluminium, and to a lesser extent iron, increase stability in some soils but not in others (Deshpande *et al.* 1968; Hamblin and Greenland 1977).

Exchangeable sodium. The exchangeable sodium percentage (ESP) is the proportion of sodium ions on the exchange sites of the clay in relation to the total cation exchange capacity (ESP = exchangeable sodium x 100/cation exchange capacity). In Australia, a soil is 'sodic' if the ESP >6 and 'highly sodic' if the ESP >15 (Northcote and Skene 1972). Many so-called sodic soils only have a sodic clayey subsoil, the topsoil being non-sodic.

Soils with sodium on the exchange complex are more susceptible to dispersion. Sodium cations expand the diffuse electrical double layer, because they are highly hydrated. On wetting, the aggregates swell and if the electrolyte concentration is low dispersion will occur. Therefore, sodic and highly sodic soils can be spontaneously dispersive, depending on the electrolyte concentration.

Non-sodic soils can be dispersive under the influence of remoulding forces such as cultivation and stock trampling (Emerson 1983; Rengasamy *et al.* 1984) or when prolonged rain lowers electrolyte concentrations near the surface. Many medium and fine-textured agricultural soils in WA have non-sodic surfaces, but are dispersive and hardsetting due to their low EC and excessive cultivation or cultivation when the soil was wetter than the lower plastic limit (Howell 1987b).

Other exchangeable cations. Calcium on clay exchange sites is beneficial to structure and helps to flocculate the soil. Magnesium is not as effective as calcium in alleviating the effect of sodium. Although Mg is also a divalent ion, it can have a deleterious effect on soil stability (Bakker and Emerson 1973; Aylmore and Sills 1982). The important consideration is not the absolute amount of exchangeable magnesium, but rather the ratio of exchangeable Ca to exchangeable Mg. For a given ESP, lower Ca:Mg ratios (i.e. <1.0) exacerbate dispersion (Rengasamy *et al.* 1986; Mullins *et al.* 1990).

Exchangeable aluminium is only present in significant quantities in highly acid soils. With their high charge, Al ions are thought to reduce dispersion by compressing the diffuse electrical double layer (Emerson and Bakker 1973; Emerson 1983).

Low electrolyte. In the presence or absence of sodicity, soils can be dispersive due to the low salt concentration in the soil solution. Soils require a minimum concentration of salt (or electrolyte) to remain stable or flocculated (Quirk and Schofield 1955). This does not infer that the soil is saline, because all soils contain some salts.

Organic matter. The role of organic matter (OM) as an important soil conditioner is widely accepted, even though it is chemically complex and the mechanisms by which it stabilises soil aggregates are not understood fully (Tisdall and Oades 1982). It is cheap to produce and can be grown on-site to ameliorate problems of soil structure. There is a useful increase in organic matter under a productive pasture phase, but after two or more cultivations the OM is mineralised and the OM content reverts rapidly to pre-pasture concentrations (e.g. Stoneman 1973).

There appears to be an upper limit of OM in the agricultural soils of WA depending on the soil texture and climate. In general, the upper limit is higher in finer textured soils than coarse-textured soils, because OM binds strongly to clay minerals (Tisdall and Oades 1982). The upper limit for fine-textured soils in WA is about 2 to 2.5% organic carbon (OC). There also is a lower limit corresponding to that fraction which is highly stable and resistant to physical and microbial breakdown, making it effectively inert. The lower limit appears to be about 0.9% OC for fine-textured soils, but would be lower for coarse-textured soils (Howell 1987c). These values may change under no-till systems or with better pastures. The bulk organic matter is a broad indicator of soil stability, but is of limited use for detecting changes in stability with changes in management (Jarvis 1987).

(iii) The effect of management practices

Most uncultivated surface soils in WA are stable, but are likely to degrade if they are cultivated when wet. Pasture is seen as beneficial as it promotes the accumulation of organic matter and aggregate stability, but pugging of the soil by grazing animals can cause degradation.

On many soils, a combination of management practices have contributed to the present condition. Four of the main effects are listed below and described in Table 3.2.2. Their relative importance depends on soil type and management history.

- Breakdown and remoulding of aggregates after cultivation of wet soils
- Loss of organic matter
- Stock trampling wet soils and surface compaction
- Mixing of dispersive subsoils with surface soil during cultivation

Table 3.2.2 The main effects of management practices on soil structure.

Management practice	Effect on soil structure
Cultivation of wet soils	Excessive cultivation and poorly timed cultivations are the major causes of soil structure decline. Many soils disperse if remoulded, either because they are being cultivated above the lower plastic limit or because stock are pugging wet soils. Without remoulding, there are few soils in south-western Australia which will disperse simply on wetting, i.e. are spontaneously dispersive (Howell 1987c). The effect of excess tillage on structural condition has been demonstrated in many trials comparing conventional cultivation with reduced or no-till sowing (e.g. Hamblin 1984).
Loss of organic matter	There is generally a decline in soil organic matter (OM) after clearing (Stoneman 1962) and following cultivation (Tisdall and Oades 1982). Cultivation increases the rate of mineralisation by exposing previously inaccessible OM to micro-organisms and oxidation. Cultivation, even with tined implements, dilutes OM through the soil. In an undisturbed soil or after a long pasture phase, OM accumulates in the surface 0-3 cm (Hamblin 1984), where it has the largest effect on stability.
Stock trampling wet soils	Trampling can cause two problems. At low to medium soil water contents compaction by hooves predominates, decreasing the volume of large pores (Beckman and Smith 1974). Under wetter conditions compaction can still occur, but the soil can also be progressively weakened by the repeated remoulding action of hooves as rainwater is incorporated and bonds between particles are disturbed (Mullins and Fraser 1980). Soils under pasture are usually pedal, although often only a weak crumb structure, as deterioration can occur through poor grazing management. The severity of structural decline through animal trampling depends on factors such as soil texture, structural condition, the length of time grazing occurs under wet conditions, stocking rate, pasture species and vigour. For example, removing stock when the soil was close to the lower plastic limit helped to reduce structural decline, which ultimately improved workability in subsequent cropping years (Proffitt *et al.* 1995a, b). Reduced stocking rates and deferred grazing have also been shown to help reduce the degree of structural damage to fragile surface soils (Proffitt *et al.* 1993, 1995c).
Mixing of dispersive subsoils	Mixing dispersive subsoil clays with surface layers is an important cause of structural decline in permeability contrast soils, especially in the Great Southern. Before clearing, many of these soils had a shallow (<10 cm), coarse-textured A1 horizon overlying a clayey subsoil. Mixing the surface soil with the subsoil has occurred because of deep cultivation and/or removal of part or all of the A1 horizon through erosion (wind erosion and/or sheet erosion). For example, in the Salmon Gums district, the difference in surface texture between Circle Valley loamy sand and Circle Valley sandy loam was caused by the mixing of a very shallow A1 horizon with the clayey subsoil in the latter. Before clearing, all Circle Valley soils had a similar surface texture (Burvill 1988). Mixing clayey subsoils with the A1 horizon increases the clay content, but it is the properties of the clay that cause most problems. The clayey subsoils contain high concentrations of exchangeable sodium and/or magnesium. For example, on undisturbed sites in the Jerramungup district, the ESP typically increases rapidly from 2-10 in the top 10 cm (clayey sand texture), to highly sodic (ESP 20-30) clayey soils (20 to 60% clay) at a depth of 20 cm (T. Overheu unpublished data). In general, stability depends on the amount of subsoil incorporation, decreasing as more subsoil is mixed with the surface soil. Mixing of subsoil with surface soil is often highly variable, so the surface condition reflects this variation.

Stock trampling wet soils cause (a) smearing and (b) compaction (T. Proffitt).

ASSESSING SUSCEPTIBILITY TO STRUCTURAL DECLINE

Two methods for assessing the susceptibility to soil structure decline are described. The first method was developed for duplex soils from the WA wheatbelt; the second is a general guide to the inherent stability.

(i) Identifying potentially dispersive soils: Using SAR, EC and clay dispersion test

A method has been developed to determine whether a soil is 'potentially dispersive' (i.e. dispersive on remoulding) using results from a number of field trials in WA. The test is recommended for duplex soils with a sandy loam or finer surface texture from the WA wheatbelt, which have not been treated with gypsum (Howell *et al.* 1988). A different relationship has been developed for the red-brown earths of north-eastern Victoria (Rengasamy *et al.* 1984, 1987).

The method is based on relationships between sodium adsorption ratio (SAR), electrical conductivity (EC) and clay dispersion. The conditions under which a soil is potentially dispersive can be identified by a unique line called the *critical concentration line.* This line (represented by Equation 3.2.1) separates conditions under which the clay is dispersive from conditions under which it is non-dispersive (Dellar *et al.* 1988; Howell *et al.* 1988). Soils to the left of this line were dispersive and to the right, non-dispersive (Figure 3.2.2).

$$EC_D = 0.91 + 0.12\ SAR_D \qquad (3.2.1)$$

Where,

EC_D is electrical conductivity measured in 1:0.75 soil:water extract (dS/m).

SAR_D is the sodium adsorption ratio. $SAR_D = Na/(Ca + Mg)^{1/2}$, where Ca, Mg and Na are measured in the supernatant of a 1:0.75 soil:water extract.

There is a close relationship between SAR and ESP for hardsetting wheatbelt soils, SAR being a more straightforward test (Equations 3.2.2, 3.2.3, Dellar *et al.* 1988).

Low sulphate soils ($SO_4^{2-} \leq 4$ me/L)

$$ESP = 1.4\ SAR + 0.8 \quad R^2 = 0.95 \qquad (3.2.2)$$

High sulphate soils ($SO_4^{2-} \geq 4$ me/L)

$$ESP = 2.2\ SAR + 0.9 \quad R^2 = 0.94 \qquad (3.2.3)$$

In the absence of information on the sulphate concentration in the 1:0.75 extract, it can be assumed that gypsum-treated soils have more than 4 me SO_4^{2-}/L for the first few years after application.

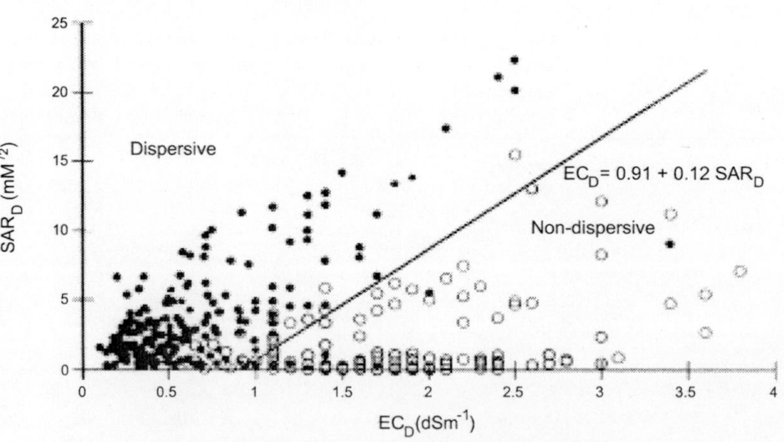

Figure 3.2.2 *The dispersive behaviour of some WA wheatbelt soils in relation to sodium adsorption ratio (SAR) and electrical conductivity (EC) measured in a 1:0.75 soil to water extract (from Figure 1, Dellar and Proffitt 1994).*

Method (from Dellar *et al.* 1988)

- Collect soil samples from the 0 to 5 cm layer
- Air dry soil samples and pass through a 2 mm sieve
- Prepare a 1:0.75 soil to water extract by adding 80 g of soil to 60 mL of distilled water and leave to stand overnight
- Gently stir (four to five revolutions) the suspension with a spatula and stand for one hour
- Decant the supernatant, stand for one hour then note whether it is dispersed or flocculated
- Measure the EC of the supernatant.

INTERPRETING THE RESULTS

Dispersed clay in the supernatant suggests that the soil has the potential to develop unfavourable physical characteristics, however it does not necessarily mean it is currently in poor condition.

Dispersed soils with EC >100 mS/m are most likely to be sodic. Their management may be different to non-sodic dispersive soils.

(ii) Estimating stability from laboratory data

Soil stability must be measured as it cannot be determined visually. It is useful to differentiate between soil stability and present condition. Soil stability is an inherent property, largely determined by soil chemistry; present condition is a reflection of both stability and management.

It is well established that exchangeable sodium, and to a lesser extent exchangeable magnesium, increase susceptibility to dispersion, but this does not relate directly to soil stability. For instance a highly sodic surface soil (ESP >15) can be structurally stable, while soils with an ESP as low as 2 can be highly dispersive (Cochrane *et al.* 1994). The inherent stability of a soil depends on the net effect of a number of properties.

When laboratory data are available, a *soil stability score* can be used as a general guide to stability. The soil stability score is the sum of the individual scores for four soil properties: sodicity, soil organic matter, calcium:magnesium ratio and electrical conductivity (Table 3.2.3).

Soils with similar stability scores do not necessarily behave in the same way, because behaviour also depends on susceptibility to slaking or dispersion. For instance, a red earth (10% clay) with a very low organic carbon, low electrical conductivity and a soil stability score of -2 would be susceptible to slaking and behave differently to a sodic clay with the same score.

Table 3.2.3 Determining the soil stability score for soils with a sandy loam or finer surface texture.

Soil property	Unit of measure and soil stability score			
Sodicity	**ESP**			
	<6	**6-15**		**>15**
Stability score	0	-2		-3
Soil organic matter	**Organic carbon %**			
	<0.8	**0.8-1.5**	**1.5-2.5**	**>2.5**
Stability score	-1	0	+1	+2
Exchangeable Ca and Mg	**Ca:Mg ratio**			
	<1	**1-3**		**>3**
Stability score	-1	0		+1
Electrical conductivity	**ECe* (mS/m)**			
	<50	**50-100**	**100-150**	**>150**
Stability score	-2	-1	+1	+2

* *To convert EC(1:5) to ECe refer to Section 5.3.*

Soil stability score

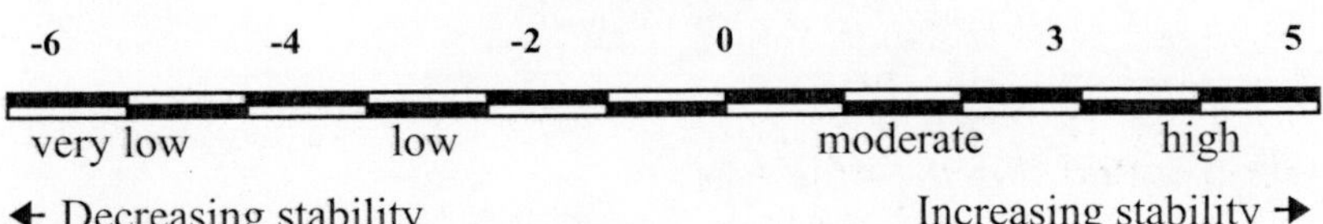

Example: *The soil stability score is equal to the sum of the scores for sodicity, organic matter, Ca:Mg and electrical conductivity. A non-sodic soil with an ESP <6, 1.2% organic carbon, a Ca:Mg ratio of 0.5 and an ECe of 130 mS/m has a overall soil stability score of 0 (i.e. 0+0-1+1 = 0), which corresponds to a low to moderate stability.*

ASSESSING PADDOCK STATUS

It is difficult to quantify the difference between a well structured and poorly structured soil. Many methods have been used, including hydraulic properties such as infiltration and hydraulic conductivity (disc permeameter, rainfall simulator), soil density, porosity and strength, aggregate stability (water stable aggregates), modulus of rupture and image analyses of resin impregnated soil cores. Many of these measurements are time consuming and need a large number of replicates to allow for spatial variability.

To help determine appropriate management options, qualitative or semi-quantitative measurements are generally satisfactory. The emphasis in this section is placed on field observations and straightforward measurements.

If field observations indicate a soil is partially degraded to degraded, use the Emerson aggregate test to determine whether the instability is predominantly caused by slaking or dispersion. If a soil's degraded condition has been predominantly caused by dispersion, the modified modulus of rupture test or paddock test strips can be used to determine whether it is likely to respond to gypsum.

(i) Field observations

Indicators of a structurally degraded surface include a surface crust, hardset surface and ponded, cloudy rain water. Visible features of the profile and observations on how the soil behaves when cultivated are useful indicators of the soil condition (Table 3.2.4). However, field observations can give misleading results as soil structure is dynamic and short-term management could mask unfavourable attributes. Field assessments should be made after harvest or early in the growing season about a month after seeding.

(ii) Identifying the degradation processes: Emerson aggregate test

The *Emerson aggregate test* (Emerson 1967) is straightforward, quick, useful for screening large numbers of soils and the results are reproducible. It is useful for broadly defining the stability of soils and for differentiating between the processes of slaking and dispersion. Unfortunately the test is not very sensitive to changes in soil condition caused by management.

The test assesses how aggregates breakdown in water and classifies a soil into eight categories (Figure 3.2.3). It involves placing soil aggregates into distilled water and observing their behaviour, after one hour and then again after 24 hours. If the aggregates slake but are

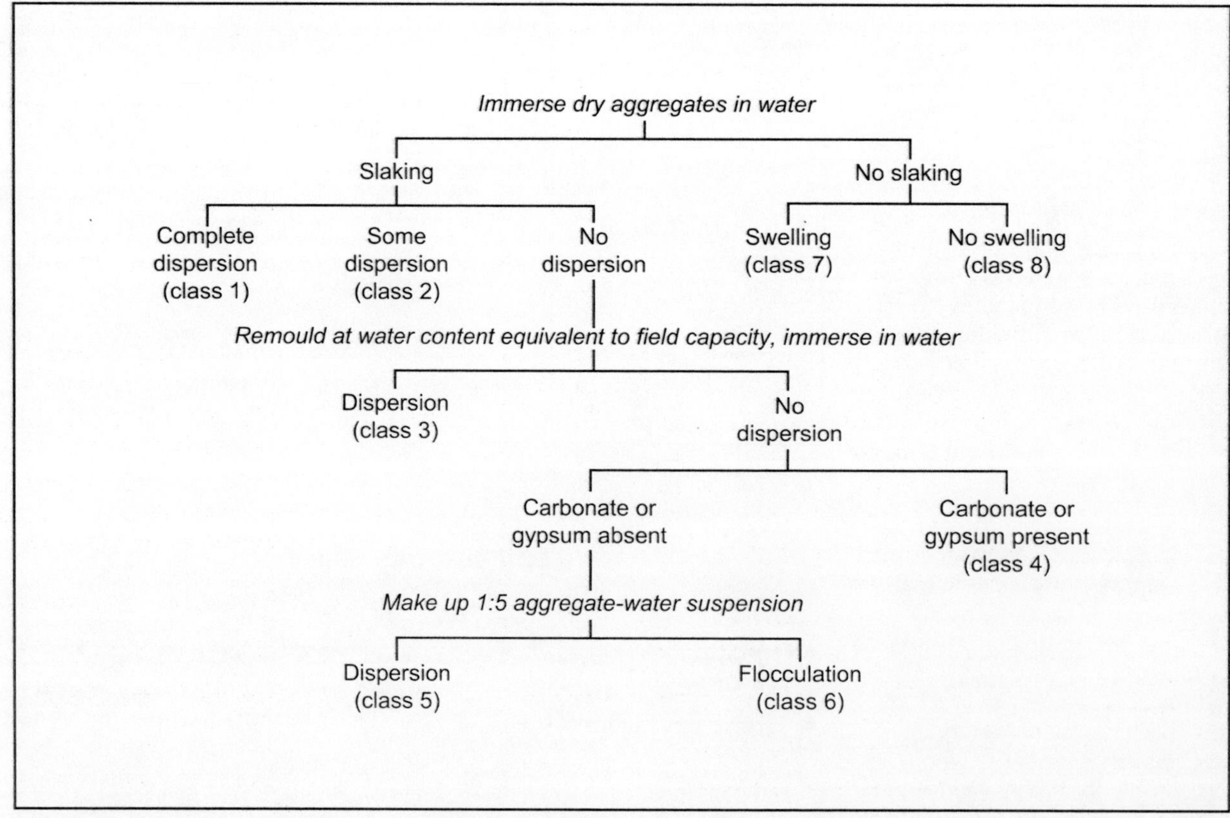

Figure 3.2.3 *Scheme for determining the Emerson aggregate test class (from Figure 1, Emerson 1967).*

not dispersive then they are remoulded at a water content approximating upper storage limit (i.e. field capacity) and again placed in distilled water and assessed for dispersion (Emerson 1967). The remoulding is designed to simulate the cultivation of wet soils. The method is fully described in Emerson (1967) with a modified version in Emerson (1991).

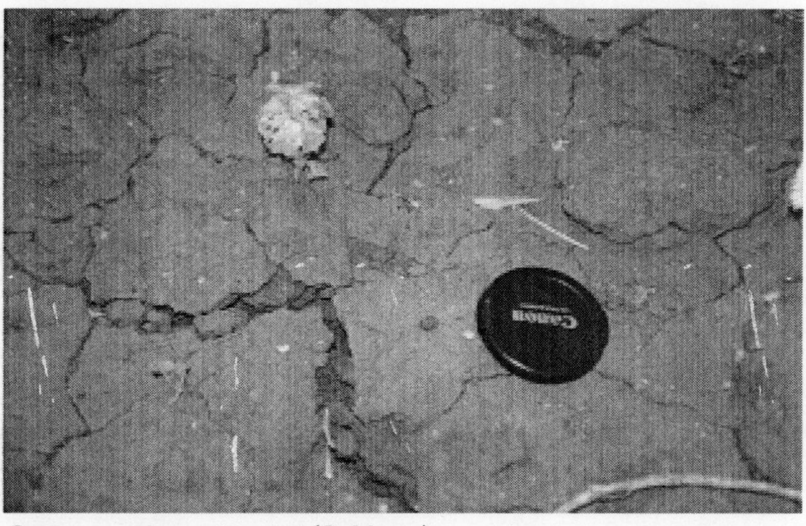

Surface cracking when dry (G. Moore).

The Emerson aggregate test is a simple way of identifying four important soil groups with respect to their behaviour when cultivated: soils which are spontaneously dispersive to varying degrees (classes 1 and 2): soils which are

Table 3.2.4 Assessing the present condition of surface soil from field observations.

Type of observation	Observations of soil behaviour	Condition of surface
Surface soil when dry	Surface is soft or crumbly and a sharpened pencil can be pushed into the soil easily.	Good
	Crust evident, can be lifted off underlying soil.	Partially degraded
	Whole topsoil is hard like concrete, i.e. hardset (dispersion and slaking). A sharpened pencil cannot be pushed into the soil.	Degraded
	The depth of penetration of a mattock or pick falling under its own weight can be used as a guide to soil strength. The mattock will penetrate a soil in good condition, but will not penetrate a degraded soil.	Good to degraded
	Dry surface forms widely spaced, deep cracks. (Not to be confused with cracking clays which contain shrink-swell properties.)	Partially degraded
Rainfall infiltration	Water infiltrates readily. Steady state infiltration measured with a ring infiltrometer* of 30-70 mm/h. Soils with excellent structure may have an infiltration >70 mm/h.	Good
	Water ponds readily on surface after rain, cloudy water may be present indicating clay dispersion. Steady state infiltration of 10-30 mm/h.	Partially degraded
	Ponded water becomes cloudy and remains present for days to weeks. Steady state infiltration is <10 mm/h.	Degraded
Aggregation and porosity under pasture	Loose or strongly aggregated soil indicating a large number of pores.	Good
	Weakly structured or massive (pores visible to the naked eye and evenly distributed).	Good condition
	Dense soil packed tightly (no pores or few visible to naked eye).	Partially degraded to degraded
Management observations	Crop growth is patchy, characteristically with undulating growth or seedling emergence.	Degraded
	Soil becomes boggy/slippery after light rainfall making machinery operations difficult.	Degraded
	Seed-beds are cloddy and difficult to produce under nearly all conditions. The clods reflect the size of tillage implement used.	Degraded
	Seed-bed breaks down rapidly and 'melts' with rain.	Degraded

** A straightforward method for measuring infiltration on the farm is described in Geeves* et al. *(1994) and Hunt and Gilkes (1992).*

potentially dispersive, i.e. they can disperse if remoulded when wet (class 3); soils which slake but are non-dispersive (classes 4, 5, 6); and soils which have a high inherent stability (classes 7, 8). Class 1 soils are highly unstable and invariably sodic to highly sodic (Emerson 1967), but are rare in south-western Australia where most hardsetting soils only disperse on remoulding (i.e. class 3, Howell 1987c).

(iii) Monitoring paddock condition: Modified modulus of rupture test

The modulus of rupture test measures the dry strength of a compacted soil brick (Richards 1953; Aylmore and Sills 1982). For WA wheatbelt soils with textures from sandy loam to clay loam, and where kaolinite is the dominant clay mineral, there was a good correlation between modulus of rupture (MOR), soil behaviour and crop yields (Aylmore and Cochrane 1987).

This test has been modified into a straightforward and reliable way to quantify the condition of hardsetting soils (Cochrane 1989). The modified test involves packing soil loosely into a plastic mould, subjecting it to a wetting and drying cycle, then measuring the strength of the dry blocks using a hand-held penetrometer. The greater the strength of the dry soil, the less stable the soil.

The modified test can also be used to estimate the likely yield response to gypsum, by comparing the dry strength of a soil wet with distilled water (i.e. dispersion) with the dry strength of the same soil wet with a weak salt solution (i.e. no dispersion). The salt solution simulates the addition of gypsum. Cochrane (1989) calibrated results against 35 gypsum field trials to estimate the likely yield response to gypsum application.

The modified MOR test is suitable for:

- degraded, hardsetting soils
- identifying soils in which structure is already a problem, or may become so in the future
- identifying soils which are likely to respond to gypsum
- providing an objective measure for monitoring changes in soil stability with changes in management.

The test is sensitive enough to detect the small changes in structural stability that occur from year to year. For example, Dellar and Proffitt (1994) used it on 46 paired sites which had different tillage treatments. The test detected a lower soil strength on the direct-drilled sites, even though the amount of dispersed clay was similar for each site.

The test is less suitable for:

- soils with well developed macro-aggregate structure, because the aggregates are destroyed when the soils are sieved (2 mm diameter)
- soils that shrink and swell on wetting and drying, as they are likely to produce cracked or deformed blocks
- subsoils, because they are not subject to the same degree of wetting and drying as the topsoil, and so different criteria would be necessary to interpret the results.

The method is described in Cochrane (1989), Hunt and Gilkes (1992), with a slightly modified version in Geeves *et al.* (1995). The most critical steps are: sample collection, sample preparation (especially sieving, when it is important to gently crush, not grind, the sample) and drying the blocks (three to seven days if air drying).

Interpreting the results

The average strength of blocks wet with distilled water indicates their structural stability; the higher the result the less stable the soil. Soils with a high stability give modified MOR readings near zero, while values greater than 2.5 using a large tip (2.5 cm diameter) on the penetrometer, or 0.7 using a small tip (0.6 cm) indicate low stability.

The strength of blocks formed using the modified MOR test is usually lower than the same soils measured in the field. The effect on plant growth depends on the size of the hard blocks of soil, not just their strength.

Response to gypsum. The likely response to gypsum can be estimated from Equation 3.2.4 (Cochrane 1989):

$$\text{\% increase in grain yield} = \{BS_w - (BS_{ss} \times 0.6)\} \times 5.5^* \quad (3.2.4)$$

where,

BS_w is the average strength (kg/cm^2) for soil blocks wet with distilled water

BS_{ss} is the average strength (kg/cm^2) for soil blocks wet with a salt solution

* Multiply by 5.5 if strength measured with a large tip on penetrometer, or by 20 for small tip.

Estimating modulus of rupture. Modulus of rupture can be estimated from Equations 3.2.5a-b; using the results from bricks formed by rapid wetting with distilled water (H. Cochrane cited in Geeves *et al.* 1995).

$$MOR = 21.4BSL + 10.5 \quad R^2 = 0.928 \quad (3.2.5a)$$

$$MOR = 75.0BSS + 11.1 \quad R^2 = 0.934 \quad (3.2.5b)$$

where,

MOR is modulus of rupture (kPa)

BSL is strength recorded with the large penetrometer tip (kg/cm^2)

BSS is strength recorded with the small penetrometer tip (kg/cm^2)

Modulus of rupture ranges from zero for coarse-textured soils to more than 500 kPa in highly degraded soils. A critical value for differentiating between problem and non-problem soils appears to be 60 kPa (Cochrane and Aylmore 1997).

A soil in poor condition following cultivation (G. Moore).

MANAGEMENT OPTIONS

Where soil structure decline is a problem or could become a problem, there are essentially three broad management strategies:

(i) Reduce the impact of management practices

(ii) Increase the stability of the soil

- Identify the degradation processes involved
- Increase the organic matter or use an ameliorant such as gypsum

(iii) Improve the aggregate structure of the soil

- Limited cultivation may be necessary to create some aggregates if the soil is degraded

Reducing the impact of management practices is the dominant strategy to maintain a soil in good condition. In some situations a combination of all three is required to restore soil to good condition. The appropriate combination of these strategies will vary with the present condition of the soil and its properties.

The ways of assessing and managing soil structure decline are summarised in Figure 3.2.4 and expanded below.

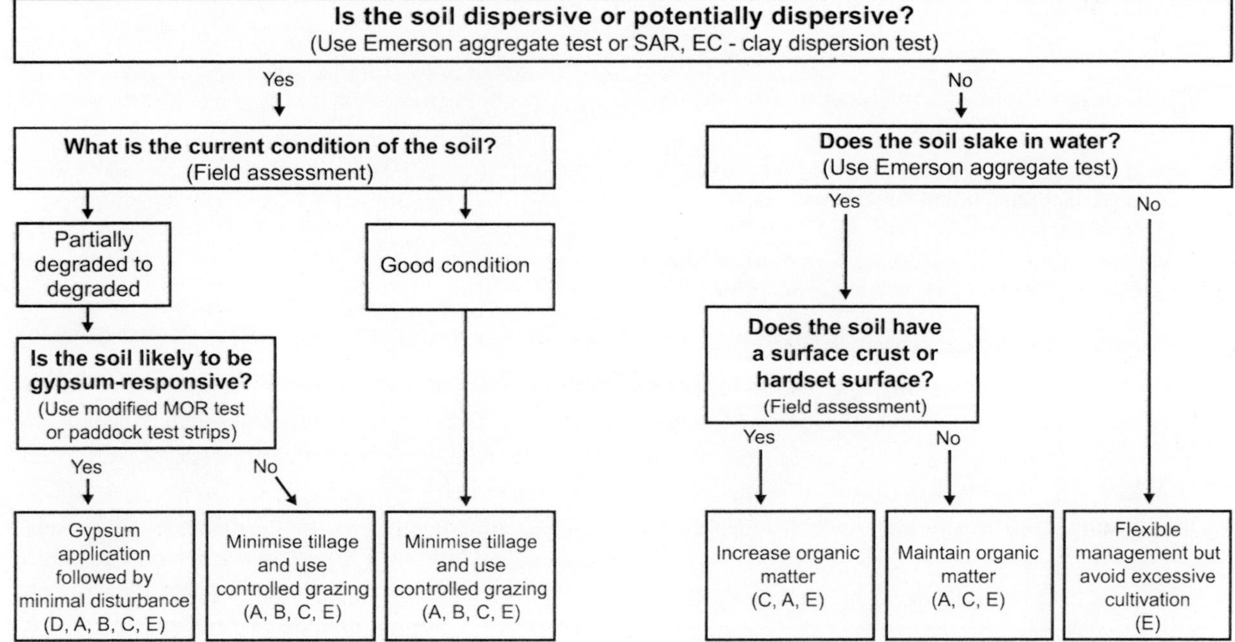

Figure 3.2.4 *Summary of assessment and management of soil structure decline.*

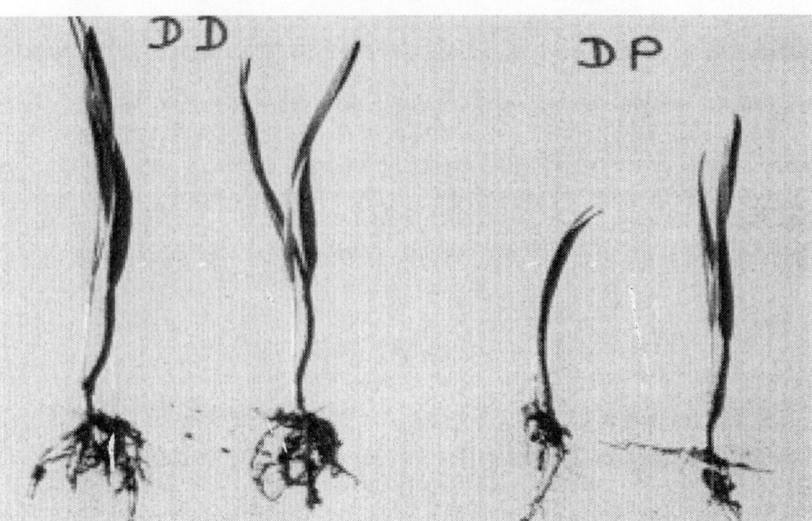

The effect of tillage on crop growth 1 month after seeding: DD - direct-drill, DP - multiple tillage (T. Proffitt).

Ways to maintain or improve soil structure (Letters refer to Figure 3.2.4)

(A) Tillage

Where tillage is necessary, apply the following guidelines:

- Only till when the soil moisture content is right. In particular, do not till at high soil water contents, i.e. if moisture content is greater than the lower plastic limit (see box) then the soil is too wet.
- Do not invert the soil, and avoid deep cultivations which bring up clayey subsoil. On permeability contrast soils with a very shallow topsoil, ensure subsoil is not mixed with topsoil.
- Avoid more than one cultivation before seeding.
- Cultivate early and use herbicides where necessary instead of repeating cultivations.
- Avoid closely spaced cultivations, as this destroys the macro-aggregates and tends to increase the rate of structural decline through age-hardening (i.e. hardening due to cementation of micro-aggregates in tilled soils, Dexter *et al.* 1988).

(B) Control surface damage by stock

- Defer grazing on susceptible soils until pasture is well established. This reduces the number of days the soil is at risk.
- Use lower stocking rates on susceptible paddocks early in the season when the soil is wet and pasture cover is low.
- Use rotational/controlled grazing; remove stock at risky times when the soil is wet (i.e. close to the lower plastic limit) and only return them when the soil has dried to below the lower plastic limit.

> *The key is to avoid all unnecessary tillage operations. The highest priority is to reduce tillage to a minimum, adopting either a direct-drill or no-till system.*

(C) Increase organic matter

Maintain high organic matter, particularly at the surface. Pastures can be very beneficial in increasing organic matter levels, microbial activity and improving aggregate stability. However, remember:

- To have a beneficial effect on soil structure, the pasture must be well managed, productive and not overgrazed.
- Groundcover should be sufficient to ensure the surface is protected from raindrop impact.
- Control grazing as in (B).

(D) Apply gypsum

Relevant to dispersive and potentially dispersive soils in degraded condition which are gypsum-responsive according to the modified MOR test or paddock test strips.

- Apply gypsum at 2.5 t/ha (rates up to 5 t/ha may be necessary on highly sodic soils), before the break of season. Surface application is adequate, although in areas of 'bare' soil the gypsum should be scratched in to avoid loss through wind erosion. Sow a cereal crop in the first year.
- Apply gypsum as part of a system which minimises soil disturbance (i.e. direct-drill or no-till sowing) to maintain the stability. Refer to (A) and below.
- Consider nitrogen nutrition, because of the higher yield potential following gypsum application.

PLASTIC LIMITS

Lower plastic limit is the moisture content when soil changes from being brittle to plastic. It is the minimum moisture content at which the soil can be moulded and the maximum moisture content at which the soil is friable. In most soils it is at or near the upper storage limit.

Upper plastic limit (or liquid limit) is the moisture content at which the soil changes from a plastic solid to a viscous liquid.

The values of the lower and upper plastic limits (also called *Atterberg limits*) vary with clay content and clay mineralogy. Coarse-textured soils are non-plastic, consequently the limits described above are not applicable.

Test for lower plastic limit. A simple test to determine the lower plastic limit is the moisture content at which the soil begins to break apart and crumble when rolled by hand into rods or threads 3 mm in diameter and 75 mm in length. If thinner rods can be formed, the soil is too wet and will easily smear and compact when cultivated. If the soil will not form rods about 3 mm in diameter it is drier than the lower plastic limit (Hicks 1991). With soils that crumble and are too dry to form rods, cultivation will pulverise existing aggregates and create dust.

This test appears to be unsatisfactory for coherent soils (may be hardsetting) with a clayey sand to sandy loam texture (i.e. ~8 to 15% clay).

Implications for management

Tillage. For coherent soils, the lower plastic limit is the optimum moisture content for tillage operations. Cultivating the soil when it is wetter than the lower plastic limit can cause major structural degradation.

Compaction. Soil wetter than the lower plastic limit, but not saturated, is easily compacted.

Soil workability. The closer the lower plastic limit is to the upper storage limit (field capacity) the greater the workability. A small difference between the two means the soil will be suitable for tillage soon after free drainage has ceased. A large difference indicates the soil has to dry for a considerable time before it can be cultivated safely (Kirby 1997).

The effect of gypsum and tillage on soil condition. (a) Gypsum application, plus ten years continuous direct-drilling. (b) No gypsum application. Two cultivations prior to seeding for ten years (M. Howell).

(E) Retain stubble

The benefits of stubble retention are linked to protection of the soil surface from wind erosion, raindrop impact and water erosion. In WA there is no conclusive evidence that stubble retention has a positive effect on soil organic carbon (Perry *et al.* 1992).

- Spread stubble, or rake if necessary, to reduce quantities to a manageable level.
- Do not cultivate to reduce stubble levels. The additional cultivation will have a more adverse effect on soil structure than the benefits of stubble incorporation.

Gypsum option

Gypsum (hydrated calcium sulphate, $CaSO_4.2H_2O$) is commonly used to improve soil physical conditions.

When used strategically, gypsum can play an important role in stabilising soil. The gypsum does not create good structure, but stabilises the present structure.

Table 3.2.5 Factors which can influence yield response on gypsum-responsive soils.

Nitrogen	There is a strong nitrogen by gypsum interaction in many trials. The gypsum-treated plots have a higher nitrogen requirement simply because of a higher yield potential. It is probable that yield responses in some gypsum trials were limited by nitrogen nutrition (Hamblin and Howell 1987). Correct nitrogen nutrition is essential to ensure a positive yield response to gypsum.
Seasonal conditions	Seasonal conditions, in particular the amount and distribution of rainfall have a large effect on the grain yield response. From field observations of the development of hardsetting in different seasons, Howell (1987c) concluded: "...the expression of hardsetting in degraded soils is closely related to the type of rainfall that occurs after cultivation or seeding. .....Hardsetting occurs when the soil is wet with heavy rainfall before the soil has dried out and reverted to its non-dispersive state. In the eastern wheatbelt it has been a common occurrence in the latter half of the 1980's to have a reasonably good and early break sufficient for seeding to go ahead, followed by a three or even four week dry period. Under these conditions even the most degraded soils have been observed not to develop a severely hardset surface. If the soil is spared rapid wetting in its dispersive state then much of its structural integrity may be maintained". A dry matter response to gypsum may not translate to higher grain yield if there are no finishing rains. A similar scenario is common with deep-ripping on medium to fine-textured soils in low rainfall areas.
Cultivation versus direct-drill	Gypsum is not required on many potentially dispersive soils if direct-drilling and continuous cropping are adopted. In gypsum trials, the electrolyte effect persisted longer in direct-drilled soils (Howell 1987a). Once the residual gypsum has been leached the soil is again potentially dispersive, but the formation of more stable aggregates during the reclamation period mitigates against this possibility. Poor management will ultimately result in a return to the original problems. A soil improvement package including direct-drill or no-till sowing is necessary to maintain soil condition after gypsum is applied.
Time of seeding	A major advantage of improving the condition of the surface soil is that it can be cultivated over a wider moisture range. Together with the adoption of a direct-drill or no-till system, this allows the crop to be sown earlier with a potential increase in grain yield. Increased flexibility of operations should allow earlier seeding.
Legumes	In general, there is minimal value in applying gypsum to established and actively growing pastures (Warren 1992). There have been some negative responses to gypsum from legumes, especially at high rates of application. The decrease in field pea yields when gypsum was applied at 10 t/ha in a trial at Nokanning was attributed to gypsum adversely affecting early nodulation and vigour (Howell 1987a). The response of medic pastures to gypsum is generally negligible, although on a couple of trials negative responses were obtained (Warren 1992). Gypsum application is best followed by a cereal crop in the first year.
Other limitations to productivity	Soils with hardsetting surface horizons can have other unfavourable properties which limit yields. In many wheatbelt soils, the rooting depth is limited by unfavourable chemical (alkalinity, acidity, nutrient deficiencies or toxicities) or physical properties (waterlogging, poor subsoil structure). Waterlogging is a major problem on most permeability contrast soils. Adding gypsum will improve infiltration on the surface and therefore reduce surface waterlogging. Perched watertables on top of the clayey subsoils will not be improved by gypsum application. A number of gypsum trials show reduced yields caused by waterlogging.

Gypsum is a mineral commonly found in dunes or lunettes on the south-western edge of some salt lakes in the agricultural area (Jones 1994). Its quality for agricultural purposes (Howell *et al.* 1987) can be defined in terms of three parameters:

- low sodium concentration (<2% NaCl by weight)
- fine particle size (0.1 to 0.3 mm)
- high purity (>85% calcium sulphate).

Gypsum can improve soil stability in two ways, by 'cation exchange' and 'electrolyte effects' (Loveday 1976). In many soils both mechanisms may contribute, although for WA soils it has been established that the electrolyte effect is the dominant effect (Howell *et al.* 1987, 1988).

Cation exchange

When gypsum is applied to sodic soils, calcium can replace the exchangeable sodium. The reduction in the ESP is normally considered to be a permanent change, but it does not guarantee that the soil is always stable. It can still become unstable as stability will depend on the electrolyte concentration in the soil solution and the ratio of exchangeable calcium to magnesium. In non-sodic soils there can be some replacement of exchangeable magnesium with calcium ions (Howell 1987b).

Electrolyte effect

Gypsum dissolved by rainwater increases the concentration of total salts in the soil solution and this causes the clay particles to flocculate. The change in electrolyte can be measured by the electrical conductivity. Gypsum has a low solubility (1 to 2 g/L) compared with sodium chloride (264 g/L), but a higher solubility than calcium carbonate (0.013 to 0.14 g/L). The application of about 2 t/ha of gypsum mixed into the top 5 cm of soil results in a saturated solution of calcium sulphate, which has an ECe of about 200 mS/m.

The electrolyte effect is not permanent as the gypsum is gradually leached from the surface soil. Trial results show that gypsum can be leached within two to three years, although the gypsum persisted longer where soils were direct-drilled (Howell 1987a). This emphasises the necessity to use gypsum as part of an overall 'soil management package', whereby reduced tillage maintains the soil condition and reduces need for re-application.

Factors affecting grain yield response to gypsum

If a soil test indicates the soil is likely to give positive yield response to gypsum (refer to Modified modulus of rupture test), a number of factors (Table 3.2.5) could influence the outcome. The actual grain yield response will depend on various factors, some of which can be controlled (e.g. N nutrition) and some which cannot, such as seasonal conditions.

Further reading

Mullins, C.E., MacLeod, D.A., Northcote, K.H., Tisdall, J.M. and Young, I.M. (1990). Hardsetting soils: behaviour, occurrence and management. *Advances in Soil Science* **11:** 38-108.

Sumner, M.E. and Stewart, B.A. (eds.) (1992). 'Soil crusting. Chemical and physical processes'. Advances in Soil Science, Lewis Publishers, 372 p.

So, H.B., Smith, G.D., Raine, S.R., Schafer, B.M. and Loch, R.J. (eds.) (1994). Sealing, crusting and hardsetting soils: Productivity and conservation. *In* 'Proceedings of the second international symposium on sealing, crusting and hardsetting soils'. Australian Society of Soil Science Inc. (Queensland Branch), Brisbane, Australia 7-11 February, 1994.

3.3 SOIL WATER

Geoff Moore, David Hall and Jeff Russell***

This section describes water movement and storage in soils. At the paddock scale the primary interest is the soil's ability to store and supply water to the plant, while at the catchment scale the soil-water balance is of greatest concern. It is important to maximise plant water use for economic reasons and to minimise deep drainage below the root zone, because this is responsible for the widespread secondary salinity in WA (refer to Section 5.3).

Crop yields are often closely related to the amount of water used. For instance, French and Shultz (1984) developed a simple model to predict potential grain yield and suggested that the maximum water use efficiency for cereal crops is 20 kg grain/ha/mm, assuming evaporation from the soil surface was 110 mm. In WA, the maximum achievable water use efficiency of wheat crops appears to be 15.5 to 16.0 kg grain/ha/mm, assuming evaporation from the soil surface is 100 mm (Tennant *et al.* 1991).

Soil water storage is a major factor determining yield potential in areas with a summer-dominant rainfall such as the wheat growing areas of southern Queensland (Gardner 1997). In a Mediterranean environment, where most of the rain falls during the growing season, soil water storage is often less important, depending on seasonal conditions. For example, in some seasons, regular light showers provide an even water supply to the plant which closely matches crop transpiration, so differences between soils are minimised. In other seasons, where rainfall is abnormally high or low or is unevenly distributed throughout the growing season, differences between soils will be evident. Soils with very low water storage capacity, or unfavourable chemical or physical properties that restrict root growth, will invariably limit crop yields.

Soil-water relationships and rooting conditions are inter-related and must be considered together. The large variation in the maximum rooting depth of different crops and the tolerance of plants to different soil conditions results in the soil depth/plant rooting depth being the major variable affecting how much water is available to the plants. *Soil water storage should always be related to a specific crop or to a depth interval (e.g. 0 to 1 m).*

PRINCIPLES OF WATER MOVEMENT IN SOILS

Water moves in soils along potential gradients. A gradient can be created by slope and the matric potential of the soil (e.g. movement from wetter to drier soil, or from soil with a low matric potential such as sand to one with a higher matric potential such as clay). The rate at which the water moves also depends on the soil's permeability (the ease with which water can be transmitted). The permeability of a soil to water is described by its hydraulic conductivity (K) which is specific to a given soil. Darcy's Law combines the effects of gradient and hydraulic conductivity to calculate the quantity of water (flux) flowing in a saturated system:

$$\text{Flux rate in a saturated system (mm/h)} = -K_s * (\Delta\psi/\Delta z) \qquad (3.3.1)$$

where K_s is the saturated hydraulic conductivity, $\Delta\psi$ is the change in slope or matric potential and Δz is the change in distance.

Hydraulic conductivity is highest in soils with a porous structure and where the pores are interconnected (i.e. coarse sands, gravels and structured soils). Common values for K_s are given in Table 3.3.1. In general, K_s values greater than 0.1 m/day represent freely draining conditions, while soils where K_s is less than 0.0001 m/day are almost impermeable (Cox 1988; Tennant *et al.* 1992).

Table 3.3.1 Saturated hydraulic conductivity values for a range of soil texture classes (Brouwer 1978).

Texture	Hydraulic conductivity (m/day)
Clay soils	0.01-0.2
Deep clay beds	10^{-8}-10^{-2}
Loam soils (surface)	0.1-1
Fine sand	1-5
Medium sand	5-20
Coarse sand	20-100
Gravel	100-1000
Sand and gravel mixtures	5-100
Clayey sand and gravel mixture	0.001-0.1

* *Agriculture Western Australia, Esperance.*

** *Agriculture Western Australia, Merredin.*

Equation 3.3.1 is often used to calculate flow rates in aquifers. However, within the root zone the soil is rarely saturated. To accommodate unsaturated flow, knowledge of how hydraulic conductivity is affected by changes in water content and matric potential is necessary. Hydraulic conductivity (K) diminishes exponentially as the soil dries. For unsaturated conditions in a vertical orientation, Darcy's Law has been rewritten (Rose and Stern 1965; Williams 1983) as:

Flux rate in an unsaturated
system (mm/h) $= K(\psi) * (1+\Delta\psi/\Delta z)$ (3.3.2)

Darcy's Law allows calculation of the flux of water through the soil matrix. However, in more complex systems where macro-pores are sporadically distributed across the landscape, particularly in areas prone to waterlogging, the flux will be severely underestimated. Due to their large radii, macro-pores exert little matric potential (Table 3.3.2). Water flow into macro-pores therefore occurs when the soil's matric potential is near zero (i.e. saturation). However, once saturated, the volume of water transmitted in these pores can be substantial. Nulsen (1996) gave the example that in a clay soil, one hole with a diameter of 2 mm can transmit water at the same rate as matrix flow through 1.3 m^2 of clay. The corresponding area of clay for a 4 mm hole is 21 m^2. Quantifying and including macro-pores into water flow calculations remains a challenge.

For water infiltrating into the soil from a ponded surface a solution to Darcy's Law was derived by Phillip (1969). Phillip proposed that infiltration has two phases. During the initial phase, water is drawn into the soil by capillarity or sorption. Sorptivity is also known as the 'blotting paper effect' and depends on the initial water content of the soil and the pore configuration. During the second phase, the rate of water infiltration decreases to a constant level (i.e. steady state infiltration). Phillip described infiltration with the equation:

Infiltration rate $= St^{1/2} + At$ (3.3.3)

where S is the sorptivity, A is the steady state infiltration rate from which saturated hydraulic conductivity can be derived and t is time.

In general, the theory of water movement in soils under controlled conditions is well understood as outlined above and in many soil physics texts (e.g. Hillel 1980; Jury *et al.* 1991). However, field soils are not homogenous and most hydraulic properties are highly variable from site to site. A soil profile normally consists of a number of horizons each with different physical properties, so it is generally difficult to predict its behaviour quantitatively.

SOIL WATER STORAGE

Pore space is that fraction of the soil occupied by air and water. The *matric potential* (ψ) is the potential produced by capillary and surface forces, or alternatively, it can be described as the suction with which water is held by the soil. There are a number of units for matric potential and these can be related to the equivalent pore radius (Table 3.3.2).

Soil properties which influence water retention vary depending on the matric potential. There is a strong positive correlation between clay content and the lower storage limit. Gardner (1968) found a strong correlation between the soil water content at -1500 kPa and surface area. These findings reinforce the theory that the amount of water held at high matric potentials depends on the surface area to which the water is absorbed. Colloidal solids, being either clay or organic matter, have a dominant effect on the surface area, thus increasing either will result in higher water contents at all matric potentials (Williams 1983).

Table 3.3.2 Relationship between pore size and matric potential (adapted from Hamblin 1985).

Matric potential				Equivalent pore radius	Significance
cm	Bars	kPa	MPa	(µm)	
1	0.001	0.1	1×10^{-4}	1,500	Cracks in ploughed soil or dried clays.
10	0.01	1	1×10^{-3}	150	Fine seed-bed cracks, seminal roots.
15	0.015	1.5	1.5×10^{-3}	100	First-order lateral roots.
50	0.05	6	6×10^{-3}	30	Roots cannot penetrate rigid pores.
100	0.1	10	1×10^{-2}	15	Root hairs, upper storage limit.
500	0.5	50	5×10^{-2}	3	
1,000	1.0	100	0.1		Bacteria can occupy pores.
15,000	15	1,500	1.5	0.2-0.5	Lower storage limit (wilting point).
10^9	10^6	10^8	10^5	1 nm	3 layers of water molecules on a clay surface.

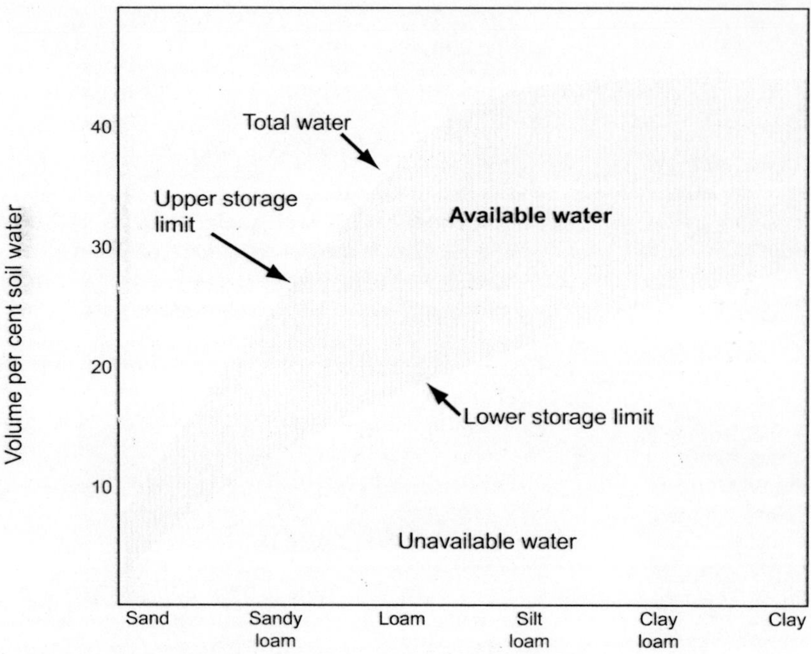

Figure 3.3.1 *The relative amounts of water available and unavailable for plant growth in soils with textures from sand to clay (from Kramer 1983).*

Correlations between soil properties and the upper storage limit are more tenuous. Water retention at low matric potentials (0 to -100 kPa) tends to be dominated by the arrangement of the solid particles and the resulting structural pore space (Childs 1969). The presence or absence of structure can have a dramatic influence on water retention at the upper storage limit (Williams *et al.* 1983). However, there are significant problems in objectively measuring structure in the field (McKenzie *et al.* 1991).

The general relationship between soil texture, the upper and lower storage limits, available water and unavailable water is illustrated in Figure 3.3.1.

SOIL WATER BALANCE

Studies of the fate of rainfall in agricultural systems generally begin with the premise that there is a balance between water inputs and outputs. This 'water balance' is written as the sum of all inputs being equal to the outputs as follows:

$$P = I + E + T + DD + L + R + \Delta S \qquad (3.3.4)$$

where P is precipitation, I is interception by the plant canopy, E is evaporation from the soil surface, T is transpiration, DD is deep drainage, L is lateral water movement, R is surface runoff and ΔS is the change in soil water storage. Other inputs including capillary rise and irrigation can be included in more complex systems.

In stable land and vegetation systems, an equilibrium exists such that the ratios of input and output parameters are relatively constant over time. Changing these ratios can have a profound effect on the type and quantity of vegetation the land can support. The evidence for this is the impact that the extensive clearing of native vegetation has had in reducing E+T and increasing DD in south-western Australia and elsewhere. In attaining a new equilibrium, periodic waterlogging, rising watertables and secondary salinity are commonly encountered. Reversing this trend requires knowledge of how to manipulate the water balance so that it more closely resembles the regime before clearing began.

Precipitation is easy to measure, but few other parameters of the water balance can be determined without considerable effort. It is for this reason that researchers generally simplify the equation to exclude parameters which are thought to be insignificant in the overall balance. For instance, capillary rise will occur only where a perched watertable is within 1 to 2 m of the surface. Canopy interception (I) is often excluded in annual crops, despite values as high as 33% of the rainfall being recorded (Leuning *et al.* 1994). Runoff (R) and lateral flow (L) are only considered important on sloping (>2%) land and where there is a conductive layer overlying an impermeable layer within the root zone, e.g. permeability contrast soils (McFarlane and Cox 1992). Generally, surface evaporation (E) and transpiration (T) are measured together as evapotranspiration (E+T).

As a consequence the water balance equation is invariably condensed to:

$$P - \Delta S = E + T \qquad (3.3.5)$$

(Capillary rise, R, L, DD and I are assumed to be negligible.)

Volumetric water content is commonly determined at regular intervals (2 to 3 weeks) either by collecting soil cores (Anderson 1992) or by neutron moisture meter techniques (Gregory *et al.* 1992a, b). The water content is then integrated over depths from the surface to below the root zone.

There have been very few soil studies in WA in which all the components of the soil water balance were measured (Table 3.3.3). For example, many studies were undertaken to measure the water use efficiency of crops at the plot scale and these concentrated on E+T and ΔS, assuming DD, R and L to be negligible. Catchment studies tend to concentrate on DD, R and L which are more easily measured over large areas.

TERMINOLOGY

- **Soil water content**

The quantity of water in soils is described using two measures; gravimetric water content (θ_g) which is the mass of water per mass of dry soil, and the volumetric water content (θ_v) which is the volume of water per volume of soil.

$$\theta_v = \theta_g \times \rho_b$$

where, ρ_b is bulk density; $\rho_b = m_s/V$, where m_s is the mass of dry soil in volume V. Dry soil is taken as oven dry soil (24 h of drying at 105°C), even though this does not remove interlayer water absorbed onto clay surfaces.

- **Upper storage limit (USL)**

The maximum water content that can be stored under irrigation or rainfall. It is defined as the water content following saturation when free drainage has effectively stopped and the water content has become relatively stable. This usually takes about 48 hours in free draining soils (less in coarse-textured soils). It is preferable to measure the USL *in situ*, but it can also be measured in the laboratory at a suction of -10 or -33 kPa depending on the soil texture (Refer to *Laboratory measurement*, below). USL replaces the term *field capacity.*

- **Lower storage limit (LSL)**

The lower water content to which a crop can extract water. Can be measured *in situ* at crop maturation or in the laboratory (-1500 kPa). The laboratory measure is usually favoured because soil evaporation can dry the soil well below LSL, especially in the surface horizons. LSL replaces the term *permanent wilting point.*

The moisture content at LSL is closely related to the specific surface area of the soil, which in turn is primarily determined by the amount and type of clay minerals present (Section 2.3). For instance, cracking clays have a large specific surface area and a high water content (20-30% v/v) at the LSL compared with coarse-textured soils which have a low water content (2-4% v/v).

- **Plant available water (PAW)**
PAW = Σ(USL - LSL)

PAW is the volume of water (mm) summed over the profile to the depth of water extraction, corresponding to the difference between the upper storage limit (USL) and lower storage limit (LSL). (Also called the *profile extractable water.*)

- **Available water capacity (AWC)**
(mm/m or % v/v or % w/w)

The AWC is the difference between the upper storage limit (USL) and lower storage limit (LSL) per unit depth (v/v) or mass (w/w). The AWC is a capacity measure (e.g. 150 mm/m) while *available water* (or available water storage) is a mass or volume measure related to water extraction by a crop or to a specified depth (e.g. 75 mm to a depth of 0.5 m). Values of AWC range from 20 mm/m in very coarse sands to more than 250 mm/m in well structured loams, with the normal range being 50-150 mm/m.

- **Moisture characteristic**

The relationship between water suction or matric potential (ψ) and water content (θ). The moisture characteristic shows the availability of water in a soil, plant available water and as each suction (or potential) corresponds with the emptying of pores of a certain diameter, then a range of pore diameters can be inferred. The slope of the moisture characteristic represents a pore size distribution curve, and this can be used to infer conductivities at different water contents.

Table 3.3.3 Summary of soil water balance components (mm) recorded in WA.

Where, P is rainfall, R is runoff, L is lateral flow, ΔS is change in soil water storage, DD is deep drainage, E is evaporation from the soil surface and T is transpiration (E + T are often measured together).

Site (Reference) Soil description	Year(s) and dates	Land use	P	R	L	ΔS	DD	E	T
		Uniform coarse-textured soils							
Wongan Hills	1966	Sub. clover	103			-30	17	116	
(Sedgley *et al.* 1981)	9/8 to 31/10	Rose clover	103			-25	21	107	
Deep loamy sand		Wheat	103			-57	9	151	
		Cereal rye	103			-61	7	157	
Swan coastal plain (Carbon *et al.* 1982)	1965-67	*Pinus pinaster*	669-701	0			11%	89%	
Deep yellow sand	1965-67								
	9/11-30/3	Ungrazed	33			-184	59	158	
	30/3-22/9	lucerne	578			124	24	430	
	22/9-13/3		90			-142	26	206	
	1967-69								
	11/10-4/4	Grazed sub.	123			10	106	7	
	4/4-10/10	clover pasture	704			7	244	453	
	10/10-18/4		99			-22	97	24	
Three Springs (Hamblin *et al.* 1988)	1985	Wheat	374				40	321	
Deep pale sand (>2.5 m)		Cereal rye					36	328	
		Narrow-leafed lupins					32	348	
Pale sand to 1.3 m	1985	Wheat	374				26	365	
		Cereal rye					30	337	
		Narrow-leafed lupins					23	400	
		Permeability contrast soils							
West Dale (Laing *et al.* 1991) Loamy sand to sandy loam (0.15-0.55 m) over sandy clay	1988	Sub. clover	497	21.5 (0.5-50)	55.2 (1.5-94)				
Three Springs (Hamblin *et al.* 1988)	1985	Wheat	374				15	313	
Pale sand (0.6 m) over clay		Cereal rye					15	350	
		Narrow-leafed lupins					13	372	
East Beverley	1990	Wheat					3.6	220	
(Gregory *et al.* 1992a, b)		Narrow-leafed lupins					7.0	220	
Loamy sand (0.3-0.5 m) over sandy clay	1991	Wheat					28	239	
		Narrow-leafed lupins					36	213	
Narrogin (Cox 1988)	1984		330	0-3.4	2.3-5.7				
	1985		356	3.7-4.8	2.6-4.0				
	1986		267	0.2-0.6	1.3-4.3				
(Gregory *et al.* 1992b)	1991		508	5	15				

Table 3.3.3 continued

Where, P is rainfall, R is runoff, L is lateral flow, ΔS is change in soil water storage, DD is deep drainage, E is evaporation from the soil surface and T is transpiration (E + T are often measured together).

Site (Reference) Soil description	Year	Land use	P	R	L	ΔS	DD	E	T
Medium to fine-textured soils									
Merredin (Rickert *et al.* 1987) Loamy sand over clay at 0.5 m	1973	Wheat	247				0	240	
Merredin (Hamblin and Tennant 1987) Loamy sand over clay at 0.5 m	1986	Wheat	247					247	
		Narrow-leafed lupins						248	
		Annual pasture						241	
Merredin (Siddique *et al.* 1990) Loamy sand over clay at 0.5 m	1987	Wheat	185					64-86	106-129
Merredin (Hamblin 1984) Red-brown sandy clay loam over medium clay	1982	Wheat: multiple tillage	197					224	
		Wheat: direct-drill						269	
		Wheat: no-till						265	
		Pasture						224	
Merredin (Yunusa *et al.* 1992) Red-brown sandy clay loam over medium clay	1989	*Medicago polymorpha*	177					140-146	

THE FOUR PROFILE HYDROLOGY GROUPS

The application of soil physical principles to predict the hydrology of a soil profile is extremely complex, because soils display considerable vertical and horizontal heterogeneity. Morphological properties, such as texture and structure, influence the hydrology of the soil profile, even though the relationships are not quantified.

There are key properties which largely control the hydrology of the four *profile hydrology groups* described in Section 1.2. Each group is discussed in relation to the principles of water movement and available water capacity.

(i) Uniform coarse-textured soils

Water movement

Uniform coarse-textured soils characteristically have high infiltration, low soil water storage and significant deep drainage. Infiltration is usually rapid (>50 mm/h) unless the soil is water repellent (which results in highly variable infiltration and redistribution or fingering; Section 3.1), or becomes hardset over summer. After rain, the wetting front gradually moves through the soil profile (Figure 3.3.2). Water movement in these soils has been studied by Carbon (1975). Typical values for hydraulic properties are summarised in Table 3.3.4.

Available water capacity

Available water capacity varies from about 20 mm/m in very coarse sands with negligible clay (<1%) to 50 to 70 mm/m for fine sands, to about 100 mm/m for a loamy fine sand. The sand size distribution, especially the fine sand fraction, is more important than the amount of clay. Conversely, coarse sand has a negative influence on AWC (e.g. Salter and Williams 1969; Hamblin and Blake 1986; Williams *et al.* 1991).

These soils are frequently deep, with few restrictions to root growth, so a deep root system develops which partly compensates for the low AWC. Crop and pasture species may approach their genetic potential in terms of maximum rooting depth if nutrients are non-limiting (Hamblin and Hamblin 1985).

The available water capacity in uniform coarse-textured soils is not strictly comparable with other soils, because water continues to drain through these soils at the pressure used to identify the upper storage limit.

With these soils the traditional concept of 'field capacity' is inappropriate, as it relies on the soil reaching a relatively steady moisture content, 24 to 48 hours after rain. It is a useful concept for soils with a bimodal or skewed pore size distribution, but not for those with a relatively uniform pore size (Hamblin 1985).

Crops and pastures growing on coarse-textured soils are able to respond to light showers following a dry spell, because most of the rain is available for plant growth. In low rainfall years, there is less water available for crops grown on fine-textured than coarse-textured soils resulting in lower yields.

(ii) Permeability contrast soils

Water movement

Infiltration can vary considerably depending on the surface texture, surface condition and existing soil moisture. For example, a coarse sand over an impermeable clay will have a high infiltration rate until the surface soil becomes saturated, after which K_s will approach zero (refer to *saturation excess runoff*, Section 7.2).

The slow permeability of the B horizon dominates the hydrology because it limits water entry into the lower part of the profile. Perched watertables above the clayey subsoil are a feature of this group.

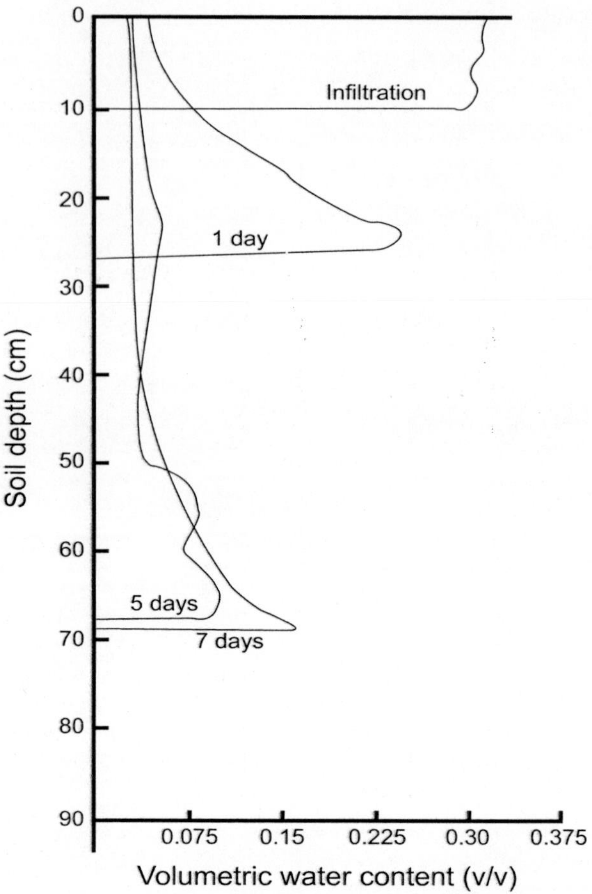

Figure 3.3.2 *Moisture profiles showing the redistribution of 37 mm of water applied to a column of Bassendean sand (from Carbon 1975).*

Table 3.3.4 Hydraulic properties of uniform coarse-textured soils from WA.

Soil properties	Three Springs (Hamblin *et al.* 1988)		Merredin (J. Russell unpublished data)		Swan Coastal Plain (Carbon *et al.* 1982)	Merredin (J. Russell unpublished data)	
Soil description	Pale deep sand		Yellow earthy sand		'Spearwood' sand (yellow deep sand)	Acid yellow earthy sand	
Soil parent material(s)	Colluvial sand over truncated laterite on sedimentary deposits		Laterite on granite		Aeolian sand over limestone	Laterite on granite	
Depth (m)	0-0.1	0.1-3.0	0-0.2	0.3-0.9	0-0.6	0-0.2	0.3-1.0
Texture (% clay)	Coarse sand	Sand	Coarse loamy sand (9)	Coarse sandy loam (18-23)	Sand (2)	Clayey sand (9)	Coarse sandy loam (15-25)
Bulk density (Mg/m^3)	1.28	1.45-1.5	1.6	1.6-1.75	-	1.5	1.5-1.6
LSL (v/v)	0.025	0.03	0.02	0.07	0.02-0.03	0.02	0.07
USL (v/v)	0.06	0.06	0.06	0.13	0.07	0.06	0.13
Available water capacity (mm/m)	35	30	40	60	40-50	40	60
K_S (m/day) (range)	-	-	2.7 (1.7-4.5)	1.5 (0.9-2.6)	~4	3.7 (2.3-4.3)	2.8 (1.4-3.7)

Water moves down through the profile under a tension gradient (matrix flow). When water infiltrating through a sand encounters a clayey layer such as a B horizon, the water will move from the sand into the clay because the clay tends to have finer pores which exert higher tensions at a given water content. However, the unsaturated hydraulic conductivity of clay subsoils is very low.

For water to move through large pores or cracks in a clayey B horizon there has to be ponding above the B horizon (i.e. saturated conditions) for example, a perched watertable. The flux of water moving through preferred pathways may be several orders of magnitude higher than the flux moving through the soil matrix. Water movement in layered soils has been studied by Hillel and Talpaz (1977). Typical values for hydraulic properties are summarised in Table 3.3.5.

Available water capacity

Permeability contrast soils generally have a low AWC (25 to 100 mm/m), especially if the subsoil is sodic and/or has a very slow permeability. The depth and physical properties of the A horizon affect AWC, although water movement and root growth into the subsoil are likely to have the biggest influence. For example, wheat roots penetrated to a depth of 175 cm on a yellow duplex soil at Merredin (Collgar soil), while at East Beverley they only penetrated to 50 cm (Dracup *et al.* 1992). The large difference in root growth can be attributed to differences in subsoil conditions (structure, presence of biopores).

Water may be stored in the A horizon as perched water, therefore the traditional concept of AWC being the difference between USL and LSL can be misleading. Consequently, calculating the soil water storage by summing the AWC for each layer may under- or over-estimate the actual plant available water (PAW) depending on the depth to a perched watertable and the tolerance of the crop or pasture to waterlogged conditions.

(iii) Cracking clays

Water movement

Water movement in swelling soil is quite different from that in rigid soils. For instance, the volumetric water content of a swelling soil in equilibrium with a watertable decreases with depth, while in a rigid soil it increases (Philip 1969). A key characteristic of these soils is that the water storage is determined by the infiltration rate rather than the moisture characteristic, i.e. 90% of water stored in the profile enters through cracks rather than through the matrix (Loveday *et al.* 1978).

Infiltration is high while cracks remain open, but is very low when they close and the water runs off or ponds on the surface. Cracking clays can wet from the 'bottom up', as water enters the soil down the cracks and then wets the matrix by capillary rise.

Available water capacity

The AWC of cracking clays can vary considerably under rain-fed conditions. The principal determinants are water entry, rooting depth and subsequent cracking pattern, which are all inter-related (Williams 1983). The surface stability and the presence of a sodic layer, or low electrolyte conditions are the main factors

Table 3.3.5 Hydraulic properties of permeability contrast soils from WA.

Soil properties	**Condingup (D. Hall unpublished data)**			**East Beverley (Tennant *et al.* 1992)**		**Mount Barker (Cox 1988)**			**Narrogin (Seow *et al.* 1988)**	
Soil description	Grey fine sand over sodic clay			Loamy sand (0.3-0.5 m) over a sandy clay		Sandy loam overlying sandy clay within 0.5 m			Permeability contrast soils - various classifications	
Soil parent material(s)	Sand sheet overlying Eocene sediments			Colluvium on truncated laterite		Truncated laterite			Truncated laterite	
Horizon	Ap	A3	B	A	B	A1	A2	B	A	B
Texture (% clay)	Fine sand	Fine sand	Sandy clay (33-50)	Weak loamy sand (2-4)	Sandy clay loam (20-30)	Sandy loam	Sandy loam	Sandy clay	Sand-loamy sand	Sandy clay loam (20-30)
Bulk density (Mg/m^3)	1.35-1.5	>1.5	>1.8	1.5-1.9	-	1.6	1.5	1.49	1.4-1.55	1.6-1.7
LSL (v/v)	0.027	0.005	0.155	0.05-0.12	0.08-0.16	-	-	-	-	-
USL (v/v)	0.073	0.053	0.378	0.18-0.26	0.20-0.26	-	-	-	-	-
Available water capacity (mm/m)	46	48	220		100-120	-	-	-	-	-
K_S (m/day) (1 stand. deviation)	0.34-0.48* (0.05-0.3)	~0.7* (0.24-0.38)	0.05-0.2* (0.037-0.096)	0.03-0.04	0.002-0.0001	0.59-0.65 (0.08-0.34)	0.50-2.9 (0.1-1.0)	0.095-0.21 (0.06-0.23)	1.6-2.1**	0.02-0.2**
S ($mm/h^{1/2}$) (1 stand. deviation)	6-11* (3.4-5)	~20* (2-4)	8-20* (3-11)	-	-	-	-	-	13-20**	-

* *Hydraulic conductivity/sorptivity measured at a tension of 15 mm.*
** *Hydraulic conductivity/sorptivity measured at a tension of 10 mm.*

Table 3.3.6 Hydraulic properties of medium to fine-textured soils from WA.

Soil properties	Merredin district (J. Russell unpublished data)							
Soil description	Red-brown clayey sand grading to an alkaline sandy clay at 0.4 m		Red-brown sandy loam over calcareous sandy clay at 0.25 m		Red-brown sandy clay loam over a highly calcareous clay at 0.1 m		Red sandy clay loam over an alkaline clay at 0.1 m.	
Soil parent material(s)	Colluvium, aeolian deposits		Alluvium, aeolian deposits		Colluvium		Colluvium, aeolian deposits	
Depth (m)	0-0.1	0.4-0.6	0-0.10	0.6-0.9	0-0.10	0.2-0.6	0-0.12	0.1-0.6
Texture (% clay)	Sandy loam (11)	Sandy clay (35-40)	Sandy loam (14)	Sandy clay loam to clay loam (15-35)	Sandy clay loam (28)	Medium clay (43-52)	Light sandy clay loam (18)	Medium clay (40-50)
Bulk density (Mg/m^3)	-	-	1.7	1.6-1.7	1.3	1.25-1.5	1.4	1.3-1.4
LSL (v/v)	0.03	0.14-0.16	0.04	0.09-0.14	0.12-0.15	0.14-0.25	0.11-0.16	0.14-0.20
USL (v/v)	0.15	0.18	0.14	0.16-0.19	0.22-0.28	0.22-0.33	0.16-0.26	0.25-0.42
Available water capacity (mm/m)	120	-	100	-	-	-	-	-
K_S (m/day) (range)	1.4 (1.1-1.6)	0.27 (0.09-0.63)	1.1 (0.9-1.2)	0.20 (0.046-0.44)	0.24	0.41-0.59	0.29 (0.09-0.7)	0.15 (0.04-0.3)

controlling water entry (Loveday 1976). Soil water storage can be increased by improving infiltration, for example, by adding gypsum. Cracking clays do not conform to the model of water movement in rigid soils as explained above, so the traditional concepts of soil water storage are inappropriate (Williams 1983). As a guide, the AWC of cracking clays is 120 to 200 mm/m (Gardner 1997). Using a 'salt profile' (McCown *et al.* 1976; Mullins 1981) may be useful for identifying the normal depth of wetting and the maximum depth of root growth.

(iv) Medium to fine-textured soils

Water movement

These soils tend to follow the classical concepts of water movement. The surface soil is often the layer of lowest permeability, because of surface crusting or a hardsetting A horizon. Soil water storage is the difference between the upper storage limit and the lower storage limit, and the traditional concept of 'field capacity' is applicable on these soils. Typical values for hydraulic properties are summarised in Table 3.3.6.

Available water capacity

In general, well structured loamy soils have the highest AWC, with values reported in the international literature of 300 to 350 mm/m; but (apedal) sandy loams in WA are more likely to have an AWC of 90-120 mm/m (C. Henderson unpublished data).

In medium to fine-textured soils, PAW is largely determined by the interaction between soil depth, soil structure and depth of root growth.

In low rainfall districts, a 'salt profile' can be used to identify a salt bulge in the subsoil which corresponds with the normal depth of wetting and the maximum depth of root growth (McCown *et al.* 1976; Mullins 1981). This technique may also be useful for cracking clays and permeability contrast soils in low rainfall districts.

ASSESSING SOIL WATER STORAGE

Methods for assessing soil water storage vary with the profile hydrology group (Table 3.3.7), because not all soils follow the classic concepts of water movement and storage, as discussed above.

(i) Direct field sampling (water ponding)

Direct field measurement of the soil water storage is considered a benchmark method (Salter and Williams 1965) and is often the simplest and most practical.

There are two methods for obtaining the USL in the field. The first involves sampling the profile in July/August following significant rainfall (12 hours for sandy soils; 24 to 48 hours for other soils), when the profile is fully wet. The second method involves artificially ponding water to wet the soil profile to saturation, then allowing it to drain to the USL (Land Resources Branch Staff 1990; Gardner 1997). The LSL can be determined by sampling the soil profile at harvest (providing there has been no significant rainfall late in the growing season), or when the plants are wilting.

(ii) *In situ* measurement (e.g. neutron moisture meter)

There are a number of techniques for measuring soil water *in situ*, including the neutron moisture meter, time domain reflectometry (see below), porous blocks, lysimeters etc. The time involved and the higher level of expertise required prohibits their routine use.

Neutron moisture meters. This is the only method for which much data are available for the agricultural area of WA. It is the preferred method for most soils (except cracking clays; Table 3.3.7), providing a satisfactory calibration is available. The method is described by Greacen (1981).

The USL corresponds with the wettest profile during the growing season, while LSL is approximately the moisture content at harvest, providing there has been a dry finish to the season. The depth of water extraction can be estimated from the intersection of the curve for the wettest and driest profiles (Figure 3.3.3). Plant available water is the difference between USL and LSL to the depth of water extraction.

There is variation between seasons, therefore the results for two to three years should be assessed. The USL should not vary with different crops, but LSL does. The effect of crop type on LSL can be assessed by comparing results for two different crops on the same plot. In general, differences in the LSL are not large and mainly relate to variation in the depth of rooting (Ratliff *et al.* 1983).

Time domain reflectometry (TDR). This measures the transit time of an electrical signal along metallic probes. The time is related to the 'dielectric' constant of the material surrounding the probe, this being about 15 to 40 times greater in water than solids (Topp *et al.* 1982; Topp and Davis 1985). TDR has been used in numerous soil water studies (e.g. Herkelrath *et al.* 1991), including WA (Gregory *et al.* 1995). In a critique, Zegelin *et al.* (1992) concluded it is best suited for coarser textured soils, as soil electrical conductivity limits its usefulness in clay soils.

(iii) Laboratory measurement

Measuring soil water storage in the laboratory is not ideal, especially for estimating the USL (Richards 1960; Salter and Williams 1965; Gardner 1971; Ritchie 1981).

The traditional method of determining the USL is to extract water from a disturbed or undisturbed soil sample using a pressure plate apparatus (Richards and Weaver 1943). An alternative is the *filter paper method* (Hamblin 1981; Greacen *et al.* 1989), which has been successfully used on a range of soils from WA (C. Henderson unpublished data).

The USL is estimated using a matric potential of -10 kPa for coarse-textured soils and -33 kPa for medium and fine-textured soils. In the laboratory, undisturbed cores should be used because the water content at the USL is largely determined by the soil structure.

The LSL can be estimated using a matric potential of -1500 kPa. The LSL is controlled by the surface area of the soil, thus the clay percentage and clay mineralogy have a large influence. As a consequence, there is a reasonable relationship between the -1500 kPa measurement on disturbed and undisturbed samples (Salter and Williams 1965).

Table 3.3.7 Options for assessing soil water storage, ranked according to precision and ease of measurement.

Profile hydrology group			
Uniform coarse-textured soils[a]	Permeability contrast soils[b]	Cracking clays[c]	Medium to fine-textured soils[d]
1. Direct field sampling (water ponding)	1. Direct field sampling (water ponding)	1. Method of Shaw and Yule (1978)	1. Direct field sampling (water ponding)
2. *In situ* (neutron moisture meter or TDR[e])	2. *In situ* measurement (neutron moisture meter or TDR - A horizon only)	2. Inference from texture	2. *In situ* measurement (neutron moisture meter or TDR - not fine-textured soils)
3. Laboratory measurement (intact cores)	3. Laboratory measurement (intact cores)		3. Laboratory measurement (intact cores)
4. Inference from texture	4. Inference from particle size		4. Inference from particle size
	5. Inference from texture		5. Inference from texture

a) Uniform coarse-textured soils. *Regression models are not recommended because these soils are not well represented in the databases used to derive the models.*

b) Permeability contrast soils. *No methods are entirely satisfactory, however field methods are preferred. There are problems with neutron moisture meters when there is a perched watertable and also when using the water ponding method (Land Resources Branch Staff 1990). Assessing soil water storage from laboratory measurements, particle size or texture are at best rough estimates of the true value.*

c) Cracking clays. *The neutron moisture meter is less suitable for swelling soils, because they exhibit little or no drainable porosity (Gardner 1997). The model developed by Shaw and Yule (1978) gave a satisfactory estimate of USL for swelling soils when tested by Ahern (1988). The method is suitable for soils with an exchangeable sodium percentage <15 to a depth of 1 m and with no salt bulge at depths <60 cm (Land Resources Branch Staff 1990). Measuring USL on a core in the laboratory using -0.33 kPa results in an over-estimation of the USL, because of the unrestrained swelling of the samples when saturated (Gardner 1985).*

d) Medium to fine-textured soils. *The reliability of inferring soil water storage from the field texture will depend on the degree of structural development.*

e) TDR *is time domain reflectometry.*

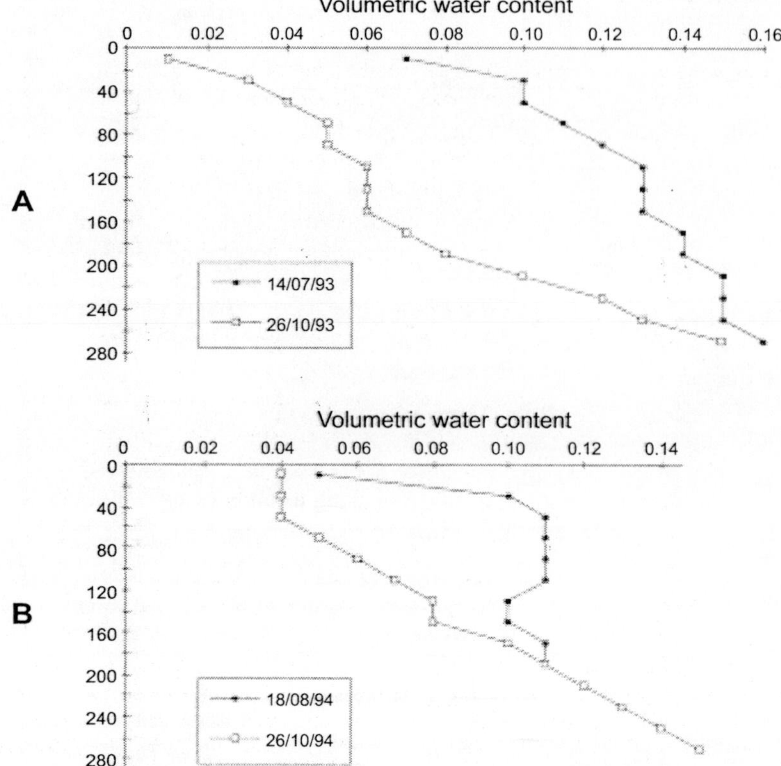

Figure 3.3.3 *The wettest and driest profiles as measured with a neutron moisture meter for wheat growing on an acid earthy sand (pH_{Ca}<4.0 at 0.4 m) during (a) 1993 and (b) 1994 (J. Russell unpublished data).*

If particle size information is available, a regression equation will generally give a more reliable estimate of the AWC than a simple texture relationship. The exception is for uniform coarse-textured soils which are not well represented in the databases for any of the models. The proliferation in the number of models available is partly because of the time and cost involved in making a large number of soil hydraulic measurements. *It may also indicate the models have a limited usefulness outside the data sets from which they were derived.*

Some of the models used to predict AWC using relatively simple parameters include:

- Shaw and Yule (1978) - developed for cracking clays (based on 14 soils from Queensland, R^2 >0.92).
- Rawls *et al.* (1982) - based on a large data set from US (2,541 soil horizons) suitable for soils >5% and <60% clay.
- Williams *et al.* (1991) - based on 186 soil horizons from eastern and southern Australia.

The above models adequately describe the data sets from which they were derived, with R^2 values from 0.7 to >0.9.

Ratliff *et al.* (1983) compared *in situ* field measurements of the upper and lower storage limits with laboratory methods (-33 and -1500 kPa) for 282 observations. Their results showed an inconsistent relationship between the USL measured in the field and the laboratory (-33 kPa). However, if the USL was obtained in the field, and LSL measured in the laboratory (-1500 kPa) the estimation of the soil water storage was good (Ratliff *et al.* 1983).

(iv) Inference from particle size

There are many models available for estimating soil water retention from physical properties (Rawls *et al.* 1991). These are usually regression equations using relatively simple parameters (texture, particle size, bulk density, organic matter). The utility of a model depends on a number of factors:

- The size of the database from which it was derived
- The statistical correlation of the model (R^2)
- The similarity of the soils in the dataset to those of interest.

Independent testing of models developed by McKeague *et al.* (1986), Hall *et al.* (1977) and Williams *et al.* (1991) found they did not satisfactorily predict the AWC of 10 Australian soil profiles (McKenzie *et al.* 1991). Cassell *et al.* (1983) describes a model based on 401 *in situ* field observations, although it requires the input of eight parameters (Note: US particle size classes are not the same as Australian classes, Chapter 10).

(v) Inference from texture

The least reliable, although frequently the only available option for estimating the AWC is to use a simple relationship with the field texture (McKenzie *et al.* 1991; Land Resources Branch Staff 1990; Table 3.3.8). To improve the estimate of AWC for coarse-textured soils the sand size fraction has been included and the values have generally been measured in the field in WA. Published values (e.g. Salter and Williams 1967, 1969; Stace *et al.* 1968) for sands and loamy sands tend to be higher than those measured from the agricultural area of WA.

CALCULATING SOIL WATER STORAGE FROM FIELD TEXTURE

Determine the effective rooting depth by measurement, observation or prediction (or use a standard depth of 0.5 or 1 m).

Sum the available water capacity for each soil layer/horizon to the required depth.

AWC (mm) = depth (m) x AWC (mm/m) x (100 - vol% coarse fragments*)/100

* If the moisture holding capacity is estimated from the particle size or texture, then any coarse fragments or gravel must be excluded. The gravel or coarse fragments are assumed to be physically inert, consequently AWC is proportionally reduced for that layer.

Example

Estimating the soil water storage of a permeability contrast soil (0 to 1 m), using a relationship between AWC and the field texture.

Layer	Depth (m)	Texture	$AWC_{est.}$ (mm/m)	Coarse fragments (%)	Soil water storage per layer (mm)
1	0-0.10	Loamy sand	75	0	(0.1 x 75) = **8**
2	0.10-0.25	Loamy sand	75	0	(0.15 x 75) = **11**
3	0.25-0.70	Sandy loam	100	30	(0.45 x 100)x 0.7 = **32**
4	0.70-1.00	Sandy clay	90	60	(0.3 x 90) x 0.4 = **11**
					TOTAL = **62**

AWC to 1 m is 62 mm. If the clayey subsoil (layer 4) was poorly structured and prevented root growth, the effective depth of root penetration for annual crops would be 0.7 m and the AWC reduced to 51 mm.

When there is a perched watertable, additional water is present. Whether this water is available for plant growth depends on the waterlogging tolerance of the crop or pasture. The PAW would be under-estimated for a tolerant pasture like kikuyu, but over-estimated for a sensitive pasture such as serradella. This demonstrates PAW is not a single fixed value for a soil.

MANAGEMENT OPTIONS

The main reasons for manipulating the soil water balance are to increase plant water use (transpiration, T), and/or to reduce deep drainage (DD). The close relationship between grain yields and water use is a major reason to increase plant water use. Options for reducing deep drainage are discussed with secondary salinity (Section 5.3).

Plant water use can be increased by increasing the plant available water (e.g. volume of soil explored), increasing soil water storage (e.g. increase organic matter), reducing soil evaporation (agronomic measures, stubble retention) and increasing or maintaining infiltration (stubble retention, managing water repellence, soil structure decline).

Increasing plant available water

Plant available water is closely linked to the depth of soil explored by the root system. By growing species or varieties more suited to the conditions, PAW can be increased without modifying soil conditions. For example, this could involve replacing shallow rooted species with deep-rooted species on uniform coarse-textured soils (Table 3.3.9). On other soils, the rooting depth may be limited by the soil's chemical or physical properties. For example, tolerance of waterlogging on permeability contrast soils or tolerance of alkalinity on some fine-textured soils (see Chapters 8 and 9).

Increasing soil water storage

Soil water storage is essentially an inherent feature of a soil, but there are some limited options to increase the amount of water available for plant growth.

In general, soil water storage can be increased by increasing the organic matter. The increase in water retention is greater at the USL than at the LSL, so there is an overall increase in PAW (Russell and Shearer 1964; Barrow 1969; Hamblin and Davies 1977; Emerson *et al.* 1994). However, the overall increases in soil water retention are small in absolute terms, particularly in coarse-textured soils (Hamblin 1985).

When the results for a range of soils were compared there was an increase in available water of 1.0 to 4.9% (gravimetric) per unit increase (1%) in organic carbon (Emerson 1995). In general, the increase in available water with carbon content decreases with the coarseness of the soil. For example, available water increased by 5 g/g of C for a silty clay compared with 2 g/g of C for a loamy sand (Emerson 1995). However, there appears to be a lower limit of organic carbon below which it does not contribute to soil water storage, due to the absence of polysaccharide gels. This lower limit is thought to be about 0.7% organic carbon (Emerson 1995), but will most likely vary with soil type and approximately corresponds to the organic carbon of an intensively cropped soil.

Reducing soil evaporation

Soil evaporation can be reduced by using a surface mulch or by increasing transpiration (T) at the expense of soil evaporation (E).

Mulches are used commonly to reduce soil evaporation. However, their effectiveness is reduced in a Mediterranean climate where rainfall is low and irregular. For example, in the late autumn period in 1984, conditions were ideal for water conservation with 82 mm of rain received during April-May. In a nil stubble treatment, 25 mm of this rainfall was stored at sowing and available for crop growth, while 39 mm was stored where 8 t/ha of stubble was on the surface. The additional 14 mm of stored moisture had the potential to increase grain yields by 210 kg/ha (Tennant 1985; Tennant *et al.* 1991). When summer rain occurs earlier there is likely to be less additional stored water under stubble. For example, in 1985 the summer rain (42 mm) was lost through evaporation, irrespective of stubble levels (Perry *et al.* 1992). In low rainfall districts, the amount of stubble available in most seasons will be about 1 to 3 t/ha, which will limit its effectiveness in reducing evaporation.

Table 3.3.8 Estimating available water capacity (AWC) from soil texture, sand size and structure.

Texture	% clay	Sand size fraction	AWC (mm/m)	
			Moderate-strong structure	Weakly structured or apedal
Sand	<5			
		Coarse to very coarse	-	~20[a]
		Medium to coarse	-	30-45[b]
		Medium	-	40-50
		Fine	-	50-70[j]
Loamy sand/	5-10	Coarse	-	50-60[f]
clayey sand		Medium	-	60-90[f]
		Fine	-	80-100[f]
Sandy loam	15-20	Coarse	110-220[i]	50-60[f]
		Medium	110-170[i]	60-100[c,d,f]
		Fine	170-220[i]	~140
Light sandy clay	15-20	Coarse	120-150	50-60[e]
loam		Medium	170-220[i]	90-100[f]
		Fine	~180	100-120
Loam	~25	-	150-240[h,i]	100-130[i]
Sandy clay loam	20-30	-	130-190[i]	100-130[g,i]
Clay loam	30-35	-	120-210[i]	~100
Sandy clay	35-40	-	130-150[i]	80-100[f,i]
Clay	>35	-	110-120[h,i]	90-140[h,i]
Clay (self-mulching)	>35	-	~210[h]	-

a *G. Luke (unpublished data)*
b *Hamblin* et al. *(1988)*
c *Hamblin and Hamblin (1985)*
d *Hamblin and Tennant (1981)*
e *S. McKeague (unpublished data)*
f *C. Henderson (unpublished data)*
g *M. Hegney (unpublished data)*
h *Williams (1983)*
i *Hollis and Jones (1987)*
j *Hamblin (1985).*

Table 3.3.9 The maximum effective rooting depth for selected annual crops and pastures.

Crop or pasture	Maximum effective rooting depth* (m) (reference)
Cereal crops	
Wheat	1.6-1.7[a, b, c]
Barley	2.1[d]
Pulses and oilseeds	
Narrow-leafed lupins	1.9[a]
Field peas	1.2-2.0[e]
Faba beans	0.8[f]
Chickpeas	1.0[g]
Canola	1.3[h]
Pastures	
Blue lupins	~1.9[c]
T. subterraneum	
ssp. *subterraneum*	0.8-1.2[i, j]
ssp. *yanninicum*	0.4-1.2[j, c]
Yellow serradella	2.5[k]

* *The maximum rooting depth of a crop is determined genetically. It can be approximated in uniform coarse-textured soils with no compacted layers or chemical limitations and a non-limiting nutrient supply (Hamblin and Hamblin 1985).*

a *Hamblin (1982)*
b *Tennant (1976)*
c *Hamblin and Hamblin (1985)*
d *Briggs (1978)*
e *Armstrong* et al. *(1994)*
f *S.P. Loss and K.H.M. Siddique unpublished data*
g *Siddique and Sedgley (1986)*
h *Suanda (1992)*
i *Humphries and Bailey (1962)*
j *Ozanne* et al. *(1965)*
k *B. Nutt unpublished data*

Stubble retention can reduce evaporation during the growing season and increase the yield potential (Perry *et al.* 1992). Increases in transpiration early in the growing season tend to be matched by a reduction in soil evaporation of a similar magnitude (D. Tennant personal communication).

Increasing infiltration, reducing runoff

Management options for increasing infiltration depend on soil texture and susceptibility to water repellence (Section 3.1) and soil structure decline (Section 3.2). Stubble retention can increase infiltration as residues may act as paths for rain to infiltrate into the soil. Stubble also helps protect the soil surface from raindrop impact which can result in crusting or hardsetting and a decline in infiltration.

Further reading

Gregory, P.J., Tennant, D., Hamblin, A.P. and Eastham, J. (1992). Components of the water balance on duplex soils in Western Australia. *Australian Journal of Experimental Agriculture* **32:** 845-55.

Miyazak, T. (1993). 'Water flow in soils'. Marcel Dekker Inc., New York, 296 p.

Williams, J. (1983). Physical properties and water relations. Soil hydrology. *In* 'Soils: an Australian viewpoint.' Division of Soils, CSIRO Melbourne/ Academic Press, London, pp. 507-530.

3.4 WATERLOGGING

Geoff Moore and Don McFarlane*

Waterlogging is widespread in the agricultural areas of Western Australia and is a major factor reducing crop yields, especially in wet years. Warnings of yield losses from waterlogged crops came in the early 1970s (Poole 1971). Later research in the Zone of Rejuvenated Drainage, initially at Mount Barker and Narrogin (Cox 1988; McFarlane *et al.* 1992) and then near Northam (Belford *et al.* 1990, 1992), quantified these losses.

The magnitude of waterlogging is difficult to measure given the large variation between seasons. Incidence is probably underestimated because perched watertables can go unnoticed unless the soil profile is examined in winter. Waterlogging early in the growing season can delay sowing with a commensurate loss in yield potential. Direct yield losses of up to 2.5 t/ha have been measured in crops (McFarlane and Wheaton 1990). Pasture production can also be significantly reduced (McFarlane *et al.* 1992). In 1988, a year with slightly above average rainfall, an estimated 32% of the cereal crops in the East Yornaning catchment were mildly to severely waterlogged. Farmers were unaware that the figure was so high (McFarlane and Wheaton 1990; McFarlane *et al.* 1992). The areas most susceptible to waterlogging in south-western Australia are illustrated in Figure 3.4.1.

PRINCIPLES OF WATERLOGGING

In well drained soils, the air-filled porosity is between 10 and 60% (i.e. 10-60% of the total soil volume is air space), allowing oxygen and carbon dioxide to diffuse freely in and out of the soil. Roots require oxygen for respiration and for most agricultural crops this is largely obtained from the soil air. This compares with a saturated soil where all the pores are filled with water and the amount of oxygen dissolved in the water is typically only about 5% of that in an aerated soil. Consequently, when a soil is saturated there is a sudden decrease in the oxygen supply and the oxygen used for respiration by the roots and micro-organisms is not replaced, because diffusion of oxygen and other gases is 10,000 times slower in water than in air (Grable 1966). While the oxygen is depleted, carbon dioxide and other gases produced in respiration and alcoholic fermentation accumulate. Gaseous exchange in aerobic and anaerobic soils is illustrated in Figure 3.4.2.

The soil does not have to be saturated for aeration to limit root growth. A critical value often used is 10% air-filled porosity, but this is only a guide and not definitive for all crops. In highly compacted soils with a low porosity, poor aeration can limit root growth when the water content is near the upper storage limit.

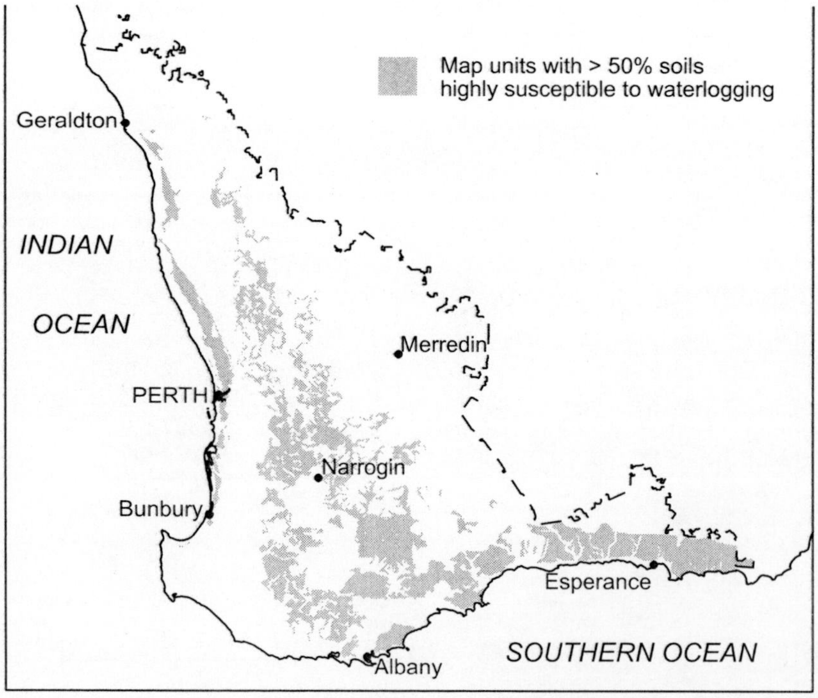

Figure 3.4.1 *The areas most susceptible to waterlogging in south-western Australia, considering rainfall, landform and soils.*

Changes in the soil environment

The key factor in waterlogged soils is the change from oxidising to reducing conditions. Aerobic micro-organisms use the oxygen present and at low oxygen concentrations the facultative and then obligate anaerobic micro-organisms begin anaerobic respiration using oxidised compounds as their substrate.

The rate at which oxygen is depleted depends on the temperature, availability of organic matter (surrogate for biomass of roots and micro-organisms), salinity and pH. Soil temperature is the major factor controlling the rate of biological activity of both roots and micro-organisms (Drew 1983). At low temperatures (<5°C), oxygen demand is low and even if the soil is waterlogged it may take many days before the supply is depleted (Drew 1983).

* *Agriculture Western Australia, Albany.*

Saturation is when all soil pores are filled with water. It may or may not be accompanied by anaerobic conditions, depending on whether there is biological activity to use the oxygen present.

Waterlogging is excess water in the root zone accompanied by anaerobic conditions. The excess water inhibits gaseous exchange with the atmosphere and biological activity uses up the available oxygen in the soil air and water. (Also called anaerobiosis, anoxia or oxygen deficiency.) Soils don't have to be saturated for gas exchange to be inhibited (i.e. waterlogged).

Inundation is water ponding on the soil surface. The effect on plant growth is severe because inundated plants cannot photosynthesise. Inundated soils are frequently also waterlogged.

Flooding is storm water flowing outside its usual channel.

However, the rate of root respiration increases exponentially with temperature up to about 35°C (Glinski and Lipiec 1990). Therefore, waterlogging causes more stress on plants under warm conditions when they are growing vigorously and have a higher oxygen demand (Letey *et al.* 1962; Trought and Drew 1982; Drew 1983). The limited data available on soil temperatures in WA suggest that for moist soils, the average temperature near the surface (5 cm) is approximately equal to the average monthly air temperature (Dracup *et al.* 1992).

With aerobic respiration, oxygen acts as an electron sink and combines with hydrogen to produce water. However, in the absence of free oxygen, different substrates can act as electron sinks and are progressively reduced according to a defined order. First, nitrate is denitrified through a series of reactions to nitrogen gas (i.e. NO_3^- to NO_2^- to NO to N_2). When nitrogen gas is lost from the system there is less nitrogen available to the plants.

Second, manganese dioxide and the insoluble ferric (Fe^{3+}) hydroxides are reduced to the soluble manganous (Mn^{2+}) and ferrous (Fe^{2+}) ions, which increases their availability to plants. However, high concentrations of Fe^{2+} and Mn^{2+} in the soil solution can actually be toxic to plants. Soil colour is largely due to iron and to a lesser extent manganese oxides. Mottles form when the soluble Fe^{2+} is oxidised to the insoluble Fe^{3+} under conditions of fluctuating watertables (Section 2.3).

The decomposition of organic matter in waterlogged soils is slower than in well drained soils, for example, plant residues accumulate in swamps and poorly drained soils. The end products are also different. Organic compounds are reduced to a number of intermediate products (e.g. acetic acid, butyric acid, ethylene, methane), compared with full oxidation to carbon dioxide and water under aerobic respiration.

Third, sulphate (SO_4^{2-}) is reduced to sulphide (S^{2-}) and occasionally hydrogen sulphide (H_2S) in very poorly drained soils. The unpleasant odour often associated with waterlogged soils is usually caused by organic compounds, rather than hydrogen sulphide. Under prolonged waterlogging (more than four weeks) the pH will often rise as the reduction reactions consume protons, however the pH usually reverts to its former

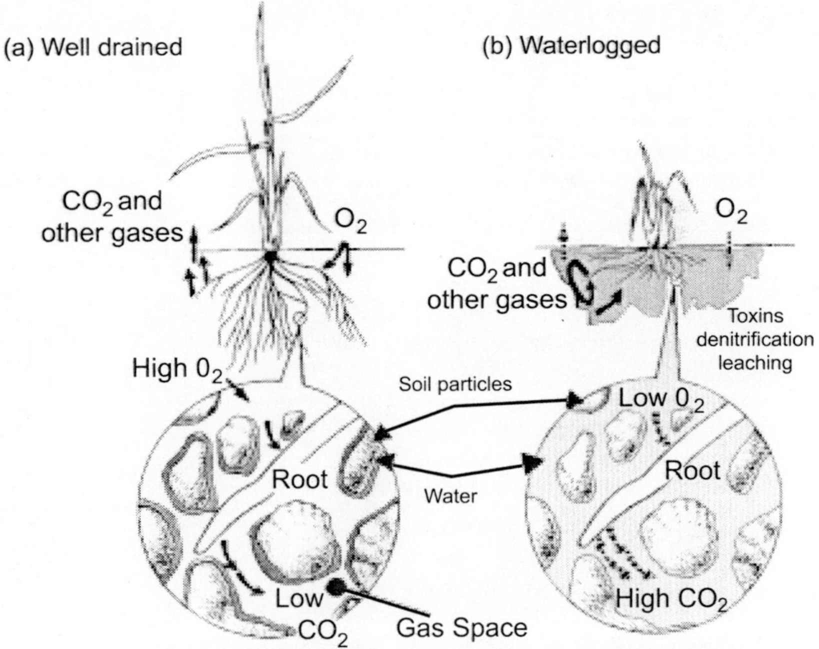

Figure 3.4.2 *In well drained soils (A) gases diffuse through the spaces between particles of soil thus supplying oxygen for respiration (energy production) of roots and allowing the products of respiration (carbon dioxide) to escape.*

During waterlogging (B) the diffusion of gases is reduced 10,000 fold. Oxygen used by roots and micro-organisms becomes deficient, while carbon dioxide and other gases such as ethylene accumulate (modified from Jackson and Drew 1984).

Waterlogging and inundation on the footslopes and valley floor in the Northam district.

level when aerobic conditions return. For further information on the chemistry of waterlogged soils refer to Ponnamperuma (1972).

Plant response to waterlogging

The effect of waterlogging on plant growth depends on the duration and intensity of waterlogging, the rate at which soil oxygen is depleted and plant factors (i.e. stage of growth, ability to adapt to waterlogged conditions and nutritional status).

For most plants, the capacity of roots to supply nutrients and water to the rest of the plant decreases in waterlogged soils. Adverse effects on root and shoot growth may originate both in the soil and in the plant.

A low oxygen supply which limits aerobic respiration is the main effect on root growth, reducing root elongation. Root tips may die, because they have a high energy requirement. The roots also take up less water. Toxins can accumulate in the soil (organic acids, high concentrations of NO_2^-, Mn^{2+}, Fe^{2+}, H_2S) and anaerobic respiration can lead to potentially harmful end products such as acetaldehyde and ethanol (Drew 1983). Stressed plants are more susceptible to root diseases such as take-all (e.g. Drew and Lynch 1980).

The shoots are not directly affected by low O_2 in the root zone, but the active uptake of nutrients is decreased because root growth is reduced, there is less energy available and less nitrate available in the soil due to denitrification. Waterlogged plants tend to translocate mobile nutrients (N, P, K) from old leaves to emerging leaves, resulting in the characteristic chlorosis and early senescence of the old leaves in cereals.

Tolerance of plants to waterlogging

Agricultural plants vary considerably in their tolerance of waterlogging (see Table 3.4.2). Tolerance is mainly related to their ability to adapt to a changed environment, dormancy during waterlogging or recovery after it ceases. Tolerance also varies with the plant's stage of development (e.g. barley; Stepniewski and Labuda 1989). All crops appear to be much more susceptible during germination and pre-emergence than later in their development (e.g. wheat; Cannell *et al.* 1980). Hence, one of the benefits of sowing crops early in paddocks susceptible to waterlogging is that they can establish quickly under warm soil conditions, before winter when the probability of waterlogging is highest. Refer to Chapters 8 and 9 for a discussion of the tolerance of individual crop and pasture species.

Some plants adapt to waterlogged conditions through one or more of the following mechanisms:

- *Shallow rooting.* The roots remain in the surface layer of soil which contains more oxygen than the deeper layers (e.g. *Trifolium subterraneum* ssp. *yanninicum*; Ozanne *et al.* 1965).
- *Nodal roots.* Nodal roots produced under waterlogged conditions are confined to the surface layers where oxygen concentrations are higher. However, shallow root systems produced in response to waterlogging can leave a plant predisposed to moisture stress in spring, after the surface layers have dried out.
- *Aerenchyma.* Monocotyledons have the ability to modify their root system to form aerenchyma (i.e. continuous gas-filled channels whereby oxygen can diffuse from the base of the stem to the root tips; Armstrong 1979). The degree to which aerenchyma can develop varies greatly between species. Wheat can form limited aerenchyma to supplement aeration in waterlogged soils. In kikuyu grass subjected to waterlogging, 70% of the root volume was occupied by aerenchyma which would almost replace normal aeration (Setter and Belford 1990). Rice is another example of a plant well adapted to waterlogged conditions because of its aerenchyma.
- *Alcoholic fermentation.* Root adaptations such as aerenchyma are not an advantage to a submerged seed at germination. However, some plants have the ability to change from aerobic to alcoholic metabolism as a survival mechanism. Alcoholic fermentation is not efficient for the plant as it only produces 1/18th of the energy of aerobic metabolism, i.e. 2 moles of adenosine triphosphate (ATP), compared with 36 moles of ATP, per mole of carbohydrate (Setter and Belford 1990).
- *Plant nutrition.* Cereal crops grown at high levels of nutrition, particularly nitrogen, can recover rapidly following a period of waterlogging stress (Barrett-Lennard *et al.* 1988; Buwalda *et al.* 1988).

Effects on soil degradation

Numerous problems are associated with waterlogging and these both reduce yields and increase the possibility of soil degradation:

- *Runoff, water erosion and flooding.* Saturated soils produce runoff, increasing the likelihood of erosion and/or flooding. Saturation-excess runoff is probably the most common runoff mechanism in areas receiving more than 400 mm rainfall (Section 7.2). Cultivated soils have low strength and have been observed to flow downslope, while rills can cause header damage at harvest. If the runoff carries particulate P into streams it can cause eutrophication. Seepage areas can be responsible for gully initiation (McFarlane and Ryder 1987; McFarlane *et al.* 1989a).
- *Soil structure decline.* Cultivating wet soils can damage their structure resulting in slaking, dispersion and hardsetting.
- *Wind erosion.* If waterlogging reduces plant growth there is an increased risk of wind erosion during summer, especially on loose, sandy soils. This is exacerbated when stock camp on these areas, causing them to become deflation basins.
- *Groundwater recharge.* Waterlogging reduces plant water use because the roots absorb less water and perched watertables can kill deep roots. Perched watertables in permeability contrast soils are thought to drain through preferred pathways, possibly old root channels, thus contributing to recharge (Engel *et al.* 1989). Ponded and perched water on fine-textured soils on valley floors infiltrates slowly and contributes to recharge. There may also be some preferential flow (McFarlane *et al.* 1989b, 1990b).
- *Salt and waterlogging.* Interactions between salinity and waterlogging can have a major impact on plant growth, because their combined effect is far greater than either factor alone. For instance, cereals can tolerate some salinity or waterlogging but have a low tolerance when both occur (John *et al.* 1977; Barrett-Lennard 1986a, b; Barrett-Lennard *et al.* 1990; refer to box).

SALINITY-WATERLOGGING INTERACTIONS

Ed Barrett-Lennard

The tolerance of agricultural plants to salinity is reduced when the plants are simultaneously stressed by waterlogging. It is now recognised that waterlogging on salt-affected land affects the growth and survival of all but the most tolerant plants. For successful revegetation, the severity of waterlogging needs to be reduced.

Waterlogging has a variety of effects on growth. Under saline conditions, waterlogging inhibits the ability of roots to screen out salt at the root surface. Increased salt concentrations in the shoots of waterlogged plants can inhibit growth and affect survival. In one example, wheat was waterlogged at a range of salt concentrations for 33 days. At all salt concentrations greater than 200 mS/m, waterlogging caused extensive leaf damage and there was no shoot growth (increase in shoot weight) after 33 days. In contrast, when plants were grown at the same salt concentrations under drained conditions, shoot growth continued even at salt concentrations of 1,200 mS/m (Barrett-Lennard 1986b).

Effects of waterlogging on wheat grown with (a) no salt, (b) salt equivalent to 4 per cent sea water, and (c) salt equivalent to 20 per cent sea water. Pots on the left are waterlogged. Pots on the right are drained (E. Barrett-Lennard).

This results in *large increases in salt uptake and in salt concentrations in the shoots* (Barrett-Lennard 1986a). For example, wheat growing in sand irrigated with solutions containing salt concentrations equivalent to 10% of that in sea water, waterlogging nearly tripled the rate of sodium uptake. After seven days, the concentration of sodium in the shoots of the waterlogged plants was double that of plants grown in drained sand (Barrett-Lennard 1986a).

The adverse effects of waterlogging under saline conditions can also affect the growth and survival of trees. With *Eucalyptus kondininensis, E. spathulata* and *E. comitae-vallis* grown in sand irrigated with solutions containing salt concentrations equivalent to 76% of sea water, 11 weeks of waterlogging killed 55 to 75% of the trees. However, fewer than 5% of the trees died at the same salt concentrations in drained pots (van der Moezel *et al.* 1988).

SITE SUSCEPTIBILITY TO WATERLOGGING

Three factors affecting susceptibility to waterlogging are:

- climate (rainfall, temperature)
- external drainage (landform, position in the landscape, runon and through-flow), and
- the soil's internal drainage (infiltration, hydraulic conductivity).

Internal drainage is generally the most important, but the combination of all three factors controls the intensity of waterlogging For example, the friable red-brown loamy earths in the Pemberton region are well drained even though the annual rainfall is >1,000 mm. In contrast, the shallow, sodic duplex soils used extensively for cereal cropping in the Scaddan area (which receives <350 mm) are not well drained. They can be cropped for two reasons. First, the low annual rainfall reduces the frequency and intensity of waterlogging, and second, cereals have a degree of waterlogging tolerance. If these soils were in a very high rainfall area (e.g. near Pemberton) they would be unsuitable for growing most crops.

Climate

The amount and distribution of rainfall are fundamental factors affecting the susceptibility of soils to waterlogging, while seasonal variations explain the large differences in waterlogging between years that are often observed. Soil temperature is closely related to the rate of root and microbial respiration as discussed. In general, the higher the temperature when waterlogging occurs, the more rapidly oxygen is depleted.

Landform

Position in the landscape is a major factor affecting *waterlogging intensity* (McFarlane *et al.* 1992). The general order of decreasing intensity is: valley floors, footslopes on concave slopes, level plains <0.5%, slopes <3%, crests and slopes 3-10% (Cox 1988).

For soils susceptible to inundation, landform controls the amount of surface runon and runoff. The location of seepage areas in uniform coarse-textured soils is controlled by the underlying geology rather than landform.

Soil hydrology

It is convenient to consider soils in terms of the four profile hydrology groups (Section 1.2) when assessing susceptibility to waterlogging (Table 3.4.1).

Uniform coarse-textured soils

These soils have high infiltration rates (unless water repellent) and there is no marked change in permeability down the profile. They are only susceptible to waterlogging if the groundwater level is high (i.e. seasonal watertable within 1 m of the surface). Where a fresh watertable is within 1 to 2 m of the surface, it can be used productively with perennial species or for farm water supplies.

Table 3.4.1 Susceptibility of soils to waterlogging in relation to the four profile hydrology groups. For susceptible sites, the factors controlling waterlogging intensity apart from climate are noted.

Susceptibility to waterlogging	Profile hydrology group			
	Uniform coarse-textured soils	**Permeability contrast soils**	**Cracking clays**	**Medium to fine-textured soils**
General susceptibility	Usually not susceptible.	>95% are susceptible under moderate to high rainfall.	Variable. Depends on site characteristics and condition of surface.	Variable. Some are very susceptible because of poor infiltration and/or low lying areas susceptible to inundation.
Requirements to be susceptible	High groundwater or seepage area (hillside seep).	Susceptible unless slope is >5%, or rainfall <300 mm.	Low lying position, slope <1-2%, poor surface condition.	Low slopes <3%, and low surface infiltration or a slowly permeable layer in the profile.
Contributing factors	High groundwater, seepage zone.	Slowly permeable subsoil, slope <1%, shallow topsoil, runon, shallow seepage, sandy topsoils which assist infiltration and inhibit evaporation.	Runon, microrelief (e.g. internal drainage to 'crabholes').	Runon, seepage from upslope, groundwater discharge.

Waterlogging is generally confined to situations such as:

- seepage areas e.g. sandplain seeps with perched aquifers (Nulsen 1985; George 1991)
- low lying areas with high regional groundwater levels
- depressions
- swamps.

The exceptions are those soils which consist of 0.8 to 1.5 m of coarse material (sand to clayey sand) overlying a clayey subsoil with a low permeability. In wet years the soil profile can fill from below and a perched watertable can saturate the root zone. In drier or lower rainfall areas the clayey subsoil acts to retain moisture within the root zone and crops will have a higher yield potential than similar soils without a clayey subsoil at depth. A marked change in soil texture at 1.2 to 1.5 m is relatively common for uniform coarse-textured soils in WA.

Permeability contrast soils

By definition these soils have a significantly lower hydraulic conductivity in the B horizon than the overlying A horizon, which predisposes them to waterlogging. Field studies have shown that they are susceptible, except where runoff or lateral seepage are extremely high e.g. on slopes >5%.

The frequency and duration of waterlogging vary considerably depending on the climate, slope, position in the landscape, permeability of the B horizon, water storage capacity of overlying horizons (a function of depth, gravel percentage, texture and structure) and slope length or distance between interceptor drains.

Waterlogging occurs in the more permeable layer (i.e. surface or subsurface) overlying the slowly permeable subsoil. The average saturated hydraulic conductivity (K_S) of clayey subsoils in the Mt. Barker area (Zone of Rejuvenated Drainage) was about 1×10^{-6} ms^{-1} (85 mm/day), although some spot measurements approached zero (Cox 1988). Frequently, the top of the clay is the least permeable part of the subsoil, so the clayey subsoil may not be saturated throughout, even when there is a perched watertable in the A horizon.

In paddocks with permeability contrast soils, there is a high degree of spatial variation in the amount of waterlogging and consequently in crop and pasture growth (Cox 1988; McFarlane *et al.* 1992; Gregory *et al.* 1992). Detailed field studies in the Zone of Rejuvenated Drainage have shown that the most important factors controlling this variability are the K_S B horizon, followed by K_S A2 horizon (which affects downslope seepage or through-flow), thickness of the A horizon (horizons thinner than 40 cm are particularly susceptible as they have a low soil water storage), slope length (slopes >50 m long accumulate seepage inflow from above), slope angle (>5% were not waterlogged) and slope shape (Cox 1988).

Cracking clays

Cracking clays can be susceptible to inundation on low lying sites and low slopes (<2%) where runoff is slow or non-existent. There is a large variation in infiltration capacity depending on whether the surface cracks are open. Some of these soils have very high infiltration rates with minimal runoff under high rainfall events but can change to very slow infiltration during a storm as the cracks close. If a cracking clay is susceptible to waterlogging, then the surface usually has the limiting permeability. If the surface soil has slow permeability, then slope and position in the landscape control the amount of surface ponding.

Medium to fine-textured soils

Drainage varies from well to poorly drained. The major influencing factors are the permeability of the surface and position in the landscape. Waterlogging will generally occur from the surface down, although mild waterlogging can occur within the soil profile.

Fine-textured soils, particularly those with a poor or unstable surface, do not allow water to infiltrate readily. In low lying areas where overland flow is restricted, surface ponding (inundation) and saturation of the topsoil often occur. If inundation is the problem, the potential to increase the infiltration capacity could be investigated (refer to Soil structure decline, Section 3.2).

ASSESSMENT OF SITE DRAINAGE

This section describes how to assess site drainage which affects the frequency, duration and intensity of waterlogging. It should be assessed on soils susceptible to waterlogging.

Site drainage is intrinsically linked to both soil and site characteristics and climate. Consequently the same soil type in a different position in the landscape or rainfall area can have markedly different drainage. On the other hand, waterlogging has a high spatial and temporal variability, but is consistent between years. The intensity of waterlogging will vary between years, but the same areas are waterlogged with the area expanding away from highly susceptible areas (McFarlane *et al.* 1992).

There are a number of ways of assessing waterlogging and categorising the results. The 'waterlogging class' can be used to describe the intensity of waterlogging (Table 3.4.2).

In general, a combination of methods results in the most reliable assessment of waterlogging intensity. First, use Table 3.4.1 to determine whether the site is susceptible. If a site is susceptible, the waterlogging class can be assessed from a combination of these methods:

(i) Observations of soil morphology

(ii) Observations of crop and pasture performance

'Probability of waterlogging' - useful for indicating the frequency of waterlogging

(iii) Field observations of perched watertables from shallow wells

'SEW_{30} index' (Sieben 1964) - an index of waterlogging intensity

(iv) Remote sensing

This list is not exhaustive and there are a number of methods for directly measuring the hydraulic properties of the soil. For instance, the saturated hydraulic conductivity (K_S) of the least permeable layer in the soil could be measured. A disc permeameter could be used to measure the hydraulic conductivity of the subsoil, but to reflect infiltration under rainfall conditions, it is preferable to use a rainfall simulator for the surface soil (Grierson and Oades 1977). Soil cores may also be used for hydraulic conductivity measurements in the laboratory, but a number of replicates are needed to overcome spatial variability (Cook 1997). Modelling the water balance of the root zone using daily rainfall records can also be used to predict waterlogging.

Hydraulic measurements are useful for comparing different treatments in a research trial, however they are not recommended for routine site assessment, because:

- There is high spatial variability of measurements. In most soils, K_S of the subsoil varies by at least 1 to 2 orders of magnitude, so a large number of replicates are necessary for confidence in the values (Cook 1997).
- The hydraulic measurements may only reflect part of the site's hydrology. On sloping sites, a number of interacting factors influence site drainage. For example, with permeability contrast soils the climate, runoff, seepage flow and moisture storage in the A horizon all contribute to the overall site drainage.
- Methods are time consuming and the benefits can be limited when compared with other options.
- Technical problems. Smearing the soil during augering will close pores. Using water with a different chemical composition to either the rainfall (when simulating surface infiltration) or soil solution (when measuring subsoil permeability) could also affect the results.

McKenzie and Jacquier (1997) have developed a 'regression tree' for predicting saturated hydraulic conductivity from soil morphology, based on 99 horizons from 36 sites in eastern Australia. They found useful predictions of Ks can be made using field texture, grade of structure, areal porosity, bulk density, dispersion index and horizon type. The models are not applicable to horizons that are organic (OC >12%), gravelly, cemented or water repellent; or for cracking clays (McKenzie and Jacquier 1997).

(i) Observations of soil morphology

In general, soil morphology alone is not a reliable indicator of waterlogging intensity, but when used in conjunction with other observations it can be very useful. Mottles, bleached A2 horizons, subsoil colour, the presence of gleyed horizons, iron precipitation along root channels and root-mats can all be used as indicators of waterlogging. Soil morphology as an indicator of waterlogging has been demonstrated in various environments (e.g. Simonson and Boersma 1972; Venneman *et al.* 1976; Coventry and Williams 1984).

The reliability of morphology as an indicator of waterlogging depends on whether the morphological features developed under the current climatic regime. For instance, in a landscape like WA which has been extensively laterised, it is important to differentiate between contemporary mottling and relict features from a past climate. For a discussion of mottles as an indicator of waterlogging for WA conditions refer to Section 2.3, *Mottles and other indicators of waterlogging. Are they are a relict feature?*

In uniform coarse-textured soils susceptible to waterlogging, observations of the soil morphology are less useful than observations of plant growth and measurements of the watertable. Waterlogging may only be indicated by an absence of coatings on the sand grains.

Soil morphology is useful for identifying poorly drained soils, but less useful for separating moderately well and imperfectly drained soils (Figure 3.4.3).

Root-mats. These are iron oxide copies of the root system (Bartlett 1961; Schwertmann 1993) and appear as fine lines of mottled soil following the pattern of root growth. Root-mats are reliable indicators of reducing conditions under the current environment, but are not common in the agricultural soils of WA.

Bleached A2 horizon. Also called an eluvial or E horizon, this is a pale horizon below the cultivation layer, usually just above a clayey subsoil of low permeability. The presence of a bleached A2 horizon indicates that nutrients have been leached out of this horizon. It is associated with fluctuating watertables (Northcote 1983).

Ferruginous gravel. The laterite and gravel which are abundant in many WA soils are a relict feature formed under a warmer, wetter climate (Section 2.2). Some profiles are largely gravel-free, but a few (<5%) soft ferruginous nodules may be observed. These may indicate current reducing conditions.

Bog iron ore. Bog iron ore (or ferricrete) is formed in wetlands that have groundwater discharging into them. The soluble Fe^{2+} in the groundwater is carried to the surface where oxidation causes iron oxides to accumulate (Richardson and Daniels 1993). It usually contains ample ferrihydrite mixed with goethite.

Table 3.4.2 Description of waterlogging classes in relation to the intensity of waterlogging, seasonal rainfall, SEW_{30} index, crop and pasture growth.

Waterlogging class *(Drainage[1])*	**Waterlogging intensity** (Perched watertable at 20 cm[(2)] or **inundation** for 3 season types)			**Average SEW_{30} Index[4]** (cm.days in average season)	**Crop and pasture tolerance to waterlogging intensity in an average season** (For non-saline soils ECe <200 mS/m[5])	
	Wet season[3]	**Average season[3]**	**Dry season[3]**		**Crops**	**Annual and perennial pasture legumes**
1	Nil	Nil	Nil	**0**		

1. *These terms approximately correspond with the* drainage classes *in McDonald* et al. *(1990).*
2. *Or equivalent intensity if watertable is shallower or deeper.*
3. *A wet season is defined as one where the June to September rainfall has a 25% probability of being exceeded in any year. An average season has a 50% probability of exceedance and a dry year a 75% chance.*
4. *SEW_{30} index is used to describe waterlogging intensity (see below).*
5. *ECe is the electrical conductivity of a saturation extract. To convert EC(1:5) to ECe refer to Section 5.3.*
6. *The approximate tolerance of lentils to waterlogging, assuming an average season. If lentils are grown under more waterlogged conditions there is likely to be a yield penalty.*

Groundwater oil slick. The 'oil slick' appearance of groundwater (iridescent film) indicates the presence of iron oxides. This can be distinguished from true oil when disturbed; the iridescent film from iron oxides fractures but an oil film remains continuous.

(ii) Observations of crop and pasture performance

Field observations of plant growth and indicator species can be used to detect waterlogging. When the waterlogging is severe or inundation occurs it is obvious. However transient waterlogging, which can reduce crop yields by up to 25%, may go undetected. Field indicators of waterlogging are:

Crops

- *Chlorosis in cereals (yellowing of older leaves).* Waterlogged crops take up fewer of the major nutrients, especially nitrogen. This means they must translocate nitrogen from older to younger leaves, resulting in the characteristic chlorosis and early senescence of the lower, old leaves in cereals. (Note: Similar symptoms can also occur from nitrogen deficiency and barley yellow mosaic virus.) Plant stress occurs well before the onset of visible symptoms. This early stress can be detected in infra-red bands using remote sensing (McFarlane *et al.* 1989a).
- *Small areas of bare ground or very poor crop growth.* These areas are usually small, however they may indicate widespread sub-clinical waterlogging in the rest of the paddock.
- *Reduced yield.* This may be difficult to detect and separate from the effects of weed competition, disease and nutritional deficiencies. Use assessment methods (i) and (iii), or (iv) to determine if reduced yield is due to waterlogging or another factor.

Pasture

- *Presence of indicator plants.* Toad rush, docks, sedges and Yorkshire fog grass indicate waterlogging.
- *Yellowing of grasses in pasture.* The older leaves of pasture grasses may show chlorosis (yellowing) similar to cereal crops.
- *Effect on pasture growth.* This varies with intensity and tolerance of the species (Table 3.4.2 and Chapter 9). In severely waterlogged paddocks, production will be reduced during most of the year. For example, severely waterlogged sites at Yornaning in 1987 produced only half the growth of mildly waterlogged sites from late June to late August (McFarlane *et al.* 1992). In mildly affected pastures there will be lower winter growth, but extra spring growth may partly or completely compensate. In low rainfall areas, waterlogging can actually increase production. However, it is not simply the total dry matter production which is important, as the marginal value of pasture in late spring is less than in early winter (June-July) which may limit the overall carrying capacity of the pasture (McFarlane *et al.* 1989a, 1992; McFarlane and Setter 1990).
- *Pasture composition.* Composition and quality can change following waterlogging if the grass is more tolerant than the legumes.

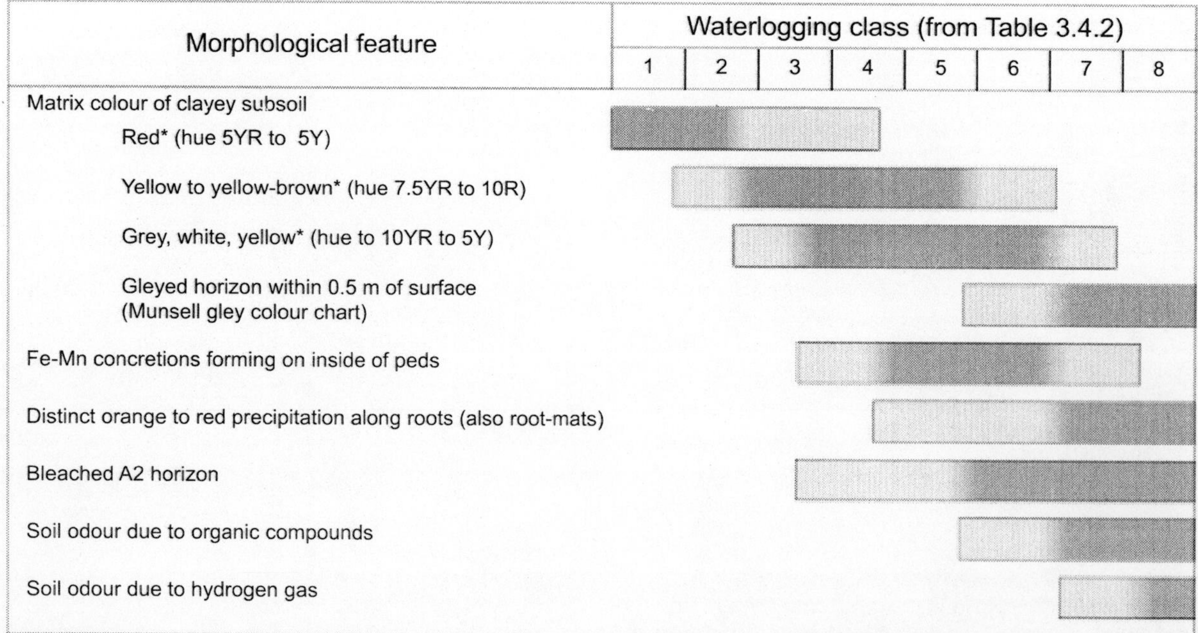

* Providing the mottles are not relict features, soils with mottled subsoils are more waterlogged than whole coloured subsoils.

Figure 3.4.3 *The relationship between soil morphology and waterlogging intensity for use with permeability contrast and medium to fine-textured soils. The intensity of shading indicates the likelihood that a particular site is in that waterlogging class(es).*

*Indicators of poor drainage in pasture; (a) Yorkshire fog grass (*Holcus lanatus*), (b) Dock (*Rumex *spp.), (c) Button grass or waterbuttons (*Cotula coronopifolia*) (G. Moore).*

(iii) Field observations of perched watertables from shallow wells

Field observations of perched watertables can be taken by digging a few holes in winter or by establishing a network of shallow observation wells. With the first method, a few holes should be dug during July-August to the top of the clayey subsoil. If there is a perched watertable, seepage will flow into the hole above the clay and cause the sides to collapse. Shallow wells are installed using slotted PVC pipe and are more accurate. They allow the depth to the free water surface to be monitored (McFarlane *et al.* 1992; Hunt and Gilkes 1993).

The high spatial variability of waterlogging (Figure 3.4.5), can limit the usefulness of observations from only one or two wells. A single well may only be a reliable guide to the degree of waterlogging within a few metres (McFarlane *et al.* 1992). To reduce the effect of this variability, knowledge of crop and pasture production can be used to site observation wells in areas of comparatively uniform growth.

To categorise the results from monitoring wells use the waterlogging categories (Table 3.4.2) or the SEW_{30} index (see below) if regular recordings are made.

PROBABILITY OF WATERLOGGING

The likelihood of waterlogging in any year at a particular site can be estimated by observing crop and pasture performance for a few years. The frequency of waterlogging is estimated by comparing field observations with rainfall records. The method assumes that rainfall during the two wettest months (usually June and July) is the critical factor affecting waterlogging (McFarlane 1985).

Field observations are used to determine in which years waterlogging does and does not occur. The critical amount of rain for waterlogging to occur is then deduced from the rainfall records. A comparison of this critical value with the rainfall percentiles for June and July from the nearest Meteorology station indicates the probability of waterlogging at that site in any year (McFarlane 1985).

Example: In an area where the average rainfall during June and July totals 150 mm, there is a 20% probability of receiving 200 mm (Figure 3.4.4). If waterlogging occurs when the rainfall is 200 mm, but not when the rainfall is 150 mm, it may be assumed that 175 mm is the critical amount. There is a 33% probability of receiving this amount of rain. There is a 33% probability of waterlogging in any year and a 10% probability of waterlogging during a cropping year if the paddock is cropped one year in three.

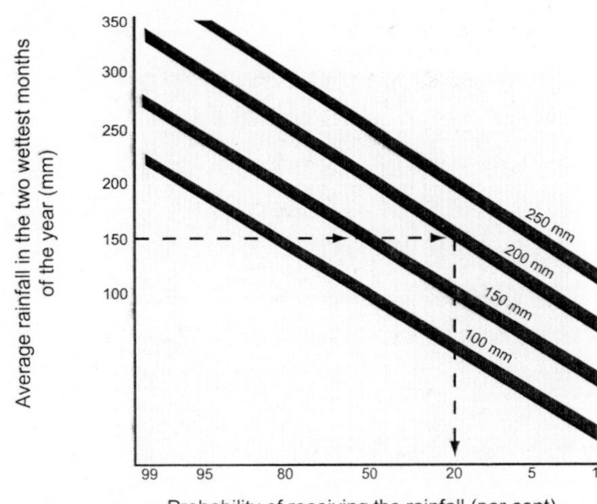

***Figure 3.4.4** The probability of receiving 100, 150, 200 or 250 mm of rainfall in the two wettest months of the year (from McFarlane 1985).*

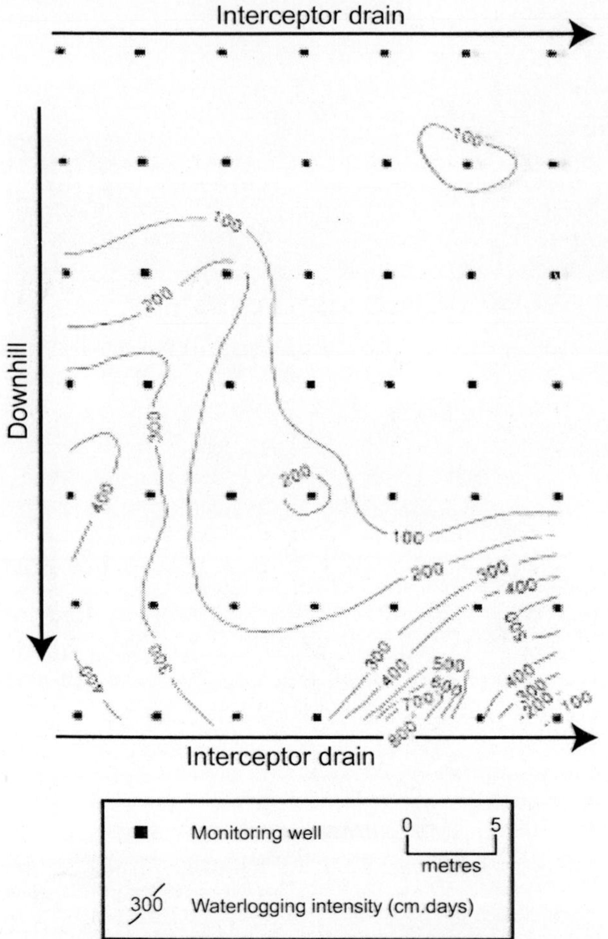

Figure 3.4.5 An example of the high spatial variability of waterlogging from the Zone of Rejuvenated Drainage (from Figure 4, McFarlane et al. *1992).*

SEW_{30} index

The SEW_{30} index (Sieben 1964) can be used to describe waterlogging intensity at experimental or monitoring sites. It was originally developed for drainage purposes and nitrogen uptake of cultivated crops in the newly reclaimed polder at Ijssel Meerl, but has subsequently been correlated with crop and pasture growth in WA (McFarlane *et al.* 1992). The index is calculated using the depth to the perched watertable and for how long it persists.

The SEW_{30} index is the sum of the daily values (in centimetres) when the watertable is within 30 cm of the soil surface. For example, if a watertable is at 10 cm for eight consecutive days, then at 20 cm for a further five days this corresponds to a waterlogging intensity of 210 cm.days

$$SEW_{30} = (30-10) \times 8 + (30-20) \times 5 = 210 \text{ cm.days}$$

Perched watertables fluctuate rapidly in response to rainfall, especially once the soil profile is wet. Small amounts of rain cause them to rise markedly (Figure 3.4.6).

Points to note when using the SEW_{30} index:

- The index assumes that waterlogging intensity increases linearly when the watertable is within 30 cm of the soil surface.
- The relationship between SEW_{30} and plant growth is not always consistent, because it is not high watertables as such that adversely affect root growth and metabolism (McFarlane *et al.* 1989a). The stage of growth and distribution of roots in relation to the watertable are also important (Dracup *et al.* 1992).
- A number of short periods of waterlogging may not have the same effect on plants as one single waterlogging event of the same intensity. This is related to the number of days before the oxygen concentration drops to a critical level (McFarlane *et al.* 1989a). However, every time the watertable falls it can remove soluble nutrients (e.g. NO_3^-).
- Soil temperature and organic matter need to be considered because they determine microbial and root respiration and thus the rate at which soil O_2 is depleted.
- The SEW_{30} index requires regular monitoring of the dip wells. Ideally, the wells are measured continuously or at regular short intervals

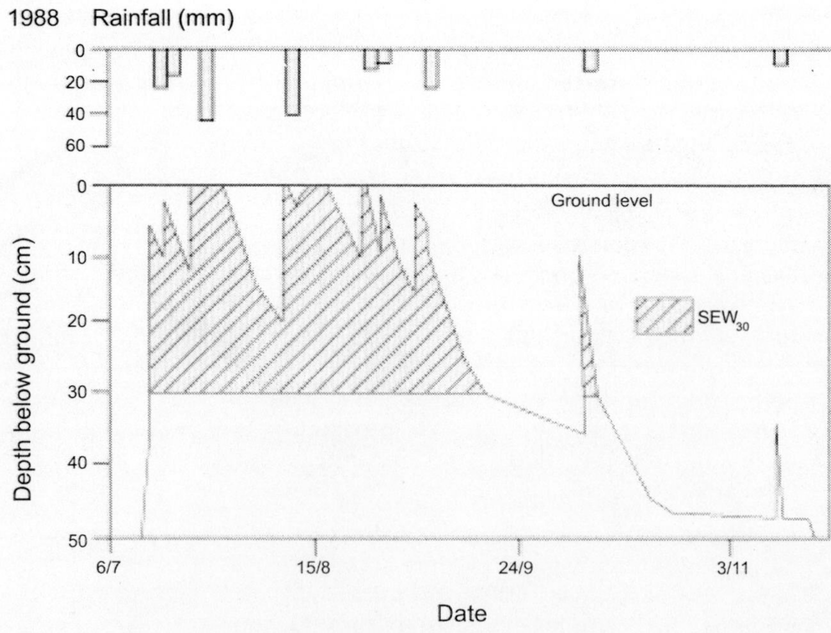

Figure 3.4.6 Perched water levels in the soil fluctuate rapidly in response to rainfall. The shaded area is the SEW_{30} index, which measures waterlogging intensity (from Cox and McFarlane 1990).

(e.g. 1 to 3 days). Non-continuous data underestimate SEW_{30}, especially high values (Cox 1988). Equation 3.4.1 can be used to correct non-continuous data measured every three days.

$$Y = 19.5 + 0.67X + 0.0014X^2 \quad (R = 0.99) \qquad (3.4.1)$$

where,

Y= the continuous SEW_{30} value (cm.days)

X = the non-continuous SEW_{30} value (cm.days)

(from Cox 1988).

- A perched watertable below 30 cm can still affect root growth. For example, if the clayey subsoil was at 50 cm, then a perched watertable at 30 cm for four weeks would have an SEW_{30} value of zero, but root growth could still be affected.

(iv) Remote sensing

LANDSAT TM digital information was used to map waterlogged crops in the East Yornaning catchment (McFarlane *et al.* 1992). The method can be used to estimate regional waterlogging and to map areas 50 m across, making it relevant at the farm scale. The spectral classifier used would be expected to vary between seasons and major landforms, although the bands would be expected to remain constant (McFarlane *et al.* 1992). The method shows considerable promise, although it has not been used widely. Detecting waterlogged pasture was unsuccessful.

Band 4, in the near infra-red, indicates the relative biomass of crops and pastures. Therefore there is no need to classify the images if the extent of poor growth (for whatever reason) is to be displayed on computer screens or shown as printed images. This method is superior to colour aerial photography, which only shows severely waterlogged areas and is often unavailable. LANDSAT TM is available at 16-day intervals, subject to cloud cover.

More recently, multidate Landsat TM imagery was used to identify waterlogged crops in successive seasons in the upper Great Southern (Wheaton *et al.* 1995). An index based on a combination of visible and near infra-red bands was derived to indicate the degree and extent of waterlogging. The index proved to be a reliable indicator where waterlogging was known to occur (B. Wheaton personal communication).

The airborne multispectoral scanner requires specialised equipment and although the results are superior to LANDSAT TM, it is more expensive.

MANAGEMENT OPTIONS

Management depends on the frequency and intensity of waterlogging. On sites which are waterlogged occasionally, changing agronomic practices to reduce the risk of yield losses may be all that is required. At the other extreme, poorly drained sites, especially those with insufficient grade to benefit from drains, are best excluded from cropping programs, or if presently uncleared should remain uncleared. To manage waterlogged or potentially waterlogged sites:

1. Identify waterlogged areas either with posts or markers in the paddock or on a farm plan or aerial photo,
2. Assess degree of waterlogging using the methods outlined previously,
3. Fence-off or manage the poorly drained areas separately (e.g. exclude from cropping program),
4. For other areas, consider drainage options and agronomic options.

Drainage options

Drainage is the best option for reducing the effect of waterlogging and inundation in many areas. Drains can reduce waterlogging intensity, control runoff and allow more flexibility in farming operations, such as earlier sowing of crops and timely applications of herbicide and fertiliser (McFarlane and Cox 1992).

The most suitable drainage system for a particular area depends on soil, site and economic factors. In WA, open drains have successfully reduced waterlogging. Interceptor drains are used on sloping sites (>1.5%) and relief drains on areas of low slope (Cox 1988; McFarlane and Cox 1990, 1992; McFarlane *et al.* 1990b). In Victoria, mole drains have been successful (e.g. MacKewan *et al.* 1992) but where they were tried in WA the subsoils have been unstable and the drains have collapsed. Slotted pipe drains have been used to reduce salinity in irrigation areas, but are not generally recommended because of their cost. They are also liable to be clogged by iron and are unable to handle storm flows (McFarlane and Cox 1992). The cost effectiveness of drains can be assessed with a cost benefit model such as DRAINS (Salerian and McFarlane 1987). When assessing interceptor drains in WA, the model showed that the greatest cost was the lost production from the land occupied by the drains.

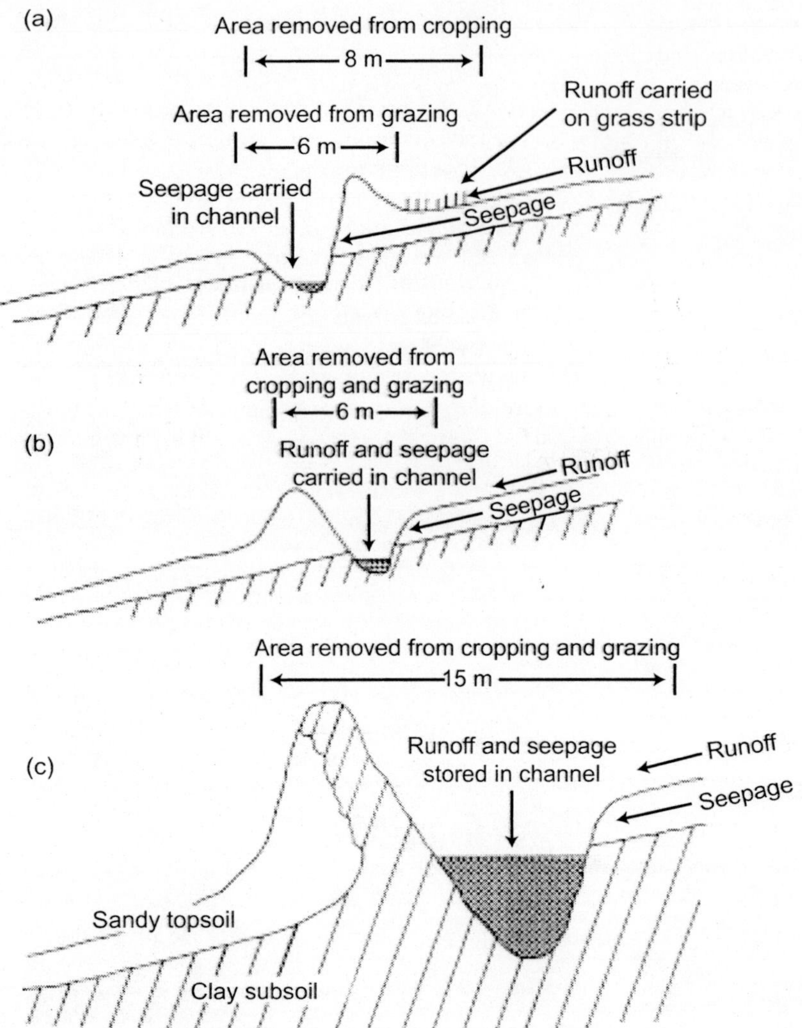

Figure 3.4.7 *Cross-sections of (a) reverse and (b) conventional bank seepage interceptor drains and (c) a WISALTS interceptor bank (from McFarlane and Cox 1990).*

Interceptor drains are suitable for permeability contrast soils on slopes greater than 1.5% if there is a safe disposal point for the water (e.g. grassed waterway, uneroded streamline). They are designed to intercept lateral flow above the clayey subsoil and to have sufficient grade to remove the water before it can seep through the bank. They are most effective when the subsurface soil has a high hydraulic conductivity to allow significant lateral flow (e.g. permeability contrast soils with a moderately deep (30 to 50 cm), coarse textured A2 horizon). There are two types of interceptor drains and an alternative, the WISALTS interceptor bank.

Conventional interceptor drains are in effect deep grade banks; the spoil is on the lower side, and the channel collects both through-flow and surface runoff (Figure 3.4.7). They are usually on a grade of about 0.4% to avoid channel erosion from storm flows. However, the upslope batter is often susceptible to rilling because the seepage face is unstable, resulting in siltation of the channel. *Reverse interceptor drains* were designed to prevent these problems (Figure 3.4.7). The overland flow is removed on a vegetated strip upslope of the earthen mound formed from the spoil from the channel. The seepage flow is collected in a channel on the lower side (Negus 1983a, b). About 8 to 10% of the land is removed from production with conventional or reverse interceptor drains and they are usually constructed with a grader except when the subsoil clay is deep.

Drain performance can be assessed by the waterlogging intensity adjacent to the drains, by the amount of water removed and their combined effect on crop yields or pasture production. The effectiveness of conventional and reverse interceptor drains was studied at Narrogin and Mt. Barker (Cox 1988). Interceptor drains have a greater effect downslope than upslope. In the wettest year, the distance waterlogging was reduced downslope was 3.5 to 5 times further than it was upslope, depending on the waterlogging intensity at the site (Cox 1988). The amount of water removed by interceptor drains varies between years and between sites: e.g. from 1 to 7% of the annual rainfall in the 450 mm annual rainfall zone (McFarlane and Cox 1990) and from 3 to 19% in a higher rainfall area at Mt Barker (Cox 1988). More rainfall is drained in wet years than in dry years, so the risk of over-drainage in dry years is considered to be low.

WISALTS interceptor banks are larger, deeper channels and either have no slope or a very low grade (0.03%). They occupy more land (up to 15%) and are constructed with a bulldozer, so they are more expensive. WISALTS banks are designed to hold both seepage and storm water in large impervious channels, allowing it to evaporate (Figure 3.4.7). They can reduce waterlogging and appear to reduce salinity, however they can also contribute to groundwater recharge. In a five year study of WISALTS banks on a hillslope, McFarlane *et al.* (1990a) found that about one-quarter of the water in the banks evaporated while the other three-quarters drained vertically from the channels, almost certainly increasing groundwater recharge and exacerbating salinity.

Relief drains are used on land with a low slope (<1.5%) and include W drains, spoon (V) drains and bedding. W drains have an excavated channel on either side of a central spoil mound (Figure 3.4.8), so water can enter from both sides. Spoon drains have the soil from the channel spread or scattered away from the channel so that they do not prevent water from flowing into the channel (Figure 3.4.8). Bedding is a common practice in some parts of the world, but has not been used extensively in WA. It consists of long raised beds from which surface waters drain into adjacent furrows. Spoon drains, W drains and bedding are effective in reducing inundation on level to very gently sloping land, but have limited effects on waterlogging (McFarlane *et al.* 1990b).

Raised permanent beds have been widely used to alleviate waterlogging on medium to fine-textured soils in irrigated agriculture (e.g. Tisdall and Hodgson 1990). The use of raised permanent beds, with traffic confined to the furrows between the beds has implications for reducing waterlogging, compaction and improving the structure of the surface soil. They are presently being tested for dryland farming in WA (Bakker and Hamilton 1997).

Agronomic options

The main agronomic options for reducing the effect of waterlogging on crop and pasture yields are summarised in Table 3.4.3. They are usually sufficient for sites which are waterlogged infrequently or can be used to complement drainage systems.

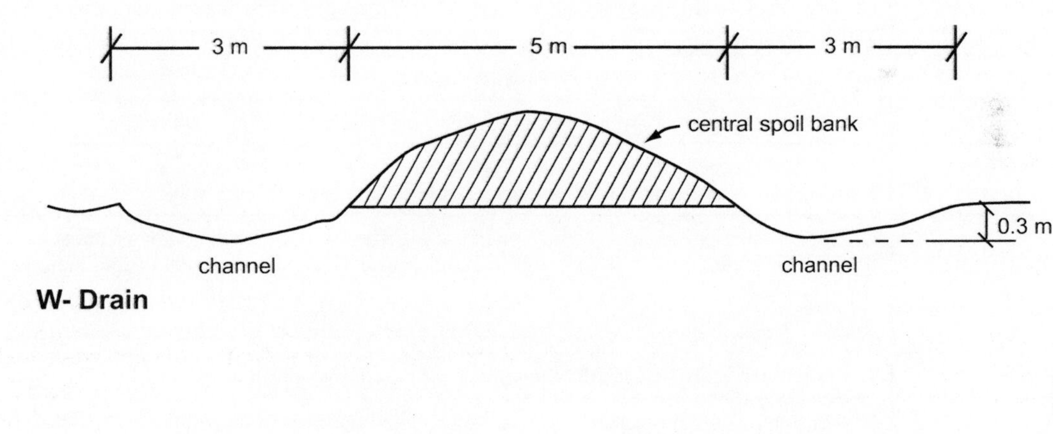

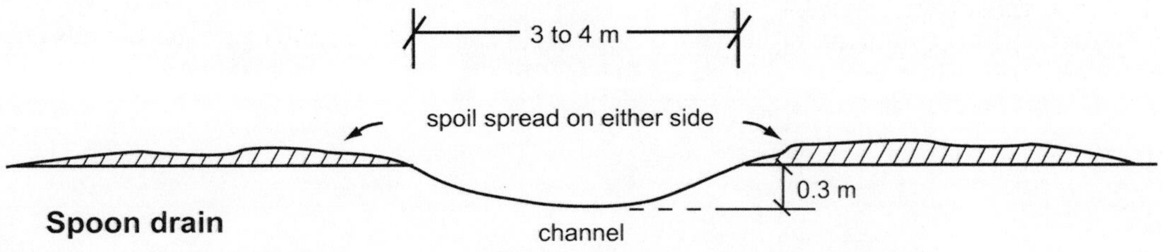

Figure 3.4.8 *Cross-sections of W and spoon drains (from McFarlane* et al. *1990b).*

Table 3.4.3 Agronomic options for reducing the effect of waterlogging on crop and pasture yields (adapted from Belford and McFarlane 1993; McFarlane *et al.* 1993).

Crop management	
Choice of crop species	Select a tolerant crop. Grain legumes and canola are generally more susceptible than cereals and faba beans. In descending order of waterlogging tolerance, the major crops are faba beans >oats >wheat >barley >canola >narrow-leafed lupins >field peas >chickpeas >lentils (also see Table 3.4.2 and Chapter 8).
Seeding	Early sowing, long season varieties and higher seeding rates are recommended for early establishment. Crops are more susceptible during germination and emergence. Increase seeding rates to insure against uneven germination and reduce dependence of cereals on tillering to produce grain, because waterlogging depresses tillering. High seeding rates also assist stressed crops to compete against weeds. Varieties which develop quickly are more likely to be at an advanced stage when waterlogging starts. Unfortunately, these varieties continue to develop quickly, flower early (increasing the risk of frost damage) and do not take advantage of the long growing season in high rainfall areas.
Fertiliser	A crop with a high nitrogen status is more tolerant of waterlogging (e.g. Watson *et al.* 1976; Barrett-Lennard *et al.* 1988; Buwalda *et al.* 1988). Aerial applications of N immediately after waterlogging can be an advantage if the previously applied N has been leached or denitrified. Ammonium is less likely to leach than nitrate, therefore ammonium forms of fertiliser (e.g. Agras, DAP) are preferable to nitrate forms (e.g. Agran) but they may increase soil acidity (Section 5.1).
Weed control	Weeds compete for water and nitrogen, so they retard recovery from waterlogging. Waterlogged crops are frequently weedy.
Disease control	Root diseases such as take-all can be more severe in waterlogged crops, therefore grass control in the previous crop or pasture year is important to reduce disease carryover. Leaf diseases are likely to be more severe in waterlogged crops because the crop is already under stress.
Pasture management	
Choice of species	Select tolerant species (refer to Table 3.4.2 and Chapter 9; Maling 1988).
Sowing	Resow early, especially on saline and waterlogged sites, because even pasture plants tolerant of salinity and waterlogging are more susceptible during early growth stages.
Grazing management	Graze the waterlogged areas lightly or not at all when waterlogged. Short pastures are more likely to be submerged during periods of inundation. Even short periods of inundation can kill some plants. Remove stock to reduce surface pugging on soils susceptible to structural decline (Section 3.2).
Weed control	Undesirable species such as dock compete for water and nitrogen, adversely affecting recovery.
Fertiliser	Nitrogen and sulphur are leached from waterlogged pastures in high rainfall districts. A strategic application of N in late-break seasons can stimulate grass and provide early feed; it also increases the likelihood that the pastures will enter a period of waterlogging with sufficient dry matter and N to continue growing.
Use of complementary species	Kikuyu has well developed aerenchyma in the roots and is dormant in winter. It is thought that oxygen leaks out of kikuyu roots, increasing the oxygen supply in the surrounding soil. This can benefit annual clover-based pastures in waterlogged soils (Setter and Belford 1990).

CHAPTER 4

PHYSICAL RESTRICTIONS TO ROOT GROWTH

4.1 HARD LAYERS IN SOILS

Paula Needham[*], Gottfried Scholz[**] and Geoff Moore

Several types of hard layers in soils can restrict crop emergence or restrict or prevent root elongation. Hard layers are hard when dry, but vary in hardness on wetting and in the degree to which they impede root growth and/or emergence. For instance, a hardsetting surface limits the timing of cultivations and sowing, can restrict crop emergence and root growth, is hard when dry, but is soft when moist. On the other hand, a red-brown hardpan is hard when both dry and moist because it contains insoluble cementing agents. If this type of hardpan is continuous it is a barrier to root growth.

Surface crust (P. Findlater).

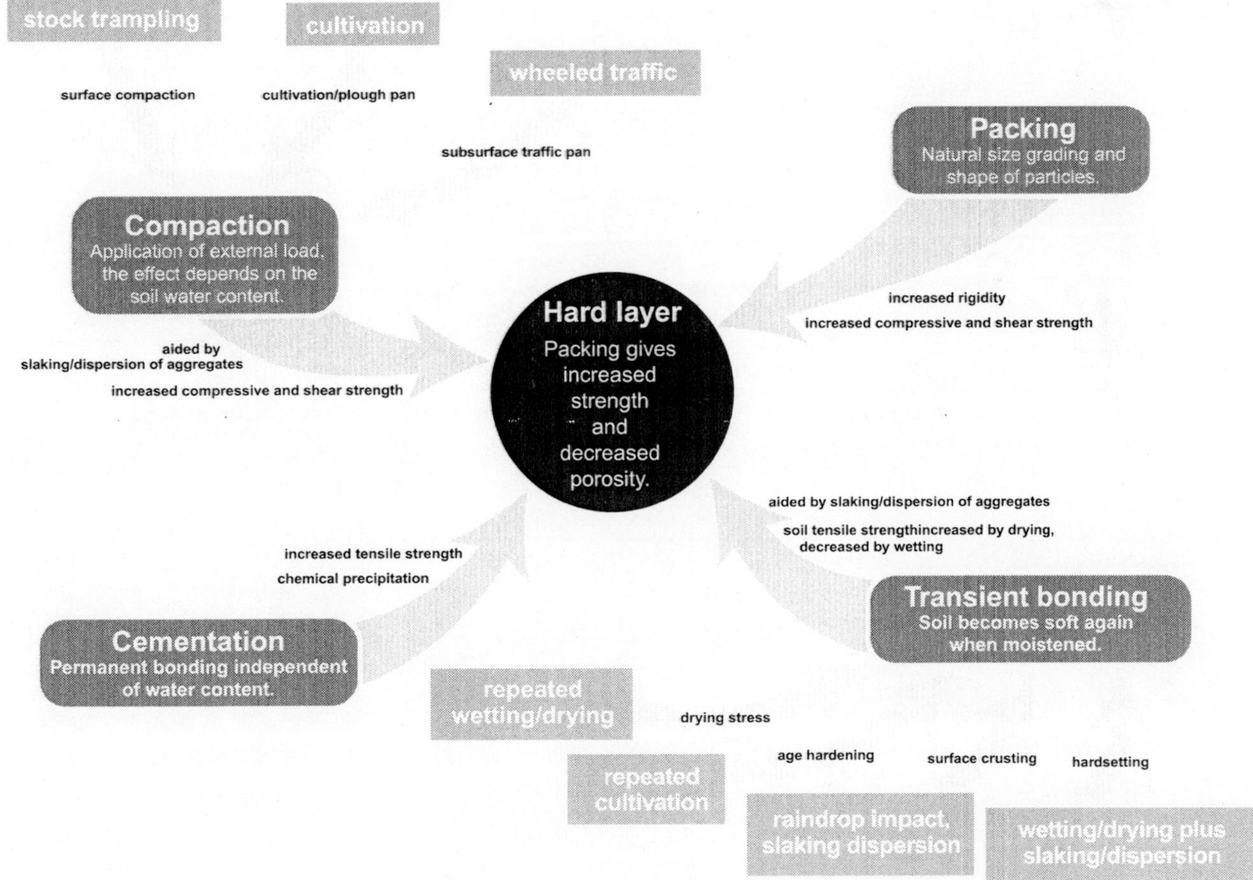

Figure 4.1.1 *A summary of the main hard layers in soils, together with the key processes responsible for their development and some of the factors affecting soil strength (B. Purdie).*

[*] Formerly Agriculture Western Australia, now Sussex, England.
[**] Formerly Agriculture Western Australia, now Scholz Environmental Consulting Perth.

The four main processes in the development of hard layers in soils: transient bonding, compaction, cementation and packing are summarised in Figure 4.1.1 (see previous page), together with the resulting hard layers formed and the key processes responsible for strength development. The processes involved in strength development (Table 4.1.1), hard layers in surface soils (Table 4.1.2) as well as the types of cemented pans (Table 4.1.3) are discussed in this section. Subsurface compaction is discussed in Section 4.2.

A hardsetting soil following cultivation. Note the massive structure (G. Moore).

Table 4.1.1 Processes involved in strength development and reduced porosity in soils.

Process	Description
Densest packing of particles (natural rigidity)	Some sands are rigid in their natural state because of the shape and size distribution of their particles. The particles are packed in a way that results in the smallest possible volume. Plant roots have difficulty penetrating these sands (Hilgard 1910; Lutz 1952).
Compaction from external forces	"Externally applied mechanical stresses will compact soils if the stress exceeds the strength of the soil" (Dexter 1988). Compaction makes the soil dense, the large pores are destroyed and the total air-filled porosity is reduced, leading to slow hydraulic conductivities. *(Refer to Subsurface compaction, Section 4.2)*
Age-hardening	Age-hardening can occur after moist soil has been completely disturbed by thorough mixing. The strength of the soil increases through micro-aggregation and flocculation of particles and also through cementation which strengthens the touching points of the particles (e.g. Utomo and Dexter 1981; Dexter *et al.* 1988). This process increases soil strength following the cultivation of moist soils and may help to explain why closely spaced cultivations can damage soil structure.
Compaction through drying (effective stress)	Through the action of water potential, *effective stresses* are generated which compact soil as it dries. The particles are pulled closer together by increasing tension between the retreating water films around them. This process increases the strength and reduces the volume of drying soils (Greacen 1960; Towner 1961; Towner and Childs 1972; Mullins and Panayiotopoulos 1984, 1989; Snyder and Miller 1985).
Transient cementation (transient bonding)	This term covers a number of transient cementing agents in soils. Cementation can involve clay bridges (very fine clay coatings) between sand particles or chemical cementation involving the progressive precipitation of slightly soluble compounds (silica, aluminosilicates, iron oxides) as very thin coatings, bridges or gels in drying soil. An important condition is that the soil is dispersive to allow the redistribution of particles. Also, it must undergo wetting and drying cycles to allow these bonds to form. The cementation is transient, being unnoticeable when the soil is wet, but rapidly increases as it dries (Mullins and Panayiotopoulos 1984; Daniel *et al.* 1988; Chartres *et al.* 1990). Transient cementation is partly responsible for the dry strength of sandplain soils (e.g. yellow deep sands, yellow sandy earths) in WA (Daniel *et al.* 1988), has been identified in some hardsetting E horizons in duplex soils (Chartres *et al.* 1990), and is thought to be widespread in hardset horizons (C. Chartres personal communication).
Cementation	Cementation is a process in which soil material hardens irreversibly when calcium carbonate, silica or iron are precipitated. The porosity decreases gradually until no water can percolate through this layer and it resembles rock. Cemented layers are an inherent feature and may occur within the soil profile or at the surface. *(Refer to the following section on Pans in soils)*

Table 4.1.2 Types and characteristics of hard layers in surface soils.

Type of hard layer (alternative names)	Description
Crust (surface hardening and sealing)	Crusting, surface hardening or sealing involves mobilisation and deposition of dispersed fine particles (silt size or smaller; Chepil 1955) at the soil surface. Crusts only affect the surface and are thinner (usually <2 cm) than hardsetting horizons. The formation of crusts is enhanced by raindrop impact and by water with low electrolyte concentration (Agassi *et al.* 1981). Soils that have massive structures or pseudo-earthy fabrics tend to surface seal or crust because they have a high proportion of dispersed silt and clay (Chepil 1955; Chen *et al.* 1980; Emerson 1984). The silt and clay form a 'wash-in-zone' at the surface (McIntyre 1958a, b). An exchangeable sodium percentage as low as 2 may be sufficient to produce seals or crusts 1 cm thick (Gal *et al.* 1984; Shainberg 1984). Apart from sands, self-mulching clays and soils with a high OM content, most bare soils are likely to crust to some extent if subject to heavy rain when air-dry (Mullins and Ley 1994). *(Refer to Soil structure decline, Section 3.2)*
	Similar processes (i.e. slaking and/or dispersion) are responsible for crusting and hardsetting and they appear to lie on a continuum from a friable condition, to crusting, to a hardsetting soil, as the organic matter decreases and sodicity and cultivation increase (Mullins* et al. *1987).
Hardset	A revised definition of hardsetting is provided by Mullins and Ley (1994): "A hardsetting horizon is one that sets to an almost homogenous mass on drying. It may have occasional cracks, typically at a spacing of >0.1 m. Air-dry hardset soil is hard and brittle, and it is not possible to push a forefinger into the profile face. Typically, it has a tensile strength of >90 kN/m^2. Soils that crust are not necessarily hardsetting since a hardsetting horizon is thicker than a crust. (In cultivated soils, the thickness of the hardsetting horizon is frequently equal to or greater than that of the cultivated layer.) Hardsetting soil is not permanently cemented and is soft when wet. Clods of a hardsetting horizon that has been cultivated will partially or totally disintegrate upon wetting. If the soil has been sufficiently wetted, it will revert to its hardset state on drying. This can happen after flood irrigation or a single intense rainfall event." Hardsetting can affect the whole surface layer (A1 or Ap horizon), while the subsurface layer (A2 or E horizon) is sometimes hardsetting. Most hardsetting soils are dominated by kaolinite and illite, and textures range from loamy sand to sandy clay, i.e. 5-40% clay (Norrish and Pickering 1977). In the natural state, topsoil exhibits a rather massive structure which is broken up during cultivation into clods (artificial aggregates). Hardsetting involves two distinct processes: first, the clods collapse during wetting because of swelling, slaking (Aylmore and Sills 1982; Emerson 1977, 1978) and/or dispersion (Cockroft and Martin 1981; Murphy *et al* 1987; Young *et al.* 1988). Second, the surface layer hardens without re-structuring during drying (Mullins *et al.* 1990). The strength on drying is due to effective stresses pulling the individual particles together, plus transient cementation. It is easy to recognise hardsetting in the field at old cuttings (road or quarry cuttings), because the hardset topsoil hangs over the subsoil with rounded edges. *(Refer to Soil structure decline, Section 3.2)*

PANS IN SOILS

There are a number of different pans in soils which may be cemented or indurated. Plough pans (also called cultivation pans) and traffic pans form as a result of farming practices, and are not an inherent feature of the soil and are not permanently cemented, although there could be some transient cementation. On the other hand, ferricrete pans and red-brown hardpans are permanently cemented and an inherent feature of the soil.

Pans can be described in terms of their:

- degree of cementation (whether they slake in water and if not, their strength);
- type (Table 4.1.3);
- continuity *(continuous, discontinuous,* or *broken);*
- structure *(massive, vesicular, concretionary, nodular, platy,* or *vermicular)*

(refer to McDonald *et al.* 1990).

Stability on wetting

Slaking in water is an easy way to identify whether pans are cemented. If a small fragment (~30 mm) of the pan is placed in water for an hour and slakes, it is called uncemented. If it does not slake, the pan is cemented.

Pans that do not slake in water will be a barrier to root growth and prevent root elongation if they are continuous. Some pans are discontinuous or fractured, and these allow limited

Table 4.1.3 Types and characteristics of pans in soils.

Type of pan (alternative names)	**Slakes in water**	**Description**	**Occurrence in south-western Australia**
Plough and traffic pans	**Yes**	Caused by repeated cultivations to the same depth (plough pan) or from compressive forces induced by heavy machinery (traffic pans). *(Refer to Subsurface compaction, Section 4.2)*	Traffic pans are very common in uniform coarse-textured soils (loamy sands to sandy loams), especially in the northern and eastern wheatbelt. The distribution of plough pans in medium to fine-textured soils is unknown.
Densipan	**Yes**	An earthy pan with a high content of very fine sand. Less stable on exposure than underlying or overlying soil horizons (Smith *et al.* 1975).	Minor.
Fragipan	**Yes** (when dry)	An earthy pan that is usually loamy. Fragments will slake in water when dry, but not when wet. More stable on exposure than overlying or underlying soil (Soil Survey Staff 1975; McDonald *et al.* 1990).	Not a common term in WA, but is similar to a hardsetting horizon.
Calcrete (carbonate pan, kankar, secondary limestone, travertine)	**No**	All calcrete pans contain $CaCO_3$, so they show a slight to strong effervescence with 1M HCl. May vary in appearance and degree of cementation.	In the north-eastern wheatbelt, sometimes close to areas of red-brown hardpan. May form a coating over a red-brown hardpan or to some extent be interleaved with it.
Ferricrete (cemented ironstone, duricrust, laterite, bog iron ore, petroferric horizons)	**No**	Formed by the cementation of discrete nodules (pisolites) and contain a high concentration of iron and aluminium oxides (goethite, haematite, bauxite). Considered to be a relict feature of a past climate (humid and wet). When part of a lateritic profile, the term *laterite* can be used. *Duricrust* is a local term (Woolnough 1918). *Bog iron ore* has a similar appearance but is less pisolitic and formed under different environmental conditions. Bog iron ore forms in swamps.	Widespread. Common on the surface or at shallow depth or on breakaways in all areas that have been subject to laterisation. Duricrust is usually higher in the landscape, corresponding to the original plateau surface. If on the valley floor, it is likely to be bog iron ore rather than duricrust. Occurs in the Darling Range, in the Zone of Rejuvenated Drainage and in the Zone of Ancient Drainage, and scattered in other areas.
Ortstein (iron organic hardpan)	**No**	A black organic pan in organic sands which forms from fluctuating watertable. The depth to the pan corresponds to the height of the wet season watertable. Contains high concentrations of organic matter e.g. the concentration of organic carbon in a pan on the Swan Coastal Plain was 10 times that in the overlying soil layer (McArthur 1991).	Common on the Swan Coastal Plain, underlying poorly drained, leached, grey to white sands of the Bassendean Association. The depth to the iron organic hardpan varies from <1 m to >2 m, but it is commonly >1 m below the surface (D. van Gool personal communication).
Red-brown hardpan	**No**	Earthy with a reddish-brown to red colour (2.5YR 4/6-4/8 or 5/6-5/8 moist) with a dense, but porous appearance and a platy structure. May also contain gravel lenses, wavy black manganiferous linings and small amounts of calcium carbonate. The principal cementing agent is silica (Teakle 1936), although iron oxides and calcium carbonate may also be present (Wright 1983).	Widespread in southern pastoral areas from north of Menzies, south to Three Springs (Teakle 1936; Northcote *et al.* 1967). Thought to be a general substrate which can underlie all soil types in an area (Wright 1983). Widespread in the north-eastern wheatbelt on broad valley floors north-east of Three Springs and Mullewa (McArthur 1991; Rogers 1996).
Silcrete	**No**	Any hard layer where the cementing agent is silica. Can form anywhere that high concentrations of silica occur in wet conditions.	Minor areas are scattered throughout the agricultural area in wet soils.
Duripan	**No**	An earthy pan which is strongly cemented by silica and does not slake in water. Duripans do not soften in water and are always brittle even when moist (Soil Survey Staff 1975).	Not a common term in WA, but probably equivalent to silcrete or red-brown hardpan.

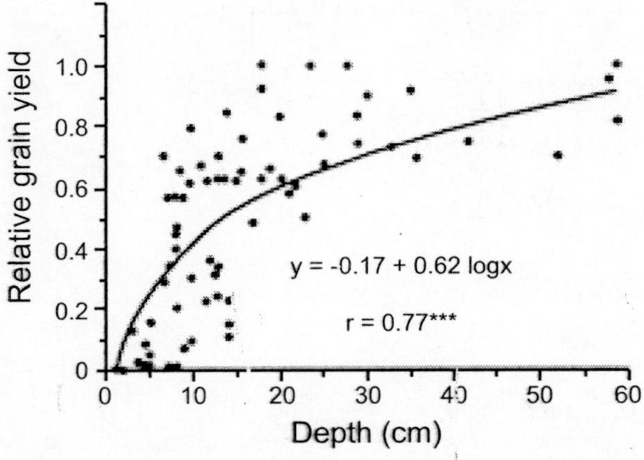

Figure 4.1.2 *The relative grain yield compared with the depth to a ferricrete pan (from Figure 1b; Sullivan 1991).*

root penetration. In general, the depth to a cemented pan can be assumed to be the depth of maximum root growth for annual crops and pastures, even though there could be some local variation due to cracks or discontinuity.

Effect on plant growth

A pan restricts the volume of soil available for root growth and indirectly, the plant available water. Their effect on plant growth will depend to a large extent on the depth and continuity of the pan. Very shallow pans can affect plant anchorage and stability. In general, if the pan is less than 30 cm below the soil surface, it will restrict crop production, but if it is deeper than 100 cm, the effect is likely to be minimal or even beneficial.

For example, cereal crops growing on shallow, gravelly loamy sands over a cemented pan (a ferricrete pan, 5 to 60 cm below the surface) in a low rainfall area in the eastern wheatbelt were severely water stressed during all stages of crop growth (Sullivan 1991). Wheat yields were reduced by at least 25% when the ferricrete was within 30 cm of the surface (Figure 4.1.2; Sullivan 1991).

For non-irrigated tree crops, a pan within a few metres of the surface can be a barrier to root growth and reduce survival during extended dry periods.

4.2 SUBSURFACE COMPACTION

Paula Needham, Geoff Moore and Gottfried Scholz***

Traffic pans are common on many of the coarse-textured soils in the agricultural area of Western Australia. Ameliorating subsurface compaction, using deep tillage can result in spectacular yield responses (Jarvis and Porritt 1985), which is unusual in a semi-arid environment. In other parts of the world significant responses are usually only reported from temperate climates or for irrigated soils (Poole 1987).

Various types of hard layers exist in soils. Some result from physical processes, others through chemical precipitation of iron, carbonates or silica. This section concentrates on the physical compaction of the subsurface soil (i.e. 10 to 60 cm below the surface). The relationship between subsurface compaction, soil structure decline and other hard layers is explained in Section 4.1.

PRINCIPLES OF SUBSURFACE COMPACTION

Soil compaction describes the rearrangement of soil particles and reduction in macro-porosity and total pore space through applied stresses. These stresses can be internal (e.g. due to increasing pore water suction) or external such as those imposed by traffic. *The main cause of subsurface compaction on tilled soils is wheeled vehicular traffic, especially heavy dual axle tractors.* Other stresses may include tillage, stock trampling and overburden pressure.

The main processes of compaction are:

- *Compression.* Packing (differential particle movement) resulting from a vertical force, for example, from wheels or animals.
- *Shearing.* Deformation of the soil mass, resulting from horizontal force, such as spinning and slipping tractor wheels or other implements.
- *Smearing.* Realignment of soil particles in a thin layer from random to parallel orientation by slipping wheels (extreme shearing).

Compression and shearing do the most damage in moist soils. They have the greatest effect on soil macro-pores, which are usually filled with air in unsaturated soils and are therefore easy to compress. As the water content increases towards saturation, the pores are filled with water rather than air. This means that water has to be squeezed out to cause compression (water being essentially incompressible at atmospheric pressure), so the soil becomes more resistant to compaction, but begins to flow and shear. On the other hand, dry soils resist compaction because of their interparticle forces, including water films, internal friction, fibres of organic matter and bonding (Hillel 1980; Towner 1983).

There are two types of pan resulting from subsurface compaction:

- *Plough pans* are caused by repeatedly tilling the soil at a constant depth for many years. Tillage implements such as tines, discs and mouldboard ploughs can smear and compact the soil immediately below their operating depth, especially at soil water contents above the lower plastic limit. Plough pans are characterised by an abrupt boundary between the tilled and compacted layers and there are often signs of smearing on the surface of the compacted layer. This often has a platy structure with a horizontal orientation, while the soil below is less dense. Plough pans form mainly in soils with medium to fine textures (i.e. sandy loams or finer).
- *Traffic pans* are layers of high strength caused by compression of the soil by traffic. They are deeper than plough pans, with the layer of maximum strength at 10 to 40 cm. Traffic pans are common in soils with a coarse to medium texture (i.e. sands to light sand clay loams).

Soil compaction is discussed below under: topsoil stresses, subsurface stresses, internal pressures, depth of compacted layers and effect on crop growth.

Compression stresses on the topsoil

The stresses exerted on a soil depend on the nature of the object imposing the pressure. Dexter and Tanner (1973) quoted the maximum pressures generated by animals and tractors as:

	Stress (MPa)
Horses and cows	0.16-0.39
Sheep/humans	0.06-0.10
Small tractors	0.03-0.10
Large tractors (2-axle)	0.1-0.2

Stock trampling is a significant cause of compaction, especially in the surface horizon of finer textured soils, but effects are confined to the upper 15 cm of the profile (Willatt and Pullar 1983; Kelly 1985).

* *Formerly Agriculture Western Australia, now Sussex, England.*

** *Formerly Agriculture Western Australia, now Scholz Environmental Consulting Perth.*

Compression stresses in the subsurface soil

Total load is the primary factor affecting soil compaction, so heavy, tractors with dual tyres cause more compaction than small tractors. Tyre type, size, shape, inflation pressure and speed also have a direct effect on surface compaction, but less effect on subsurface compaction (Porterfield and Carpenter 1986). The effect of a tractor tyre compressing the soil to depth is illustrated in Figure 4.2.1. Soil compaction caused by agricultural machinery is reviewed by Soane *et al.* (1981a, b, 1982).

The effect of traffic frequency on compaction depends on the initial condition of the soil. If it is initially loose (e.g. deep-ripped soil, newly cleared land), then most compaction occurs during the first pass (Henderson 1989b). On a soil which is consolidated, but not compacted, there is an almost linear increase in compaction with number of tractor passes (e.g. Soane *et al.* 1981b). On the uniform coarse-textured soils of the sandplains of WA (e.g. yellow deep sands), there is a relatively linear increase in compaction up to the fourth pass and then minimal effect from additional passes (Henderson 1985; Jarvis and Porritt 1985).

Compaction of the surface soil from traffic on wet soils (G. Moore).

Internal pressures

Tight packing of soils occurs when the physical force of the water menisci around particles pulls them closer together as the soil dries out. This *effective stress* increases at low moisture contents (Greacen 1960; Towner 1961). Wetting and drying phenomena can cause considerable shrinkage of sandy soils, e.g. Mullins and Panayiotopoulos (1984) found that the volume of a saturated, sand-kaolin mixture (4% clay) decreased by 17% when air dried. In WA, subsurface compaction is absent under uncleared native vegetation, so this wetting and drying is not an important process in undisturbed conditions (Henderson 1989b). On the other hand, this process appears to be partly responsible for recompaction on cleared sites when bonds have been weakened by deep tillage. This has been observed as the 'self-compaction' of yellow sandy earths at Wongan Hills even without traffic (P. Blackwell personal communication).

Transient cementation is partly responsible for the dry strength of sandplain soils (e.g. yellow deep sands, yellow sandy earths) in WA. It involves the development of thin coatings and bridges of soluble silica and fine clay between the particles in drying soils. The autumn rains are sufficient to dissolve most of the precipitated silica and reduce soil strength. The cementing agents are also readily disrupted by tillage and require at least one wetting and drying cycle to reform (Mullins and Panayiotopoulos 1984; Daniel *et al.* 1988; Henderson *et al.* 1988).

Depth of compacted layers in different soils

Plough pans are immediately below the cultivation depth, while traffic pans are usually located 10 to 60 cm below the surface. They become shallower and thinner as the soil texture becomes finer (Table 4.2.1). Depth is also influenced by types of machinery and cultivation implements which differ in the load applied and contact area. For example, light vehicles compact the soil to a depth of 30 cm, whereas heavy farm machinery can compact the soil to 60 cm (Raghavan *et al.* 1990).

Effect on crop growth

Subsurface compaction influences plant growth by:

- decreasing the soil pore size, volume and pore continuity
- decreasing root penetration and density
- reducing access to moisture and nutrients (especially N and K)
- increasing the harmful effects of pathogens resident in the topsoil (Wilson 1986).

Reduction in root growth can reduce access to water and uptake of mobile nutrients (especially nitrogen) with a corresponding decrease in yield (Bowden 1986). The problem is exacerbated in well drained, coarse-textured soils where nitrogen is readily leached beyond

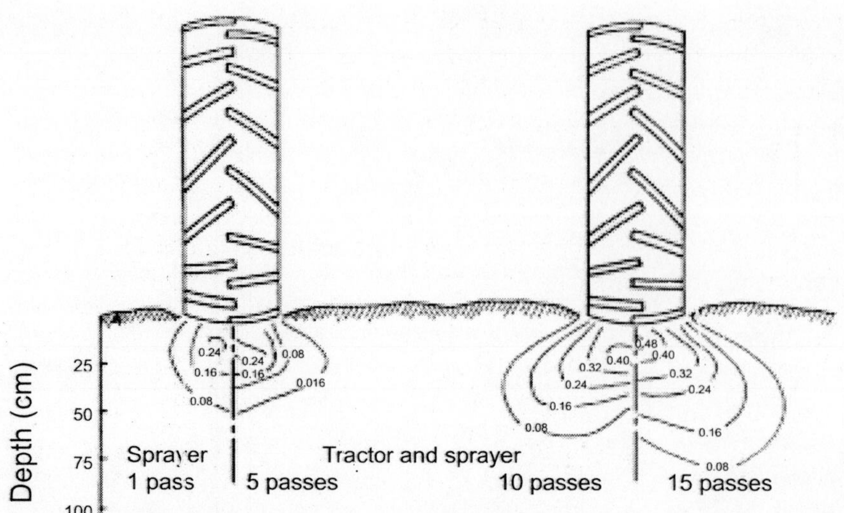

Figure 4.2.1 *Distribution of change in soil dry density in a sandy loam with different numbers of tractor passes (from Figure 3, Raghavan* et al. *1976).*

the root zone and water percolates rapidly to an inaccessible depth. In some duplex soils, a compacted layer in the A horizon does not reduce yield when moisture and nutrients in the topsoil are adequate (Belford *et al.* 1992), but may be a limitation in dry seasons when access to subsoil moisture is required.

The effect of compacted layers on crop growth is complicated by different responses from the various species and cultivars (e.g. root distribution of seedlings; Materechera *et al.* 1991). At later growth stages, the degree of root restriction can also differ markedly. For instance, the root system of narrow-leafed lupins is less restricted than that of cereals (Whiteley and Dexter 1982; Henderson 1989c). See Chapter 8 for further information.

SUSCEPTIBILITY TO SUBSURFACE COMPACTION

Most soils can be compacted if field operations occur at the wrong time, but some are relatively more susceptible. Susceptibility is reduced in strongly structured, well drained soils with high organic matter and also in dry regions where soils rarely reach the water content at which severe compaction is possible.

Two methods of assessing susceptibility to subsurface compaction are described here: the susceptibility to compaction based on texture classes, and then a compaction index which requires a detailed particle size analysis. Both methods are based on the results of Proctor compaction tests on a range of soils from WA (Henderson 1989b; Daniel *et al.* 1992). This test involves compacting the soil under standardised conditions and measuring the resulting strength with a penetrometer.

Properties affecting the susceptibility of a soil to compaction include: structure, particle size distribution, particle shape, clay mineralogy, organic matter content and moisture content (Larson *et al.* 1980; Dexter 1988). The three most important factors appear to be soil moisture (discussed previously), particle size distribution and particle shape.

Particle size distribution. Particle size distribution or grading is important in determining susceptibility to compaction. A well graded soil has a good representation of all particle sizes and is more susceptible to compaction than a narrowly graded soil.

In coarse-textured soils, many studies have shown a good correlation between particle size distribution and compactability (e.g. Panayiotopoulos and Mullins 1985; Spivey *et al.* 1986). In a range of coarse-textured soils from WA, the maximum soil strength was related to the clay content for soils with 2 to 15% clay (Henderson *et al.* 1988). An exception was a sandy soil (3% clay) with a high proportion of very coarse sand (1 to 2 mm), which had a higher maximum strength than predicted (Henderson *et al.* 1988).

Particle shape. Rounded particles compact more than angular and roughly shaped particles (shape can be checked with a 10X hand lens). Larger particles, such as coarse sand, tend to be more rounded than fine sand and so can pack more tightly (Panayiotopoulos and Mullins 1985).

Table 4.2.1 Depth of traffic pans in relation to the clay content of the soil.

Clay % (10-30 cm)	Example	Depth of maximum strength (cm)
<4	Pale deep sand	30
4-8	Yellow deep sand to loamy sand	25
8-12	Yellow sandy earth	20

Table 4.2.2 The susceptibility of soils to compaction based on texture.

Soil texture	Relative compactability	Comments
Sand (narrowly graded)*	Low	Bulk density may be high but strength and compressibility are low.
Sand (well graded)	Moderate	
Loamy sand	Moderate to high	High bulk density, high compressibility and high strength. Depends on particle shape and presence of secondary structure.
Sandy loam	Moderate to high	As above.
Sandy clay loam**	Moderate to high	As above.
Clay loam or finer	Low	Low bulk density, compressibility and soil strength.

** Sandy soils containing very low clay (<3%) and narrowly graded (i.e. fairly uniform particle size) can cause difficulties in root penetration even though there is no evidence of a compact layer, cementation or lack of moisture (Hilgard 1910; Tennant* et al. *1992). Roots that encounter these 'tight' layers tend to be thick and stubby on the ends.*

*** Soils with a mixture of particle sizes such as sandy clay loams are physically capable of compacting more than sandy soils (Panayiotopoulos and Mullins 1985), but higher levels of organic matter and/or better structural development may reduce the degree of compaction.*

Texture classes

The general susceptibility to compaction can be estimated from soil texture (Table 4.2.2).

ASSESSMENT OF PADDOCK STATUS FOR SUBSURFACE COMPACTION

Three methods are available for assessing subsurface compaction. The preferred method will vary according to the soil type and the time of year.

- Using field indicators
- Measuring penetration resistance directly
- Measuring soil bulk density.

In *uniform coarse-textured soils* (susceptible to traffic pans), a direct measurement of moist penetration resistance is the most reliable way to assess the presence of a compacted layer. The soil must be moist, preferably near the upper storage limit. Field indicators can also be used.

In *permeability contrast soils* (susceptible to traffic pans), a direct measurement of moist penetration resistance is the most reliable method. However, if the soil has a high gravel content then bulk density is the only option. Field indicators can also be used.

In *cracking clays* (susceptible to plough pans), visual assessment and/or bulk density are recommended. Variation in moisture content is likely to cause problems with measuring penetration resistance.

In *medium to fine-textured soils* (susceptible to both traffic and plough pans), all three methods could be useful, depending on the specific situation.

Field indicators

Field observations can be useful, especially for identifying plough pans in medium to fine-textured soils. Most of these observations are not exclusive to subsurface compaction and can be associated with other problems, so care should be taken when making interpretations based solely on this evidence. Observations that can be useful include:

Detecting plough pans in medium to fine-textured soils

- Land use history that has involved regular cultivation to the same depth, especially if cultivated when too wet;
- An abrupt boundary between tilled and compacted layers, with signs of smearing on the upper surfaces of the compacted layer;
- Water ponding or slow drainage with no obvious explanation;
- Evidence of anaerobic conditions (mottling) in the top 20 cm of soil;
- Hard layer observed when digging (it must be remembered that soil tends to become denser with depth even if there is no compaction and the soil underneath the tilled layer will be tighter naturally);
- Virtually all roots confined to upper 30 cm or growing horizontally below the cultivation layer (remember that it is normal for 75-80% of the roots of most crops to be in the top 30 cm);
- Platy structure evident below the cultivation layer.

AN INDEX TO CHARACTERISE SOIL COMPACTION BEHAVIOUR

Heiko Daniel *

Particular mixtures of sand, silt, and clay can be very susceptible to soil compaction and therefore achieve very high states of soil compactness. This may be desirable for road surfaces but the high soil strength can severely limit plant root growth.

For coarse-textured soils (sands to sandy loams) the characteristic particle size distribution can be related to the compressibility of a particular soil. This follows packing theories which deal with how smaller particles can fit into spaces between larger particles and the ideal proportions of particles of different sizes to achieve the most dense packing arrangements (Staple 1975).

The compaction behaviour of a large number of coarse-textured WA soils from the wheatbelt falls into two distinct groups. The first group, representing sands, shows only a very small increase in bulk density with increasing water content when applying a constant compactive effort. Typical examples are sands from near Esperance, Badgingarra and Geraldton. The second group represents soils of a loamy character with textures from loamy sand to sandy clay loam. The bulk density steadily increases with water content, at a constant compactive effort, to a maximum at the 'optimum water content', and then declines alongside the saturation line. Typical soils are Wongan Hills loamy sands and sandy loams from Merredin. Increasing compaction (compactive effort) will increase the maximum bulk density to a limiting value.

In general, a more even particle size distribution will result in a greater compressibility. For soils with increasing clay content the compressibility and maximum bulk density decrease. Sands contrast with loams in that they can normally only exist at an already closely packed state from which extensive further compaction is not possible without crushing the actual particles. Nevertheless, within the sands, those with more even particle size distribution can achieve a higher state of packing.

In an attempt to characterise this soil behaviour with a single index and thus classify the soils according to their susceptibility to compaction, a non-uniformity coefficient (C_n) was calculated from the particle size data. C_n was defined as log (d_{75}/d_{10}), where d_{75} and d_{10} are points along a particle size summation curve (<2 mm) that represent the equivalent particle diameters at 75% and 10% by weight, respectively (Daniel *et al.* 1992). It must be emphasised that to calculate this index it is necessary to undertake a detailed particle size analysis. Besides determining the clay and silt fractions, the sand fraction should be further split into five sub-fractions. This ensures a suitable particle size distribution curve. Figure 4.2.2 shows selected soils from the WA wheatbelt sorted into well graded and poorly graded soils.

This index has been applied successfully to classify soils according to their susceptibility to subsurface compaction, and also to classify the susceptibility of surface soils to compaction in relation to the impact of animal trampling. The method is suitable for coarse-textured soils (<20% clay) and can be used for surface and subsurface soils. It is not suitable for soils which have moderate to strong secondary structure (i.e. clearly defined aggregates).

Interpretation of results:

C_n value	Susceptibility to compaction
<0.5	Very low
0.5-1.0	Low
1.0-2.0	Moderate
2.0-2.5	High
>2.5	Very high

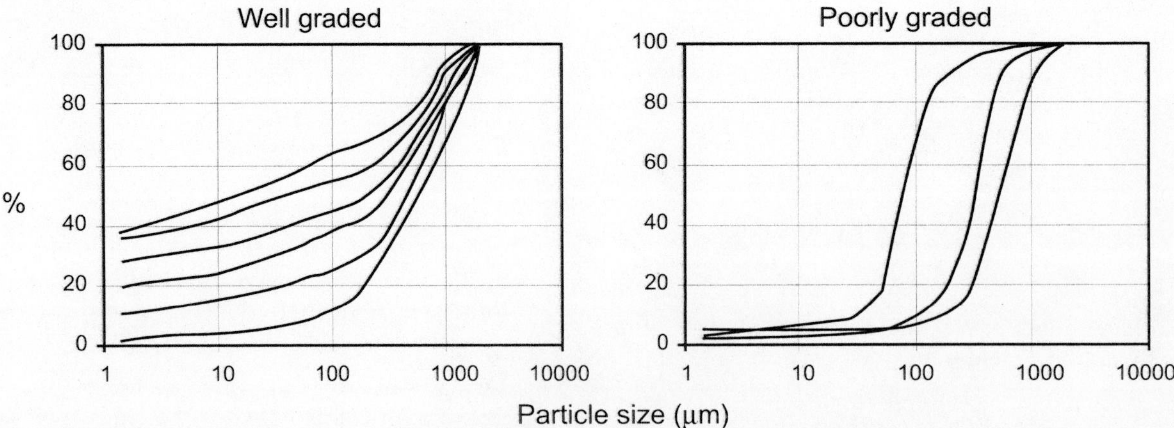

Figure 4.2.2 *Particle size summation curves of well graded and poorly graded soils. The well graded soils range from a sand from East Beverley to a sandy clay loam from York (C_n values 2 to 3). The poorly graded soils are sands from Geraldton and Badgingarra (C_n 0.3 to 1.3).*

* *University of New England, Armidale.*

Deep ripping on a uniform coarse-textured soil in the Three Springs district.

- Virtually all roots confined to upper 30 cm of soil or growing horizontally (remember that it is normal for 75-80% of the roots of most crops to be in the top 30 cm);
- If large tractors are used there may be a pattern of fluctuating crop growth across the paddock corresponding to areas which have regular traffic (the regular pattern is complicated by other management practices like topdressing and spraying).

Detecting traffic pans in soils with coarse-textured A horizons

- Land use history that has involved regular cultivation;
- Reduced yield, wilting, depressed growth of crop with no obvious explanation;
- Hard layer observed when digging (it must be remembered that soil tends to become denser with depth even if there is no compaction. Also, the soil underneath the tilled layer will be tighter naturally). Consider the over-riding effects of water content;

Measuring penetration resistance directly

Penetration resistance is an empirical measure of the degree of compaction. A penetrometer measures the force needed to push a rod, usually with a cone-shaped point, through the soil. The measurements are generally reported in kPa or MPa (100 kPa = 9.81×10^3 kg/m^2).

Measurements are quick to perform, but because they are point rather than bulk measurements, there is a large spatial variation in readings (O'Sullivan *et al.* 1987).

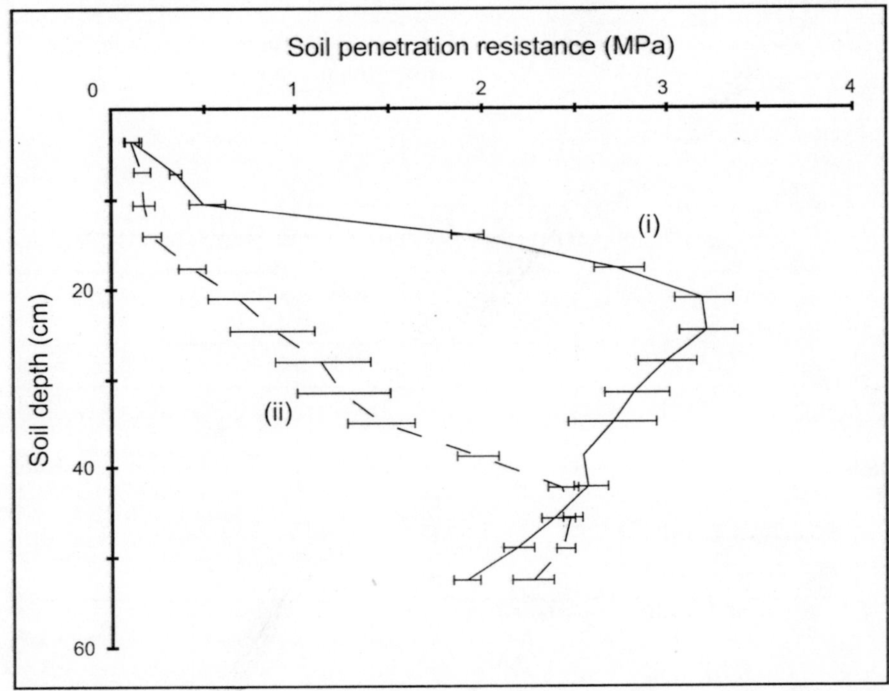

Figure 4.2.3 *A typical soil strength profile measured with a penetrometer for (i) a compacted soil and (ii) the same soil following deep-ripping (from Figure 1, Henderson 1989a).*

POINTS TO NOTE WHEN USING A PENETROMETER

Penetrometer readings are affected by many factors. These include soil particle and pore size distribution, water content and potential, particle shape and roughness and interparticle bonding (Henderson *et al.* 1988). The measurements will also vary depending on the type of instrument, and the size and type of the probe tip (Whiteley and Dexter 1981).

Penetrometer readings are greatly affected by moisture content. The soil should be at or near the upper storage limit (USL), which is one day after a soaking rain for a coarse-textured soil and two to three days after a soaking rain for most other soils. In uniform coarse-textured soils, the penetrometer readings will not be adversely affected if the soil moisture content is >70% of USL (Henderson *et al.* 1988).

Penetrometers do not necessarily estimate the pressure which has to be exerted by plant roots entering the soil. Dracup *et al.* (1992) found that lupin roots could penetrate duplex soil adequately, even though the penetrometer recorded 4 MPa, the maximum value on the scale. As the probe cannot follow the path of least resistance and use preferred pathways like a root, these anomalies are normal. The penetrometer has to exert greater pressure (two to five times more) than a root tip in order to penetrate the soil (Greacen *et al.* 1969; Whiteley *et al.* 1981). In addition, plant roots appear to be able to generate much higher pressure in compacted than in loose soil (Eavis *et al.* 1969; Whiteley *et al.* 1981).

Penetrometer values that stop root growth have not been identified conclusively as they vary with plant species, soil water content, structure and texture (Taylor *et al.* 1966; Greacen *et al.* 1969; Lowry *et al.* 1970; Henderson *et al.* 1984; Hamblin 1987; Vepraskas 1988).

This means that many measurements must be taken to allow for heterogeneity within the paddock. An advantage of penetrometers is they are more sensitive to changes in soil strength than assessments using a bulk density measure (Voorhees *et al.* 1978). Penetrometers are also useful because they give an index of soil strength (degree of compaction), especially if the readings are at a known water potential like the upper storage limit. Comparing penetrometer measurements from different locations can be difficult, because of differences between soils and water contents when assessed.

The Bush recording penetrometer is an example of a commonly used field penetrometer (Anderson *et al.* 1980). Penetrometers can also be simple home-made devices such as the hand probe described by Hunt and Gilkes (1992). Simple pocket penetrometers are also available. Note that penetrometers are not recommended for gravelly or stony soils (Anderson *et al.* 1980; O'Sullivan *et al.* 1987).

Using a penetrometer to identify a compacted layer

- The soil should be moist, preferably around the upper storage limit (i.e. field capacity).
- Check the soil profile to ensure there is no gravel present or marked changes in texture, as these may affect the soil strength. Penetrometers are not recommended for gravelly or stony soils.
- Penetrometer measurements are made at 5 cm intervals down the profile to the maximum depth of the instrument. To cope with paddock variability, a large number of measurements need to be taken (e.g. 5 to 10 replicates in a 20 x 2 m experimental plot). A deep-ripped or uncleared area with the same soil type could be used for comparison.
- A compacted layer is present if the resistance increases to a maximum then decreases again, further down the profile (Figure 4.2.3, on previous page). If a compaction pan is present, the maximum soil strength may exceed the limit of the penetrometer scale.

Table 4.2.3 Bulk density which may inhibit or prevent root growth for each soil texture class (adapted from Vepraskas 1988).

Soil texture	Bulk density[1] (g/cm^3)	
	Restrict root growth[2]	Prevent root growth[2]
Sands[3]	1.66	1.85
Loamy sands[3]	1.61	1.82
Sandy loam	1.60	1.81
Sandy clay loam	1.58	1.80
Clay loam or finer	~1.4	~1.6

1. *The values in the table assume the soil has 10% very fine sand (0.125-0.050 mm). If a soil contains appreciably more than 10%, the 'limiting' bulk density would be lower (Vepraskas 1988). The values assume that gravel is absent or <5%.*
2. *The bulk density values which restrict and prevent root growth approximate soil strengths of 1.5 and 3.0 MPa respectively (Vepraskas 1988).*
3. *Bulk density is not recommended for coarse-textured soils. A better estimate of the limiting bulk density in coarse-textured soils may be derived using a packing density model (e.g. Gupta and Larson 1979).*

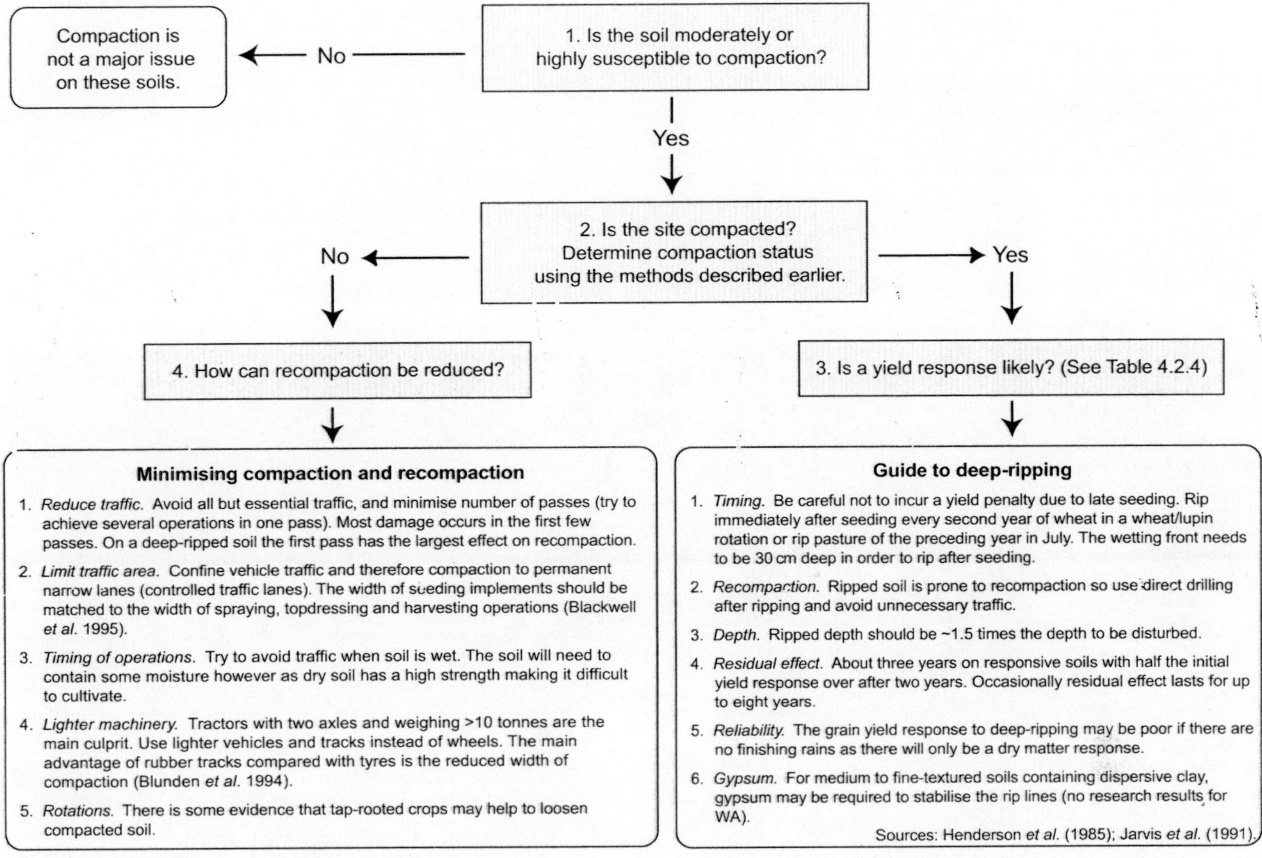

Figure 4.2.4 *A decision pathway for managing subsurface compaction.*

- The depth of compaction should be noted to ensure that subsequent treatment loosens the compacted soil.

With a hand probe, the compacted layer is detected by an increase in the pressure required to push the probe through the soil. The pressure should increase in the compacted zone and then become easier again below the compacted zone.

Bulk density

Bulk density is the oven dry weight of soil per unit volume. It can help detect plough pans in medium to fine-textured soils.

Bulk density is a useful measure of compaction status in compressible soils because it is closely correlated to soil strength, which is the critical parameter affecting plant growth.

On the other hand, some coarse-textured soils have a naturally high bulk density but a low compressibility, so bulk density is not recommended for assessing the compaction status of these soils. When subject to loads (heavy traffic, stock trampling) the increase in bulk density is likely to be small, but there can be a large increase in soil strength which adversely affects root growth and yield.

Advantages of using bulk density are the accuracy in location of the sample and the reproducibility of the measurement. Disadvantages include the slow and tedious measurement; limited value as an absolute indicator of compaction status, because it is frequently poorly correlated with either plant growth or other measures of compaction (Soane *et al.* 1981a; Cassel 1982); and a large number of samples are necessary to handle spatial variability (Erbach 1987).

In clayey soils, bulk density normally ranges from 1.2 to 1.5 g/cm^3 and in sandy soils, from 1.6 to 1.9 g/cm^3. A limiting bulk density is difficult to predict from the soil texture alone. Large variations can occur within a textural class due to differences in sand size distribution, amount of coarse sand, or the amount of clay (Henderson *et al.* 1988; Vepraskas 1988). In general, bulk density measurements are more reliable in apedal soils with few macro-pores.

Estimates of the bulk densities which may limit root growth are given in Table 4.2.3. For more information on bulk density, refer to Chapter 10.

Gravel or coarse fragments in the sample will increase the bulk density.

When determining whether there is an increase in the bulk density of the soil matrix it is necessary to allow for the gravel component if more than 5% gravel is present.

MANAGEMENT OPTIONS

The amelioration of subsurface compaction by deep-ripping can result in large yield responses, especially on uniform coarse-textured soils (e.g. Jarvis and Porritt 1985; Jarvis 1985, 1990; Jarvis *et al.* 1991). This indicates that the compacted layers were inhibiting water and nutrient uptake. The potential benefits should be compared with the cost of the treatment. The real cost of a deep-ripping program includes not only the machinery and fuel costs but the opportunity cost of delayed sowing, which in most areas results in a substantial yield penalty.

A decision pathway for managing subsurface compaction is outlined in Figure 4.2.4 (on previous page).

Table 4.2.4 Yield response to deep-ripping on different soils in Western Australia.

References

(a) Jarvis et al. *(1991)*
(b) A. Diggle personal communication
(c) Henderson (1991)
(d) Jarvis (1986)
(e) Crabtree (1989)
(f) Dracup et al. *(1992)*
(g) Belford et al. *(1992)*
(h) W. Crabtree personal communication
(i) P. Blackwell personal communication
(j) Ellington (1986)
(k) R. Jarvis personal communication
(l) H. Daniel personal communication

CHAPTER 5

CHEMICAL FACTORS AFFECTING PLANT GROWTH

5.1 SOIL ACIDITY

Geoff Moore, Perry Dolling, Bill Porter** and Linda Leonard***

Soil acidification is a natural process and is generally accelerated by agriculture. The rate of acidification varies enormously depending on the soil type, land use, productivity and management of the farming system.

In Western Australia, soil acidity has only recently become a broadscale issue because most land has been developed for a comparatively short time. This contrasts with Europe and other parts of the world where regular applications of lime have been routine practice for many decades. In WA, about 178,000 tonnes of lime were applied to agricultural soils in 1994/95, but this will need to increase substantially to maintain the current pH of the soil (Porter and McLaughlin 1993).

In the 1960s, soils supporting subterranean clover pastures were shown to be acidifying on the Swan Coastal Plain (Barrow 1964, 1969), confirming the pioneering work done with legume pastures in New South Wales (Donald and Williams 1954). The results of many early liming experiments in WA were highly variable. Some recent studies have demonstrated that acidity is already reducing yields in the wheatbelt (e.g. Dolling *et al.* 1991, 1994; Loss *et al.* 1993).

Two main groups of WA soils were inherently highly acid before they were cleared for agriculture: the yellow sandy earths in the eastern wheatbelt and the 'peaty' sands in the south-west.

Acidity is a problem throughout the soil profile. For example, acidification is evident to a depth of ~80 cm in some of the cultivated uniform coarse-textured soils that have a low pH buffering capacity (Figure 5.1.1, Dolling and Porter 1994). It is common to separate acidification of the topsoil (0 to 10 cm), from that of the underlying soil. In some cases the subsurface soil is acidifying while the topsoil pH has increased since clearing (Figure 5.1.1).

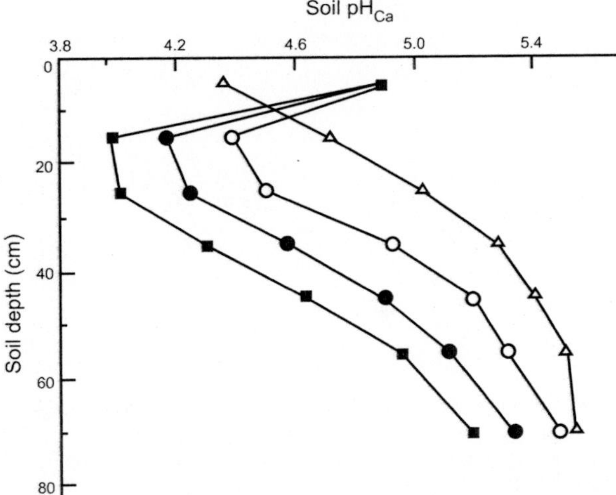

Figure 5.1.1 *Effect of depth and time since clearing on soil pH$_{Ca}$† (△ mean of uncleared sites; ○ 12, ● 50, and ■ 74 years after clearing; from Figure 1, Dolling and Porter 1994).*

Two measures are used for soil acidity. The *activity of soil acidity* refers to the concentration of hydrogen ions in the soil solution, of which pH is an indicator. *Total acidity* is the amount of acidity which needs to be neutralised with lime to increase the pH. It includes the concentration of H^+ in soil solution plus exchangeable H^+, so it varies with pH and pH buffering capacity.

PRINCIPLES OF SOIL ACIDITY

Chemistry of soil acidity

The key changes connected with acidification are an increase in the concentration of hydrogen ions, increased solubility of aluminium and manganese, and changes in the availability of several nutrients. These factors are responsible for decreased plant production on acid soils.

The chemistry of acid soils involves many complex reactions and their relative importance with regard to plant growth is still largely unknown. Current knowledge has been summarised in Coleman and Thomas (1984), Black (1984) and Ritchie (1989).

Aluminium. In general, aluminium (Al) toxicity is the major problem of acid soils in WA. Its chemistry is complex, with possibly 14 inorganic reactions in the soil solution, in addition to the reactions with organic matter (Ritchie 1989). It is difficult to identify toxic forms and the Al speciation varies with soil pH, although in general, free Al^{3+} is the major toxic form (Ritchie 1995). Comprehensive reviews are in Ritchie (1989) and Foy (1992).

The amount of toxic Al at a given pH varies between soils for several reasons, including:

- Reaction with dissolved organic matter, phosphate, fluoride or sulphate to form non-toxic complexes. For instance, chelation with organic compounds can result in Al being non-toxic to plants (Ritchie 1989).
- Ionic strength - the higher the ionic strength (adjusted for the ionic charge of the soil solution), the less active a given concentration of Al (Carr and Ritchie 1993).

* *Agriculture Western Australia, Katanning.*
** *Agriculture Western Australia, Northam.*
† *pH$_{Ca}$ is measured in a 1:5 soil:0.01M $CaCl_2$ suspension.*

- Forms of solid Al - different forms of solid Al compounds are more or less reactive at a given pH. In addition, not all soluble forms are toxic and tolerance differs between plant species and cultivars.
- Low total concentrations of Al in the soil (e.g. in pale deep sands).

The critical soil pH at which sufficient Al becomes soluble to be toxic is difficult to predict because it depends on many factors including clay mineralogy, organic matter, other cations and anions and total salts (Foy 1984). In general, Al starts to dissolve when the pH_{Ca} is lower than 5.5, while below 4.5 there is a marked increase in extractable Al (Figure 5.1.2, Wilson 1984).

Manganese. Manganese (Mn) toxicity is a problem in many acid soils in eastern Australia, but has not been reported in WA. This may be a result of the low manganese concentrations in WA soils which are acidifying to critical levels. In general, these soils have coarse-textured A horizons and a low pH buffering capacity. They are likely to contain low concentrations of Mn and may actually require the addition of Mn when sensitive crops such as narrow-leafed lupins are grown.

Hydrogen. This is unlikely to be present in toxic amounts for non-legumes unless the soil pH_{Ca} is lower than 3.4. Acidification to this degree is rare and generally confined to acid sulphate soils. Therefore the main consequence of hydrogen ions for non-leguminous plants is the reduced availability of certain nutrients.

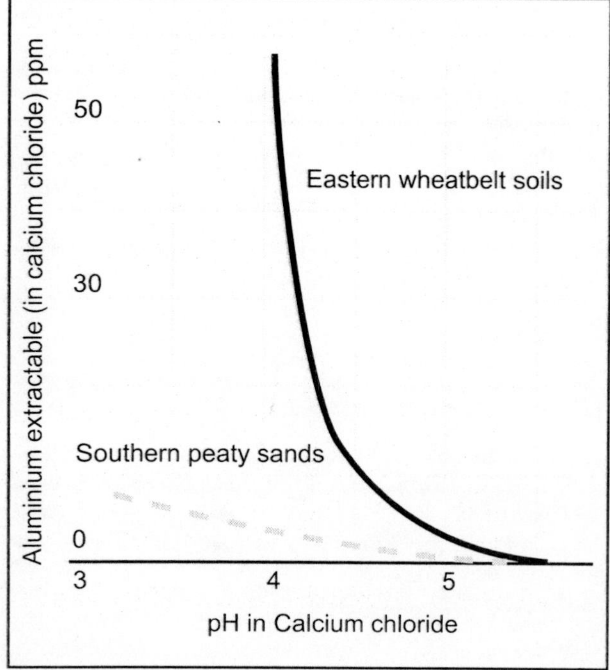

Figure 5.1.2 *The concentration of soluble aluminium increases as the soil becomes more acid (from Figure 2, Wilson 1984).*

For legumes, high concentrations of hydrogen ions affect rhizobia more than the host plant, reducing survival and nodulation (Munns 1978).

Mechanisms of soil acidification

In essence, soil acidification is the net addition of acids to an agricultural system. It is convenient to consider the acid inputs in terms of the major nutrient cycles. Carbon and nitrogen, and to a lesser extent sulphur, are the major nutrient cycles involved. The processes have been described in review papers by van Breeman *et al.* (1983), Haynes (1983) and Helyar and Porter (1989).

Carbon cycle

The role of the carbon cycle in soil acidification can be connected with two factors: product removal and increasing concentrations of organic anions in the soil.

Product removal. When plants absorb nutrients they tend to actively absorb more positively charged nutrients (e.g. NH_4^+, K^+, Ca^{2+}, Mg^{2+}, Na^+) than negatively charged nutrients (e.g. PO_4^{3-}, NO_3^-, Cl^-, SO_4^{2-}) from the soil solution. This excess of cations over anions is usually balanced by the plant excreting hydrogen ions from the roots. The plant tissue becomes alkaline, while the soil is acidified in the region of the roots (Helyar and Porter 1989).

The main source of these hydrogen ions is the production of organic acids in a process similar to photosynthesis. Some hydrogen ions are absorbed from the soil, but there is a net excretion of H^+ from the plant. If plant tissue is not removed from the area the net effect is zero, although there can be a redistribution of H^+ within the soil profile. However, removal of grain, hay, silage, meat or wool leaves a net excess of hydrogen ions in the soil, causing acidification.

The term *ash alkalinity* is used to describe the amount of alkalinity in plant material following ashing at very high temperature (Jarvis and Robson 1983; Slattery *et al.* 1991). Using this technique, the equivalent amount of lime required to neutralise the soil has been measured (Table 5.1.1). The actual quantities vary depending on nutrient uptake. In general, dicotyledons tend to be more alkaline than monocotyledons (grasses).

Stock tend to redistribute alkalinity within paddocks when large quantities of dung and urine are deposited in camps. These factors contribute to the spatial variability of soil acidity, both within and between paddocks.

Changes in organic anion content. The organic anions produced by the plant, if not removed as produce, are returned to the soil as organic matter. Increases in soil acidity have been attributed to organic matter (Williams and Donald 1957; Williams 1980), but this is not a universal relationship (Jarvis and Robson 1983). It depends on the initial pH of the soil and the dissociation constants of the weak organic acids (Helyar 1976; Ritchie and Dolling 1985; Helyar and Porter 1989).

Measurements of the accumulation of organic anions should allow for the net accumulation of anions during a certain time as well as changes in dissociation/association of the organic anions due to changes in soil pH during that time (Helyar and Porter 1989). The effect of an accumulation of organic anions can be estimated from changes in soil organic matter (~1.7 x organic carbon) over a given period and the initial soil pH (Table 5.1.2; Helyar and Porter 1989).

Nitrogen cycle

The nitrogen (N) cycle (Figure 5.1.3) is frequently implicated in acidification, but overall it is chemically neutral. The N cycle results in no net change in acidity, *providing the cycle is completed* (Helyar 1976). If nitrogen enters the soil in one form and leaves in another, there can be a net addition or removal of H^+ (Figure 5.1.3).

In broadscale agriculture, most N enters the nutrient pool through the fixation of atmospheric N by legumes. If the highly soluble nitrate ion is leached below the root zone, the cation leached with it to maintain electrical neutrality is not usually H^+, but another cation (e.g. Na^+, K^+, or Ca^{2+}). The net effect of nitrate leaching is to leave H^+ in the leached zone. Similarly, urea fertilisers cause net additions of H^+ if nitrate is leached.

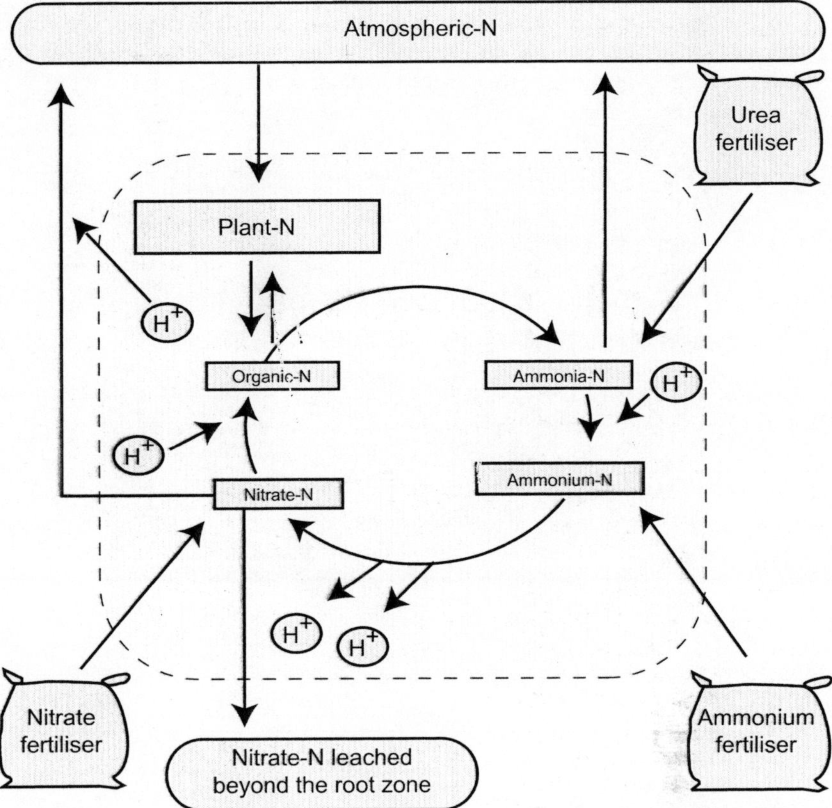

Figure 5.1.3 *The nitrogen cycle in relation to acidification of a soil-plant system. The effect of the N cycle on soil acidity depends on the form in which N enters the system and the form in which it leaves or accumulates (from Porter and McLaughlin 1993).*

Table 5.1.1 The equivalent weight of lime* ($CaCO_3$) required to replace the alkalinity exported in various farm products (from Slattery *et al.* 1991).

Product removed	$CaCO_3$ equivalent (kg/t)
Cereal grains	
wheat	9
barley	8
triticale	7
Cereal whole tops	20
Lupins	
grain	20
whole tops	60
Lucerne hay	60
Hay (mixed grasses)	30
Subterranean clover (whole plant)	40
Sheep**	
dung	25
urine	9
lambs	3
wool (6 kg/sheep)	0.4

* *The samples used for this data are from Rutherglen Research Institute in Victoria, and physical and environmental differences in growing conditions may alter the ash alkalinity.*

** *Assumes a stocking rate of 5 sheep/ha (units are kg $CaCO_3$/ha/year and values from north-eastern Victoria).*

Adding fertilisers that contain N in the form of ammonium (e.g. DAP, MAP) is acidifying even when nitrate is not leached (Figure 5.1.3, Table 5.1.3). The role of ammonium-forming fertilisers in lowering pH has been well established (e.g. Mason 1980). In a NSW study of acidification rates, the highest rate was measured under kikuyu pasture with a high input of ammonium fertiliser (Helyar *et al.* 1990).

Fertilisers

The main fertilisers that contribute to acidification (Table 5.1.3) are the nitrogenous fertilisers discussed above, plus elemental sulphur (S).

Elemental S has to be converted to sulphate ions by microbes before plants can absorb it, and this conversion releases H^+. The sulphate ions are soluble and thus readily leached. When sulphate ions leach, they usually do so in association with cations other than hydrogen, thereby acidifying the soil in the leached zone. The contribution of the S cycle to acidification is likely to be small in comparison with the N and C cycles (Helyar and Porter 1989).

Adding single, double or triple superphosphate fertilisers, which contain sulphate-S, does not contribute directly to soil acidity. These fertilisers only contribute by increasing the productivity of the farming system, so they increase the amount of N that can leach or the amount of produce that can be removed (Williams 1980; Helyar and Porter 1989).

Net change in pH. The combined effects of the nutrient cycles (especially N and C) result in a net change in pH down the profile. This is illustrated in Figure 5.1.4.

Table 5.1.2 The lime equivalent to neutralise the acidity associated with organic anions at different soil organic matter and pH levels*. The effect of a change in pH, on the acidity associated with organic anions is estimated from the initial soil pH and the change in organic matter (from Table II; Helyar and Porter 1989).

Soil organic matter (%)	$CaCO_3$ equivalent (t/ha.10 cm)		
	pH_{Ca}		
	4.0	6.0	8.0
1.0	0.54	0.93	1.32
2.0	1.08	1.86	2.63
5.0	2.71	4.66	6.58

* *Soil bulk density is assumed to be 1.37 t/m³.*

Table 5.1.3 The amount of acidity generated by various fertilisers, expressed in terms of the lime equivalent needed to neutralise their addition (adapted from Cregan and Helyar 1986).

Fertiliser	Lime required to neutralise fertiliser addition (kg $CaCO_3$/kg nutrient N, P, S) When the amount of nutrient leached is:	
	0%	100%
Nitrogen fertilisers		
Ammonium sulphate (Agras #1, MAP)	3.6 (/kg N)	7.1 (/kg N)
Ammonium nitrate (Agran)	0	3.6 (/kg N)
Urea	0	3.6 (/kg N)
DAP	1.8 (/kg N)	5.4 (/kg N)
Potassium nitrate[1]	-3.6[1] (/kg N)	0
Sodium nitrate	-3.6 (/kg N)	0
Phosphorus fertilisers		
Single, double, triple superphosphate	0	-
Dicalcium phosphate[2]	-1.6 (/kg P)	-
Rock phosphate (15.4% P)[3]	-1.6 (/kg P)	-
Sulphur fertilisers		
Elemental sulphur	3.1 (/kg S)	-
Gypsum	0	-
Potassium sulphate	0	-

1. *Negative values indicate a liming effect by the fertiliser.*
2. *Theoretical value only, actual effect will be very small due to incomplete dissolution of fertiliser in soil.*
3. *As for 2, but does not include liming effect of free $CaCO_3$ or structural CO_3 in the crystal lattice.*

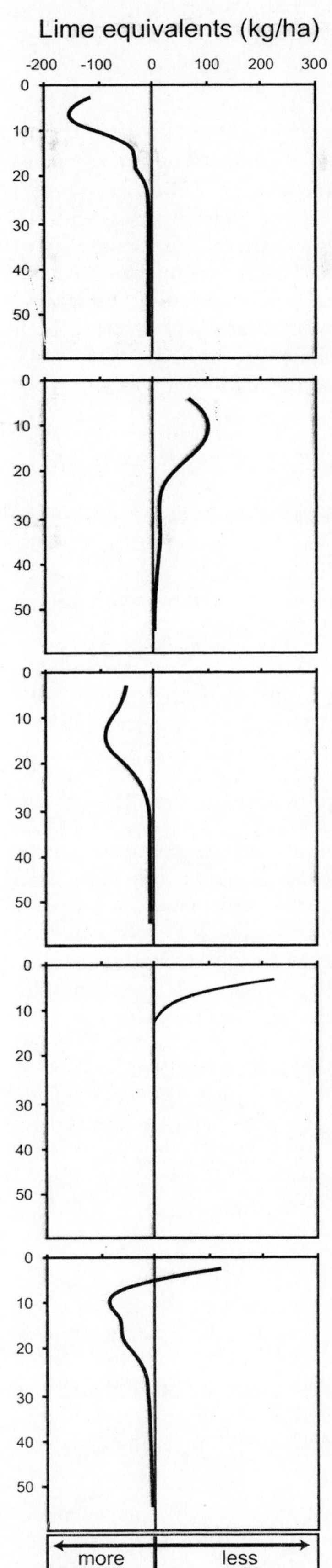

Nitrogen transformations in the soil

Most nitrogen enters the soil in the surface layers of the soil. There much of it is converted to nitrate, and leaches from that layer. The transformations from plant N or ammonium-N are acidifying. The leaching of nitrate is not acidifying, but by removing nitrate from a soil layer, it removes the potential for the uptake of nitrate from that layer by plants (see next point).

Nitrate uptake by plants

In absorbing the nitrate ion, a plant excretes a base (usually OH^-) to maintain electrical neutrality (charge balance) inside the plant. The pattern of nitrate uptake will be determined by the balance between the rate of leaching of the nitrate and the rate of growth of roots.

Non-N nutrient uptake by plants

The absorption of positively charged nutrients (cations: K, Ca, Mg, Mn, Cu, Zn) by plants is accompanied by the excretion of acid (H^+), and the absorption of negatively charged nutrients (anions: P, S, Mo) is accompanied by the excretion of base (OH^-). Under normal conditions, plants excrete more acid than base, leading to a more acid soil. The shape of the profile depends on the amounts and position of the different nutrients absorbed.

Plant material deposition on the soil surface

A result of the excretion of acids to balance an excess of uptake of cations over anions is that plant tissue contains an excess of alkalinity. The amount of alkalinity is equal to the amount of acid excreted into the soil. Most of this alkalinity ends up in the plant tops. Some is removed in produce, but the bulk is deposited onto the soil surface when the plant senesces and drops it leaves.

Net acidification profile

The net effect of the processes described above is a soil acidification profile which reflects shape of pH profiles commonly seen in agricultural soils. The shape of the final acidification profile can be modified by the manipulation of the above processes.

Figure 5.1.4 *The soil pH profile is the net result of processes that contribute acids or bases (from Figure 3.8, Leonard 1996). Please note, in this example an arbitrary set of profiles of the most important processes are presented.*

Effects of soil acidity

Soil acidity has a number of effects on plant growth including direct toxicities and nutrient deficiencies. In WA, the major toxicity in acid soils is due to aluminium. Deficiencies of molybdenum, nitrogen, sulphur, phosphorus, calcium and magnesium can also occur.

Aluminium toxicity. High concentrations of soluble aluminium in soils (e.g. >10 µg/g measured in 1:5 0.005M KCl) reduce root elongation directly (Foy 1974; Carr and Ritchie 1993). Toxic concentrations of Al result in stubby, brittle roots with thickened root tips and laterals and an overall reduction in fine branching of the root system (Foy 1984). Al toxicity can reduce root growth in both the topsoil and subsurface soil. The effect of Al toxicity in the subsurface soil will often be seen as symptoms of drought stress, resulting from the reduced root elongation and branching.

> Aluminium toxicity is often described as a chemical hardpan, because the effects on root elongation are similar to those caused by subsurface compaction.

Nodulation failure in legumes. Specific problems with nodulation can occur in slightly acid to moderately acid soils. This is predominantly associated with topsoil acidity, where most of the root nodules occur. In acid soils, the activity of *Rhizobium* declines, reducing populations. This reduces the likelihood of successful nodulation during early plant growth, which is when the roots of the host plant can be infected.

A number of components are essential for successful nodulation and efficient nitrogen fixation. In different circumstances each of these factors could be the most 'acid sensitive' and be responsible for reduced legume production (Howieson and Ewing 1984). These include:

- growth and survival of rhizobia
- nodule formation
- nodule function
- host plant growth.

Species of *Rhizobium* species vary in their tolerance of acid soil but nodulation is usually more sensitive to low pH than plant root growth (see review by Coventry and Evans 1989).

Nutrient deficiencies. The supply of most nutrients is altered in highly acid soils. Decreases in the availability of molybdenum and to a lesser extent N, S, P, Ca and Mg can occur, although this varies between soils. Increases in availability of Mn, Fe and Zn can also occur (Porter and Yeates 1984).

Molybdenum (Mo) supply is reduced when pH_{Ca} is lower than 4.2, because the Mo is strongly adsorbed by the soil, and fertiliser Mo has a low residual value. Regular applications of Mo are required for cereal crops, either as a seed dressing or by using seed with adequate Mo in it, to supply the plant requirements (Section 6.8).

The rates of both nitrification ($NH_4^+ \rightarrow NO_3^-$) and nitrogen mineralisation (organic-N$\rightarrow NH_4^+$) are lower in acid soils because soil flora (fungi, bacteria, viruses) generally do not grow rapidly under these conditions. The reduced mineralisation can reduce availability of N, S and P to plants. Conversely, liming raises the pH, causing a large increase in microbial activity, increasing mineralisation of organic matter which releases N, P and other nutrient elements (e.g for N, Barrow 1964).

INHERENTLY ACID SOILS IN WESTERN AUSTRALIA

The acid yellow earthy sands in the eastern and north-eastern wheatbelt and the dark grey 'peaty' sands in the south-west were sufficiently acid prior to clearing to affect agricultural production.

Some soils in the south-eastern wheatbelt have highly acid subsoils underlying neutral to alkaline clays. Their extent is minor and the highly acid horizons may be below the depth of root growth.

Acid yellow sandy earths

The acid yellow sands and sandy earths are frequently called 'wodjil' soils because of their native vegetation (*Acacia* spp.), but this was only one of a range of vegetation types. There are approximately 1 million ha of yellow sandy earths in the eastern and north-eastern wheatbelt (Zone of Ancient Drainage) but not all of them were highly acid before clearing. The subsurface soil is usually more acid than the topsoil (Porter and Wilson 1984). In a survey of 36 sites, 33% had a subsurface (15 to 30 cm) pH_w lower than 5.0 while 45% were greater than 6.0 (Porter and Wilson 1984). A complicating factor is that those soils which were not highly acid before clearing are now acidifying under agriculture and approaching critical levels.

Aluminium toxicity has been established as the cause of yield loss (Carr *et al.* 1991). No relationship between any morphological properties or landscape position and Al toxicity has been established. Consequently, the only reliable way of identifying the highly toxic soils is to measure the pH_{Ca} and if it is lower than 4.5, then measure the concentration of Al (Carr *et al.* 1991; refer to *Measuring Al* later in this chapter).

Carr and Ritchie (1993) report significant chemical differences even for soils of similar pH. It is possible that toxicity is related to the ionic strength of the soil solution, with an increase in strength reducing the toxicity of the Al (Adams and Lund 1966; Carr and Ritchie 1993).

Dark grey peaty sands

Many of the black to dark grey peaty sands on the Swan Coastal Plain and along the south coast are naturally acid. These occur in wet or swampy areas and the native vegetation is commonly tea-tree, bottlebrush or kangaroo grass (Fitzpatrick 1958).

The soils were valued for market gardening and summer cropping in the 1930s, but many crops failed. A number of surveys were conducted to help identify the problems (e.g. Teakle and Southern 1937a, b). These found a number of different types of peat (e.g. marl, fibrous) and sandy soils with organically stained A horizons.

Small areas of the peats had a very acid reaction (pH_w 1.3 to 3.4) and are almost certainly *acid sulphate soils* (Teakle and Southern 1937a, b). The area of acid sulphate soils in WA is very restricted. Acid sulphate soils contain pyrite (iron sulphide) which is oxidised through a combination of chemical and microbiological processes to form sulphuric acid when the soil is drained (Black 1984):

$$4FeS_2 + 15O_2 + 2H_2O \rightarrow 2Fe_2(SO_4)_3 + 2H_2SO_4$$

The area of peaty sands and organic sands with high acidity (pH_{Ca} 3.5 to 4.3) is about 100,000 ha (Porter *et al.* 1980). The area of true peats (which by definition contain at least 12 to 18% organic carbon depending on texture) is considerably smaller.

In comparison to mineral soils, the peaty sands contain low levels of extractable aluminium (Wilson 1984). The role of lime (2 t/ha) in improving the establishment of subterranean clover pastures on these soils is well established (Fitzpatrick 1958). Rock phosphate fertilisers are also very effective (Fitzpatrick 1961; Bolland and Gilkes 1990; and refer to Section 6.3).

If these soils are highly acid (pH_{Ca} <4.3) and have been under subterranean clover pasture for extended periods, a thin organic layer will develop on the surface 0 to 3 cm. The organic layer has a higher pH than the underlying soil and enables the rhizobia to function and fix atmospheric nitrogen. There is a high concentration of roots in this layer (Yeates 1988). If these soils are cultivated the organic layer is diluted and the clover may fail to regenerate. Subsequent attempts to reseed are likely to be unsuccessful unless lime is applied (Yeates *et al.* 1984; Yeates 1988).

ASSESSING SUSCEPTIBILITY TO SUBSURFACE ACIDIFICATION

All soils acidify with time, even calcareous soils, although these may take hundreds of years to reach a critical level. *Susceptibility to acidification can be considered as the time before the acidity develops to a critical level.*

Subsurface acidity refers to acidification of the soil below the normal depth of cultivation. Susceptibility to subsurface acidification (defined here as acidification at 10 to 20 cm) can be quantified using equation 5.1.1, adapted from Helyar and Porter (1989), which estimates the time before reaching a critical pH where production losses are likely. This equation can be applied where the subsurface pH_{Ca} is currently above 4.5.

$$\text{Time} = \frac{(pH - pH_{crit}) * pHBC * (1 - (gravel\%/100))}{AR} \quad (5.1.1)$$

where:

Time is the number of years until pH is reduced to a level at which production losses are likely;

pH is the current pH_{Ca} (measured in 1:5 0.01M $CaCl_2$) for the 10 to 20 cm depth;

pH_{crit} is the critical pH_{Ca} below which production losses are likely;

pHBC is pH buffering capacity (kg lime/ha for a 1 unit pH change over a 10 cm depth interval);

AR is the acidification rate (kg lime/ha.10 cm/year).

(Note: All measurements refer to the 10 to 20 cm layer of soil.)

Results from equation 5.1.1 are only estimates and should be categorised as follows:

Category	Description
High	<15 years; the subsurface soil is either currently below, or is likely to acidify below, the critical pH within 15 years.
Moderate	15-30 years; the subsurface soil is likely to acidify below the critical pH within 15 to 30 years.
Low	>30 years; the subsurface soil is unlikely to acidify below the critical pH within 30 years.

Equation 5.1.1 is not suitable for soils containing calcium carbonate ($CaCO_3$). On these soils, the measure of acidification is the rate of disappearance of $CaCO_3$, rather than the product of the change in pH and the buffer capacity (Helyar and Porter 1989).

pH buffer capacity (pHBC)

This is the ability of the soil to resist changes in pH after the addition of an acid or a base. It can be defined as the rate of acid or alkali (e.g. lime) addition per unit change in soil pH. The pHBC can be measured in the laboratory or inferred from other soil properties.

The pHBC is not constant, but varies with pH. Under alkaline conditions (pH_{Ca} >7.5), the pHBC is high because of precipitation-dissolution reactions involving carbonate minerals (Jenny 1980). Similarly, at low pH (pH_{Ca} <4.5), soils may be strongly buffered by reactions involving aluminium hydrous oxides. Between pH_{Ca} 4.5 to 7.5 the soil is less strongly buffered. The pHBC may be considered approximately constant between pH 4.5 and 6.0 (Magdoff and Bartlett 1985). Factors that increase the pHBC include increasing the ionic strength of the soil solution, and increasing the organic matter, clay and carbonate minerals.

Laboratory measurement

The pHBC is measured by titrating the soil with an acid or base using a number of different techniques (Helyar and Porter 1989). The standard method in WA involves the addition of dilute acid or alkali to a 1:5 soil:water suspension and measuring the change in pH after a standard equilibration period. A number of points are measured by varying the amount of acid or alkali and the pHBC is calculated from a regression curve (or slope) of the line when pH is plotted against the amount of acid or alkali added (Dolling and Porter 1994).

The pHBC measured using this method can be interpreted as follows:

Rating	pHBC value (cmol H^+/kg/pH unit)
Low	<1
Moderate	1-2
High	>2

Inferring pHBC from soil properties

The pHBC can be estimated from organic carbon, exchangeable Al and percentage clay using equations 5.1.2a-d, which were derived from measurements on 42 soil profiles from south-western Australia. The range of values used to calculate each linear regression is summarised. The method for measuring pHBC was a variation on that outlined above, as the samples were shaken for 17 hours and pH was measured in 1:5 0.005M KCl (C. Raphael and W. Bowden unpublished data).

For surface soil (0 to 10 cm):

$$pHBC = 0.48 + 0.54OC\% \quad R^2 = 0.91 \qquad (5.1.2a)$$

(OC is 0.13-8.04%; clay is 1.8-15.4%)

For subsurface soil (10 to 20 cm):

$$pHBC = 0.42 + 0.73OC\% + 0.40Al \qquad (5.1.2b)$$
$$R^2 = 0.82$$

(OC is 0.09-2.52%; Al is 0-2.33 me%)

For any depth:

$$pHBC = 0.66 + 0.45OC\% + 0.63Al \qquad (5.1.2c)$$
$$R^2 = 0.77$$

(OC is 0.05-8.04 %; Al is 0-2.09 me%)

$$pHBC = 0.33 + 0.49OC\% + 0.02clay\% + 0.57Al \quad (5.1.2d)$$
$$R^2 = 0.86$$

(OC is 0.05-8.04%; clay is 1.4-58.8 %; Al is 0-2.09 me%)

where:

pHBC is pH buffering capacity (cmol H^+/kg/pH unit).

OC% is the % of organic carbon (Walkley and Black).

Al is the exchangeable aluminium (me%) with extraction varying with the pH_w (refer to Chapter 10).

Clay mineralogy affects pHBC and the equations above may under-estimate the pHBC of soils containing appreciable amounts of illite and smectite. Kaolinite is the dominant clay mineral in south-western Australia. For instance, the pHBC values estimated for soils from NSW with similar clay contents were considerably higher (Helyar *et al.* 1990). This may be related to the clay mineralogy or to the different method used for measuring pHBC.

To convert pHBC from cmol H^+/kg/pH unit to kg lime/ha/pH unit.10 cm layer so that it can be used in equation 5.1.1:

pHBC (cmol H^+/kg/pH unit) x 500/BD
= pHBC (kg lime/ha/pH unit.10 cm layer) (5.1.3)

where:

BD is the bulk density (t/m^3).

Example. For a soil with a pHBC of 0.8 cmol H^+/kg/pH unit and a BD of 1.4 t/m^3 the calculation is:

pHBC = 0.8 x 500/1.4
= 286 kg lime/ha/pH unit.10 cm layer

Acidification rate

Acidification rate is a function of productivity, land use and soil type (Table 5.1.4). There are few measured rates for farming systems in WA, however some recent measurements deduced from field surveys and trial sites estimate between 0 and 13 kg $CaCO_3$/ha.10 cm/year (Dolling and Porter 1994; Dolling *et al.* 1994; Dolling 1995). Research in Victoria (Coventry and Slattery 1991) and NSW (Ridley *et al.* 1990a, b; Helyar *et al.* 1990) shows that acidification rate increases with increasing rainfall. In general, legumes especially lupins, cause more acidification than other species (e.g. Dolling *et al.* 1994; Tang 1997).

EXAMPLE: CALCULATING RISK OF SUBSURFACE ACIDIFICATION

To calculate the time before subsurface acidification develops to a critical level on a red sandy earth under a continuous wheat:lupin rotation in a 400 mm rainfall area. The pH_{Ca} at a depth of 10-20 cm is presently 4.9, texture is sandy loam (15% clay), organic carbon is 0.5%, Al is 0.18 me%, bulk density of 1.45 t/m^3 with 10% quartz gravel.

pH_{crit} is 4.5 for a lupin:wheat rotation (Table 5.1.5).

pHBC is 337 kg lime/ha per pH unit. 10 cm layer calculated from equation 5.1.2d.

Mean acidification rate is 7 kg $CaCO_3$/ha. 10 cm/year from Table 5.1.4.

Using equation 5.1.1:

Time = ((4.9-4.5)*337*(1-0.10))/7 = 17 years

Subsurface acidification would be expected to reach a critical level within 17 years. The risk is therefore categorised as *moderate*.

The influence of soil type on acidification rate has not been considered so values in Table 5.1.4 are an over-simplification because nitrate leaching and acid accretion are related to soil permeability. However, the effect of the soil on the current pH and the pHBC is included and when data become available the acidification rate could be related to specific soil types.

Critical subsurface pH

The critical subsurface pH depends on the most acid-sensitive species grown in the rotation (refer to Table 5.1.5). For instance, a critical subsurface pH_{Ca} of 4.5 can be assumed for a cereal:lupin rotation, because cereals are the more sensitive component. Field trials with barley identified yield losses of 20 to 30% due to Al toxicity when the topsoil pH_{Ca} was 4.2 to 4.3, while at pH_{Ca} 4.5 and above there appeared to be no significant yield reduction (Dolling *et al.* 1991). Critical subsurface pH can vary with:

- *Soil type.* Some soils supply higher or lower concentrations of toxic Al, even though their pH is the same, e.g. peaty sands and grey sands have lower concentrations of extractable Al than most soils (Figure 5.1.2).
- *Climate, season.* The longer the topsoil stays moist at the end of the growing season, or the less nitrogen leached early in the season, the less reliance crops and pastures have on a deep root system. Reduced root growth caused by subsurface acidity will thus be less significant.
- *Species or variety.* Sensitivity varies widely between species and genotypes (Table 5.1.5).
- *Nutrient availability.* Acidification may result in molybdenum deficiency, depending on the fertiliser history.
- *Yield.* The lower the actual yield relative to the potential yield, the less effect of soil acidity.

Table 5.1.4 Subsurface acidification rates (10-20 cm layer) versus rainfall and rotation for use in equation 5.1.1.

Annual rainfall (mm)	Rotation	Acidification rate (10-20 cm layer) (kg $CaCO_3$/ha.10 cm/year)
<350	-	~2*
350-450	4 pasture:wheat	4 ± 2**
	Lupin:wheat	7 ± 3**
450-600	Continuous pasture	7 ± 3**
>600	-	~10*

* *Estimate only*

** *Acidification rates (mean plus range) from five field surveys by P. Dolling on duplex soils (20 to 50 cm of sand to clayey sand overlying a clayey subsoil) and yellow deep sands. These values may underestimate acidification, but are the best currently available.*

SITE ASSESSMENT OF SOIL ACIDITY

The acidity of a paddock will vary with the soil, land use history, productivity and time since clearing. The pH measured in a 1:5 soil:0.01M $CaCl_2$ suspension is the accepted standard for monitoring acidity in WA. A number of alternative methods are available, but there are advantages in using dilute calcium chloride (refer to box; *Measurement of soil pH*).

An alternative method for diagnosing soil acidity is to use plant symptoms, but most crops affected by acidity display a range of symptoms. For example, in severe cases of Al toxicity, wheat crops show symptoms of nitrogen deficiency early in the growing season and drought stress late in the season. These symptoms reflect the restricted root growth in the subsurface soil.

Soils with pH_{Ca} <4.5 commonly contain high levels of soluble Al, but pH alone does not identify Al toxicity. Consequently assessment of acid soils may be improved if Al is also measured. A laboratory test for measuring toxic Al is available, although at present it is only suitable for the subsurface soil of acid yellow sandy earths in the eastern wheatbelt.

pH measurement and soil sampling

Surface soil. The depth of sampling is currently standardised at 0 to 10 cm, which is also the standard depth for testing for fertiliser requirements. Sampling with a standard 'pogo' is adequate. In the future, this depth may be refined because there can be a marked variation within the top 10 cm, especially in soils with a surface (0 to 2 cm) accumulation of organic matter from an extended pasture phase.

The critical pH ranges (Table 5.1.5) are based on sampling in summer and autumn, although soil can be collected at other times because the temporal change in pH_{Ca} between seasons is less than 0.2 pH units.

When sampling a paddock or trial, first identify any major soil type changes. If there are obvious changes these areas should be identified on a paddock or trial plan and sampled separately. Within each 'area' or paddock collect at least two bulked samples each containing 20 cores (fewer samples can be taken in paddocks of <10 ha).

Subsurface soil. The standard depth for assessing subsurface acidity is 10 to 20 cm, but in some sandy soils this will result in missing the most acid layer. On uniform coarse-textured soils the lowest pH can occur between 10 and 40 cm (Dolling and Porter 1994; Whitten 1995), therefore a deeper sample should be taken at 20 to 30 cm or 30 to 40 cm to quantify the acidity. Recommended densities are two bulked samples each of 20 subsamples, or a minimum of one bulked sample of 5 to 10 subsamples.

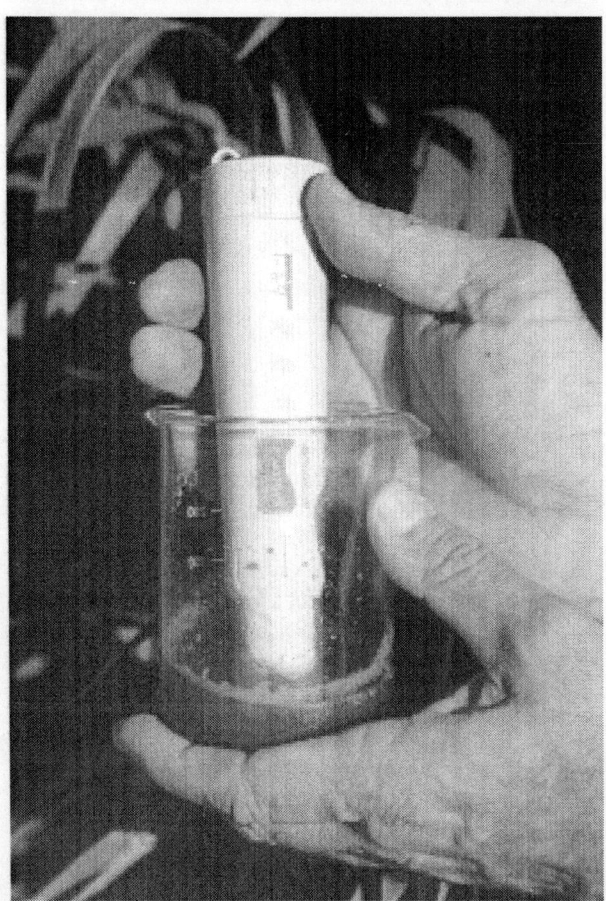

Measuring pH with a portable meter.

Measuring pH. Soil samples can be analysed in a laboratory under controlled conditions or using a portable pH meter. A portable meter should be calibrated each day using standard buffer solutions. A soil for which laboratory results are available should be included as a check in each batch.

Interpreting the results. Surface pH can be related to crop production (Table 5.1.5). The reliability of the relationship depends essentially on toxic Al, plant tolerance and soil pH. Information on pasture growth and pH is covered in Chapter 9.

For most plant species, critical levels of subsurface acidity have yet to be defined. As a first approximation, the critical pH ranges for the surface soil (Table 5.1.5) can be used.

Measuring aluminium

The primary concern with soil acidity is the concentration of toxic Al, but its measurement in most soils is complicated. A laboratory test was developed for the subsurface of yellow sandplain soils in the Zone of Ancient Drainage by Carr *et al.* (1991) plus a

Table 5.1.5 Critical surface soil pH for major crops. At the lower end of the range there is a high risk of production losses (10-30%), while at the upper end the risk is small.

Crop species (*cultivars*)	**Critical pH range** surface soil (0-10 cm) (1:5 0.01M $CaCl_2$)
Tolerant wheats	4.0 - 4.3[1]
Sensitive wheats (e.g. *Aroona, Wilgoyne, Cranbrook*)	4.2 - 4.5[2]
Barley	4.3 - 4.5[2]
Oats	4.0 - 4.3[1]
Triticale	4.0 - 4.3[1]
Narrow-leafed lupins	4.0 - 4.3[3]
Canola	4.3 - 4.5[4]
Faba beans	5.0 - 5.4[4]
Chickpeas	4.6 - 4.9[5]
Field peas	4.2 - 4.5[5]

1. *Estimate,*
2. *Dolling* et al. *(1991),*
3. *J. Howieson personal communication,*
4. *W.M. Porter personal communication,*
5. *I. Pritchard personal communication.*

MEASUREMENT OF SOIL pH

Soil pH is an important diagnostic measurement, although interpreting the results can be difficult. Complications arise from soil heterogeneity, seasonal variation, methodology, laboratory error and the ionic strength of the soil.

(i) Ideal measurement

The ideal pH measurement would cause minimum disturbance to the distribution of ions between soil surfaces and the soil solution (Schofield and Taylor 1955). The extractant should have the same ionic strength as the soil solution and consist of the most common ions in the solution. The ionic strength of soils from the agricultural area of WA ranges from 0.001 to 0.0168M, with a mean of 0.0048M according to Dolling and Ritchie (1985), or 0.0072M according to Dellar and Lambert (1992). Alternatively, the pH of the soil solution could be measured. The pH of the soil solution is not a routine laboratory measurement, although it is used in the United States. For 17 soils from WA, Dellar and Lambert (1992) developed the following relationship between pH of the soil solution (pH_{ss}) and pH_{Ca}:

$$pH_{ss} = 0.94\ pH_{Ca} + 0.25\ (r = 0.825,\ P < 0.001)$$

(ii) pH_w compared with pH_{Ca}

The pH measured in a 1:5 soil:water suspension was the standard technique for many years before pH_{Ca} was recommended. The advantages of pH_{Ca} compared with pH_w include:

- less seasonal variation (Slattery and Ronnfeldt 1992)
- more reliable relationship with plant growth
- similar ionic strength to the soil solution (for some soils)
- less change after long-term storage (Slattery and Burnett 1992).

The temporal variation in the pH_w can be partly explained by seasonal fluctuations in the ionic strength (Slattery and Ronnfeldt 1992).

The relationship between pH_{Ca} and pH_w was thought to be linear (Conyers and Davey 1988) although results from NSW (Little 1992) and Queensland (Aitken and Moody 1991) favour a non-linear relationship. The relationship changes when pH_{Ca} <4.5 and >6.5 (Figure 5.1.5). As a general rule, to convert from pH_w to pH_{Ca} subtract 0.8; however the actual difference between the two measurements ranges from 0.6 to 1.2, and in extreme cases from zero to 2.0.

(iii) Field kit pH

The pH measured with the CSIRO Inoculo Field Test kit is an approximation for pH_w. The field kit pH is only accurate to half a pH unit. It can be used to identify strongly acid soils, but cannot identify the critical pH required for acidity assessments. It is best restricted to the rough classification of soils.

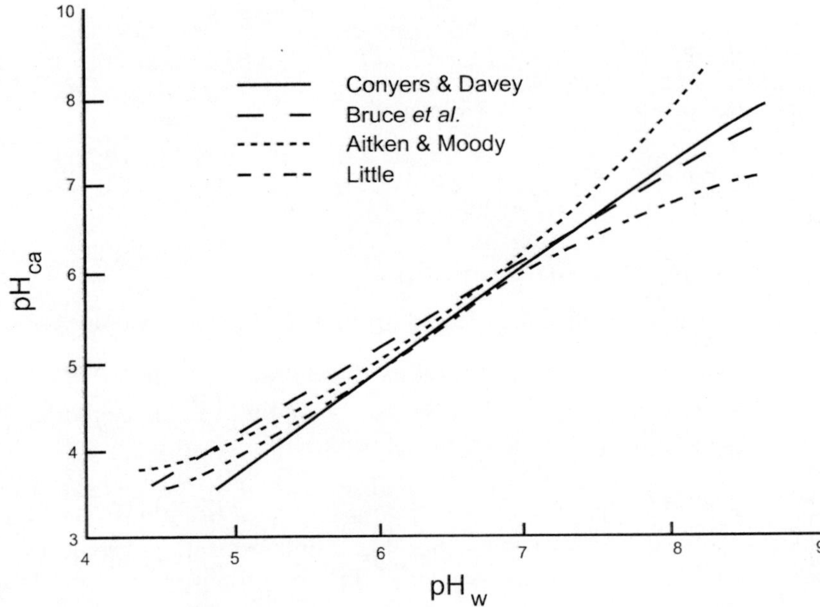

Figure 5.1.5 *Relationship between pH_w and pH_{Ca} as established by various workers (from Figure 1; Little 1992).*

simplified version of the same test (Barton and Carr 1996). The tests are not suitable for surface soils because organic matter interferes with the analysis.

- *Laboratory test*

The ratio of Al:Na in a 1:5 0.005M KCl extract from the 15 to 25 cm layer was the best predictor of grain yield when results from several regions and with different fertiliser practices were considered (Carr *et al.* 1991). In a separate experiment, the Al extracted by KCl was able to explain 97% of the variation in the fresh weight of wheat roots grown in soils containing toxic Al under controlled conditions (Carr and Ritchie 1993).

Spreading lime.

- *The aluminium quick test*

This is a simplified version of the laboratory method based on a visual estimation of the 0.005M KCl extractable Al using a colorimetric reagent, pyrocatechol violet (PCV). When PCV is added to the soil extract it turns yellow if the concentration of Al is low and a deep blue if it is high. The Al in the soil extract is estimated by visually comparing the sample with standards of known concentration (Barton and Carr 1996). The electrical conductivity (EC) of a 1:5 soil:water suspension is also required. When the Al/EC ratio is greater than 0.4, the yield potential is likely to be less than 900 kg/ha as a result of toxicity (Barton and Carr 1996).

- *Other tests for soil aluminium*

For a soil test to be useful it needs to be able to distinguish between toxic and non-toxic forms of Al. Two common tests are *exchangeable Al* (using 1M KCl) and *extractable Al* (using 1M ammonium acetate). Failure of these tests to predict Al toxicity reliably, or to consistently extract Al from biologically active pools over a range of soils, limits their overall usefulness (Ritchie 1989).

MANAGEMENT OPTIONS

Where soil acidity is a problem or will become a problem there are five management strategies:

(i) amelioration - raise the pH using lime;

(ii) reduce the rate at which soils are acidifying;

(iii) increase the acid tolerance of the plants being grown;

(iv) do nothing and accept decreasing yields;

(v) add extra nutrients (e.g. N, P, K).

Options for managing surface acidity in relation to current pH are summarised in Table 5.1.6. For a more comprehensive description refer to *Soil acidity, A reference manual* (Leonard 1996).

Methods for managing subsurface acidification

- Apply lime to the surface of coarse-textured soils and rely on leaching to reach the subsurface layer.

  Surface-applied lime will ameliorate subsurface acidity on sandy soils to a depth of 30 cm providing the topsoil is not highly acidic or does not have a high pH buffering capacity (Mason *et al.* 1994; Whitten 1995). The lime should be applied before subsurface acidity begins to affect plant growth (Sandison 1995).

- Mix lime through the subsurface layer (experimental).

- Place lime in a thin layer immediately below the surface and rely on leaching to reach the subsurface layer (experimental).

Applying lime

The amount of lime required to increase the pH of a soil by a given amount varies with the pH and pH buffering capacity. A number of methods have been developed to calculate the lime requirement of a soil, but they have not been verified in WA.

In WA, lime is graded according to its neutralising value (i.e. compared with pure $CaCO_3$) and particle size. Particles larger than 2 mm are ineffective because of their slow dissolution in soil. 'A' grade lime has a neutralising value of >75% and more than 80% of the particles pass through a 0.6 mm sieve; 'B' grade lime has a neutralising value of >50% and more than 60% of the particles pass through a sieve.

Lime amendment

Amount of lime. Apply 1 to 2 t/ha of good quality lime (neutralising value >75%). Quality may be offset against cartage if a lower quality supply is available nearby, as cartage is frequently a major component of the overall cost.

Application method. Topdress and then incorporate to a depth of 10 cm with one or two passes of a tined or disc implement. Incorporation is important because lime has a low solubility.

Effect on topsoil pH. Factors affecting pH include lime quality (neutralising value, particle size), degree of mixing through the soil, depth of incorporation, pH buffering capacity, initial pH (if pH_{Ca} <4.2 then there will probably be high concentrations of Al in solution which require extra lime to precipitate, but with little change in pH) and rainfall as lime only reacts in moist soil (Leonard 1996).

As a guide. For a soil with a sandy loam topsoil, where the pH_{Ca} is between 4.2 to 5.0, with low organic carbon (<1.5%) in a medium rainfall area (>325 mm), the pH (0 to 10 cm) will increase by about 0.5 to 0.7 units in the first year.

Effect on subsurface pH. Lime incorporated into the top 10 cm can affect the subsurface pH with time. Changes in pH will be smaller (0.1 to 0.5 units), and it may take four to seven years or longer for the lime to move down the profile (Whitten 1995).

Table 5.1.6 Overview of management options for surface acidification in relation to current pH.

Current status **pH_{Ca}** (surface soil 0-10 cm)	**Management options**
>6.0	No action required.
5.5-6.0	**Monitor** paddock every 5-10 years.
5.0-5.5	**Monitor** regularly every 3-5 years. Apply lime to avoid future production losses of sensitive species. Minimise acidification (Table 5.1.7).
4.5-5.0	**Monitor** regularly every 2-3 years. Apply lime to improve production of sensitive species and future production losses of tolerant species. Minimise acidification (Table 5.1.7).
4.2-4.5	Apply lime to improve production of most species. Use tolerant species/cultivars (crops in Table 5.1.5; pastures in Chapter 9). Minimise acidification (Table 5.1.7). Add molybdenum if required.
<4.2	Apply lime to improve production of all species. Use highly tolerant species/cultivars (crops in Table 5.1.5; pastures in Chapter 9). Minimise acidification (Table 5.1.7). Apply molybdenum regularly (Section 6.9). On cultivated permanent subterranean clover pastures growing on 'peaty sands', lime will have to be applied to raise the pH_{Ca} to >4.2, but if not cultivated it is unnecessary to apply lime unless pH_{Ca} <3.6.

Table 5.1.7 Strategies for reducing the rate of acidification* by managing the nutrient cycles (adapted from Leonard 1996).

Nutrient cycle	Management strategy
Carbon	Feed hay back onto paddocks from which it was cut to reduce the net export of alkalinity. Manage grazing to minimise the concentration of dung and urine in stock camps. Select species with low ash alkalinity (Table 5.1.1).
Nitrogen	Replace annual pastures with perennials, because perennials grow in autumn and thus reduce nitrate leaching. Sow crops early to reduce leaching. Apply some N fertiliser after crops are established. Use fertilisers which have a lower lime neutralising value (Table 5.1.3). Reduce the number of cultivations because they stimulate the mineralisation of organic matter and conversion of organic N to nitrate. Retain crop stubble. Reduce legumes in the rotation (but this is likely to have adverse effects on overall production and profitability).
Sulphur	Minimise the use of elemental sulphur as fertiliser.

* *These options should only be employed where they will lead to unchanged or higher production. A slower rate of acidification is unlikely to result in large savings in lime purchases, because the long-term cost of lime is generally low for farming systems in WA.*

Further reading

Helyar, K.R. (1976). Nitrogen cycling and soil acidification. *Journal of the Australian Institute of Agricultural Science* **42:** 217-221.

Helyar, K.R. and Porter, W.M. (1989). Soil acidification, its measurement and the processes involved. *In* 'Soil acidity and plant growth' (ed. A.D. Robson). Academic Press, pp. 61-101.

5.2 SOIL ALKALINITY AND SOIL SODICITY

Gottfried Scholz and Geoff Moore*

The term *soil alkalinity* refers to soils that are alkaline (pH_w >7.5 measured in a 1:5 soil:water suspension) in one or more layers. *Soil sodicity* refers to soils that have high levels of exchangeable sodium. On the other hand, *soil salinity* refers to soils that have a high concentration of water soluble salts. It is usually measured by the electrical conductivity of a 1:5 soil:water suspension (EC 1:5), or the conductivity of a saturation extract, ECe (refer to *Soil salinity*, Section 5.3).

In the agricultural area of Western Australia, alkaline soils occur along the coast as a narrow strip of calcareous coastal dunes, in the low rainfall districts (Zone of Ancient Drainage, Salmon Gums Mallee Zone) and in soils formed on basic rocks weathered *in situ* (Section 2.2). A soil with an alkaline to strongly alkaline reaction can have a number of nutrient deficiencies including phosphorus, nitrogen, copper, zinc, manganese and iron.

Cochrane *et al.* (1994) estimated that 26% of WA soils are sodic, but it seems that many are only sodic in the subsoil. For instance, McArthur (1991) identified 152 reference sites throughout the agricultural area and found that 57% of the soils were sodic but only 21% had sodic topsoils. The effect of sodicity on surface structure is discussed in Section 3.2, *Soil structure decline*. Sodicity can also have a direct toxic effect on plants through the release of Na from the exchange sites to form sodium hydroxide.

Alkaline soils are commonly sodic as well, but sodic soils are not necessarily alkaline and may have be acidic, neutral or alkaline.

SALTS IN WESTERN AUSTRALIAN SOILS

In alkaline soils, carbonate and bicarbonate salts are frequently present (Table 5.2.1). On the other hand, in soils with a neutral or acidic reaction, sodium chloride (NaCl) is the dominant salt.

ALKALINITY

Soils are termed alkaline if the pH_w value is above 7.5, but a further distinction is often made between *calcareous alkaline soils* (pH_w 7.5-8.4) and *alkaline sodic soils* with pH_w >8.4 (Rengasamy and Olsson 1991).

Alkalinity of soils is caused by carbonates of calcium and/or sodium. If the pH_w is above 10 then either sodium-rich clays or sodium carbonate are present. A high pH value is caused mainly by the hydrolysis of salts of weak acids and strong bases.

Where evaporation exceeds rainfall, various minerals precipitate in soils including readily soluble salts. If HCO_3^- is concentrated in the soil, the pH has a tendency to rise. The total concentration of $CO_3^{2-} + HCO_3^- + OH^- - H^+$ is the alkalinity concentration, which gives the best characterisation of an alkaline soil. If this concentration is more than 1.5 meq/L the pH value in a well aerated soil rises above 8.5 (Bolt and Bruggenwert 1978).

High alkalinity leads invariably to sodicity in soils (sodium enrichment on the exchange sites of clays), but not all sodic soils are alkaline.

The more common salts found in alkaline soils are discussed below:

Calcium carbonate ($CaCO_3$)

This is the most common of the alkaline salts found in WA soils. It can occur as nodules, be finely divided in the fine earth fraction, or occur as cemented pans (calcrete).

In calcareous soils, the pH_w varies between 7.0 and 8.4 under aerobic conditions, but under anaerobic conditions it drops by 0.5 to 1 pH unit (Bolt and Bruggenwert 1978). Calcium carbonate is a common mineral in semi-arid and arid soils. Some shelly beach sands and soils derived from them also have a high content of $CaCO_3$.

Calcium carbonate is non-toxic to plants, although it may affect nutrient uptake. Calcite (pure $CaCO_3$) has a low solubility in water and a saturated aqueous solution has a pH_w of 8.3 to 8.4 (Table 5.2.1) in equilibrium with atmospheric CO_2. The pH of the soil solution will depend mainly on the ambient CO_2 partial pressure (affecting solubility of $CaCO_3$) rather than on the absolute amount of $CaCO_3$ present in the soil. The partial pressure of CO_2 is several times higher in the soil than in the atmosphere, therefore the pH of non-sodic calcareous soils will always be lower than 8.4.

The presence of calcium carbonate can be detected by adding a few drops of dilute acid (1M HCl) to the soil. When calcium carbonate is present, effervescence occurs because CO_2 is released. This test can also be used to estimate the amount of $CaCO_3$ in the soil (Table 5.2.2).

* *Formerly Agriculture Western Australia, now Scholz Environmental Consulting Perth.*

Table 5.2.1 The properties of salts in Western Australian soils (adapted from Kovda 1973; Szabolcs 1989).

Iron deficiency (or chlorosis) in plants is frequently associated with $CaCO_3$ in the soil. It is thought to be related to:

- the amount of $CaCO_3$ (especially clay sized)
- the amount and types of Fe oxides
- the presence of bicarbonate ions
- soil aeration
- clay content (Loeppert and Hallmark 1985).

The presence of $CaCO_3$ in the soil does not infer plants will necessarily be iron deficient, only that deficiency is a possibility. The sensitivity of dicotyledonous plants to iron deficiency depends on their ability to modify the soil environment adjacent to the roots (rhizosphere), through rhizosphere acidification or the release of Fe^{3+} reductants.

Narrow-leafed lupins (*L. angustifolius*) are sensitive to $CaCO_3$ and can suffer from iron deficiency on calcareous soils. The reason for lupins sensitivity to iron deficiency is unknown and is despite their ability to acidify the rhizosphere and possessing mechanisms to enhance iron uptake (White 1990, Section 8.4).

Sodium carbonate (Na_2CO_3), and sodium bicarbonate ($NaHCO_3$)

The presence of sodium carbonate in the soil is highly significant because it is very toxic even at low concentrations. It is highly soluble (Table 5.2.1) and therefore comparatively mobile. Sodium bicarbonate differs from sodium carbonate in that it is non-toxic to plants but is strongly alkaline in solution, resulting in nutrient imbalances in the soil.

Acidic rocks such as granites with high contents of sodium-rich feldspars may weather to produce alkaline soils. Minerals such as nahcolite ($NaHCO_3$) and trona (hydrous Na_2CO_3) are formed in areas where evaporation exceeds precipitation. Rock powder from granite has an alkaline reaction and raises the pH in acid soils. Most granites in WA are sodium-rich and spreading the rock powder not only increases the pH, but the soil is also enriched with sodium, especially in the clayey horizons. In semi-arid and humid regions, this could be a problem because sodium could increase the dispersiveness of the clays and cause further alkalinisation of subsoils.

Table 5.2.2 Estimating the calcium carbonate content of soil from the degree of effervescence (Soils from the Katanning region of WA; H. Percy unpublished data).

Reaction of fine earth fraction with 1M HCl*	Category	Calcium carbonate (%)		
		Mean	Range	No. of samples
No audible or visible effervescence	Non-calcareous	0	0-2	46
Slightly audible but no visible effervescence	Slightly calcareous	1.3	<1-7	88
Audible and slightly visible effervescence	Moderately calcareous	3.3	1-12	26
Moderate visible effervescence	Highly calcareous	8.8	3-19	20
Strong visible effervescence ('jumps up')	Very highly calcareous	12.7	5–28	19

* *Using two or three drops of 1M HCl.*

Sodium carbonate is highly soluble and therefore high concentrations of Na^+ and HCO_3^- are needed (and hence very high pH values) before it is precipitated in soils.

The presence of sodium carbonate in a soil can be inferred from a high pH value and by analysing the concentrations of water soluble sodium (cation) and carbonate and bicarbonate (anions). The pH has to be measured directly in a freshly-made saturation extract. If measurement is delayed by a day the pH value is lower and misinterpretation of the results can occur. Measuring alkalinity is the preferred way to characterise the alkalinity status of soils.

SODICITY

In Australia, soils are called *sodic* if they have an exchangeable sodium percentage (ESP) of 6 to 15 and *highly sodic* if their ESP is more than 15 (Northcote and Skene 1972).

$$ESP = \frac{\text{exchangeable sodium x 100}}{\text{cation exchange capacity}}$$

Sodium adsorption ratio (SAR) may be used as an alternative, particularly in saline soils. It is normally measured in a saturation extract.

$SAR = [Na^+]/(0.5([Ca^{2+}] + [Mg^{2+}]))^{0.5}$
concentrations in me/L

A relationship between ESP and SAR for hardsetting wheatbelt soils is described in equations 3.2.2 and 3.2.3.

High sodium saturation generally leads to high pH values, but where sodic soils are neutral or acidic, additional salts are present which suppress alkaline hydrolysis. For example, a sodium-saturated clay dissociates in water giving some free Na^+ and a negative charge on the clay which is immediately hydrolysed giving free OH^-, in a reaction called *alkaline hydrolysis.* In soils with a high concentration of sodium in solution (e.g. saline soils) this reaction is suppressed. Calcium-saturated clays are less alkaline, because calcium ions do not dissociate from the clay as readily as sodium, while magnesium-saturated clays are in between (Leeper and Uren 1993).

Calcium carbonate (finely divided and in nodular form) is not readily soluble in alkaline sodic soils which contain substantial amounts of carbonate and have pH_w of 8.0 and above. Soil water that has a high carbonate concentration may show a high sodium adsorption ratio (SAR) even if the total salt content is not very high.

There is a strong link between sodicity and alkalinity. It can be reliably inferred that a soil with an alkaline reaction (pH_w >8.5) is sodic, but not all sodic soils are alkaline (Gupta and Abrol 1990). The relationship between pH and the estimated ESP is listed in Table 5.2.3 for soils with an alkaline reaction.

Table 5.2.3 The relationship between pH and the approximate exchangeable sodium percentage (adapted from Abrol *et al.* 1988; Gupta and Abrol 1990).

pH measurement		Approximate exchangeable sodium percentage* (ESP)
Saturated soil paste	1:5 soil:water	
<8.0	<8.4	pH not a reliable guide
8.0-8.2	8.4-9.1	5-15
8.2-8.4	9.1-9.5	15-30
8.4-8.8	9.5-9.8	30-50
8.8-9.0	9.8-10.0	50-70
>9.0	>10.0	>70

* *This relationship is for soils from the Indian subcontinent. A similar relationship exists for Australian soils, however the respective ESP values may be lower.*

Table 5.2.4 Relative tolerance of plants to exchangeable sodium. Within the stated sodicity range, relative crop yields are 50% of the maximum.

ESP range	Plants
<10	Citrus, deciduous fruits[c]
10-15	Peas[a]
15-30	Clover[c], Oats[b]
30-50	Wheat[a]
50-60	Barley[a]
>60	Tall wheatgrass[c], Rhodes grass[b]

a - Gupta and Abrol (1990)
b - Abrol (1982)
c - Pearson (1960).

In well aerated, alkaline sodic soils, the carbonate and bicarbonate concentrations are high in the soil solution keeping the concentrations of calcium and magnesium very low. In some duplex soils, magnesium is enriched in the exchange sites of the clays together with sodium. The ESP can be less than 15, but the magnesium saturation very high. Such soils have very poor physical behaviour.

Soils that become saline through secondary salinisation invariably become sodic with time and may also become alkaline.

A high ESP has a detrimental effect on soil physical properties and can also have a direct toxic effect on plants through the release of Na from the exchange sites forming sodium hydroxide. The ESPs that affect plant growth vary with crop species (Table 5.2.4).

ASSESSING THE PROPERTIES OF ALKALINE SOILS

Alkaline soils should be grouped with caution, as they can have markedly different properties. The cause of the alkalinity could be the hydrolysis of sodium-rich clays, or high concentrations of $CaCO_3$ or Na_2CO_3. In each case, the soils can have multiple nutrient deficiencies (e.g. phosphorus, zinc, iron, manganese).

The pH_w gives better differentiation of the alkalinity than pH measured in a 1:5 soil:0.01M $CaCl_2$ suspension. If the pH is measured in 0.01M $CaCl_2$ the corresponding pH_w can be estimated from the relationship described in Section 5.1.

A method is described (Table 5.2.5) for assessing the properties of alkaline soils, which relies on the relationships between pH_w, $CaCO_3$, Na_2CO_3 and sodicity as outlined earlier. This method requires a minimum amount of data (pH_w, ECe), but is not definitive. A more definitive method is to use *alkalinity concentration* as described previously, but this requires more complex laboratory analysis.

Table 5.2.5 Predicting the possible limitations of alkaline soils from pH_w and the electrial conductivity of a saturation extract (ECe). The major limiting factors are highlighted. The letters (A to G) refer to the key below which provides additional information on the interpretation.

pH_w	Electrical conductivity of a saturation extract (ECe) (mS/m)		
	<200	200-400	> 400
7.5-8.4	A. **Slightly alkaline to alkaline** (Check for $CaCO_3$)	D. **Marginally saline (Cl^-)** (especially for sensitive crops or if soil has limited aeration.) **Slightly alkaline to alkaline** (Check for $CaCO_3$)	G. **Saline (Cl^-)** Slightly alkaline to alkaline
8.4-9.5	B. **Alkaline** (Check for $CaCO_3$) Sodic to strongly sodic	E. **Saline (Na_2CO_3, $NaHCO_3$)** Alkaline (Check for $CaCO_3$) Sodic to strongly sodic (ESP 10-30)	G. **Saline (Cl^-)** Alkaline Sodic to strongly sodic (ESP 10-30)
>9.5	C. **Strongly alkaline** (alkaline hydrolysis) **Strongly sodic (ESP>30)**	F. **Saline (Na_2CO_3, $NaHCO_3$)** Strongly alkaline Strongly sodic (ESP>30)	G. **Saline (Cl^-)** Strongly alkaline Strongly sodic (ESP>30)

Note: There are exceptions to the general principles outlined, e.g. if gypsum ($CaSO_4.2H_2O$) is present in large quantities, it is unlikely that Na_2CO_3 will also occur (Szabolcs 1989).

The possible limitations of an alkaline soil can be inferred from the pH_w (1:5 soil:water) and the ECe (Table 5.2.5). The effects on crop yields are complex; because often the whole soil profile is not uniformly affected by alkalinity, e.g. only the subsoil is alkaline, or the degree of alkalinity varies between horizons.

Key for groups identified in Table 5.2.5

A - These soils are slightly alkaline to alkaline. Test for calcium carbonate with 1M HCl and estimate the % $CaCO_3$ from Table 5.2.2.

The growth of cereals, field peas, canola, faba beans and chickpeas will not be affected by the alkalinity. Narrow-leafed lupins (*L. angustifolius*) are sensitive to $CaCO_3$, which induces iron deficiency.

B - These soils are alkaline because of exchangeable Na which is capable of alkaline hydrolysis (sodic to strongly sodic). Sodicity will induce poor physical conditions such as limited infiltration, hardsetting, crusting, waterlogging and low aeration. Calcium carbonate may also be present.

The growth of cereals, field peas or chickpeas will not be directly affected by alkalinity, but may be affected by the poor physical condition and there may be nutrient deficiencies. This soil (or horizon) will prevent the root growth of narrow-leafed lupins.

C - These soils are strongly alkaline (alkaline hydrolysis), so deficiencies of phosphorus, zinc and copper are possible. The soil is invariably highly sodic with an ESP of 30 to 60 or higher. If pH_w is >10, then the ESP could be even higher. The high ESP will result in poor physical conditions and can have a direct toxic effect on plant growth (refer to *Sodicity*). It is highly likely that Na_2CO_3 will be present, although the low electrolyte concentration (ECe) may not suggest this.

The growth of most agricultural plants will be severely restricted.

D - The salt content will affect sensitive crops (e.g. *L. angustifolius*). If the horizon is sodic and is subject to poor aeration or waterlogging, then the salinity will affect a wide range of crops (Section 3.4). These soils are slightly alkaline to alkaline. Test for $CaCO_3$ with 1M HCl and estimate the percentage of $CaCO_3$ from Table 5.2.2.

E - The soil is likely to be sodic (ESP 10 to 30) and Na_2CO_3 or $NaHCO_3$ could be present. The salinity level will definitely affect sensitive crops (e.g. *L. angustifolius*), but is likely to coincide with poor aeration, so the salt will affect a wide range of crops (Section 3.4).

F - There is a high likelihood that Na_2CO_3 is present. This can have a direct toxic effect due to the strongly alkaline reaction (through alkaline hydrolysis). Deficiencies of phosphorus, zinc and copper are possible. The soil is invariably highly sodic, with an ESP in the range of 30 to 60 or higher. If pH_w is >10, then the ESP could be even higher. The high ESP will result in poor physical conditions and can have a direct toxic effect on plant growth (refer to *Sodicity*).

G - This is a saline soil, so the osmotic effect dominates. If the subsoil is saline then the depth to that horizon will be the available rooting depth. If the salinity is near the surface, the soil should be managed as a saline soil (refer to Section 5.3).

MANAGEMENT OPTIONS FOR ALKALINE AND SODIC SOILS

There are limited options for minimising the effects of alkalinity:

- Use tolerant crops and pastures
- Reduce the sodicity
- Reduce the alkalinity (lower the pH).

Tolerant crops and pastures

The selection of tolerant species is usually the only practical option available to the land manager. The tolerance of common crop and pasture species to alkaline conditions is summarised in Chapters 8 and 9 respectively.

Reducing sodicity

Soil sodicity can be reduced by ameliorating with gypsum. If the soil is in a degraded condition it may be feasible to reduce the sodicity of the surface by treating with gypsum (*Soil structure decline*, Section 3.2). The amelioration of sodic layers in the subsoil has received scant investigation.

Reducing alkalinity

The pH of soil can be lowered by applying elemental sulphur, aluminium sulphate or sulphuric acid, but these methods have not been tested in WA and are unlikely to be economic for low input farming systems.

In general, agriculture is an acidifying process through product removal and the use of nitrogenous fertilisers (*Soil acidity*, Section 5.1), although the time scale over which alkaline soils are neutralised could be hundreds of years.

Further reading

Naidu, R., Sumner, M.E. and Rengasamy, P.(eds.) (1995). 'Australian sodic soils. Distribution, properties and management'. 1st National Conference and Workshop on Sodic soils, Adelaide, South Australia, November 9-13th, 1992. CSIRO Publications, Australia.

Szabolcs, I. (1989). 'Salt-affected soils'. CRC Press, Boca Raton, Florida.

5.3 SOIL SALINITY

Geoff Moore

Salinisation by rising groundwater has caused widespread degradation on valley floors in the agricultural area of WA (G. Moore).

Soil salinisation following clearing has been a major disaster in the agricultural area of WA. It has led to large tracts of once productive agricultural land, especially those on the valley floors, becoming saline and has made many of the major rivers in the south-west too salty for irrigation and human consumption.

It is useful to differentiate between primary and secondary (dryland) salinity, because of the different mechanisms involved. Primary salinity refers to soils and landscapes which were saline in their virgin or uncleared state. Secondary salinity is the development of salinity after clearing and is associated with the presence of saline groundwater. Large areas of productive land have been degraded by secondary salinity in south-western Australia, eastern Australia (e.g. Bradd *et al.* 1997) and in North America. In WA, the area affected is more than 1.8 million ha, or 9.4% of the cleared farmland (George *et al.* 1997).

Predicting where secondary salinity may develop is extremely difficult, because it involves a complex interaction between the geology, geomorphology, degree of clearing, drainage patterns, climate, land use history and management. This section is confined to the assessment of the current salinity status of soils.

Sodium chloride (NaCl) is the major salt in soils in the medium and high rainfall zones of Western Australia where secondary salinity predominates. In low rainfall areas (<350 mm/annum), primary salinity is again mainly due to NaCl, but other salts causing high pH values may be present. (If the pH_w >8.5 refer to *Soil alkalinity and soil sodicity*; Section 5.2).

SALINITY AND PLANTS

Salt-affected plants usually appear normal, although they are stunted and may have darker green leaves. There is a general stunting of plant growth, because as the salt concentration increases above a threshold level both the growth rate and ultimate size of the plants decrease (Maas and Hoffman 1977). Shoot growth is frequently suppressed more than root growth.

A plant exposed to a high concentration of salts in the root zone will respond almost immediately by reducing the rate of leaf expansion. This short-term response is due to the increased osmotic potential hindering water uptake by the roots. In the long-term, there is a build-up of salt in the leaves, especially the old leaves, resulting in necrosis. Net effects on the plant depend on the rate of leaf production compared with leaf necrosis (Munns and Termaat 1986).

There is considerable variation in the salt tolerance of agricultural plants and crops. In general, crops tolerate salinity up to a threshold level but above this, yields decrease approximately linearly with increasing salt concentrations (Table 5.3.1; Maas and Hoffman 1977).

The field response of crops to salinity is not necessarily straightforward. The data presented in Table 5.3.1, are from controlled experiments with a uniform concentration of salt throughout the soil profile. In the field, the distribution of soluble salts is usually highly variable (see Figure 5.3.2). The plant response is likely to be related to the weighted-mean salinity concentration, based on the amount of water absorbed at each depth and its salt content (Bernstein and Francois 1973). Therefore, plants can probably withstand higher salt concentrations than those reported above, if part of the root zone has access to water with a low salt concentration.

Salt tolerance also varies with:

- Stage of growth. Sensitivity varies from one growth stage to the next, e.g. barley and wheat are more sensitive during emergence and early seedling growth than during germination and grain development (Ayers *et al.* 1952).
- Varietal differences.

- Soil water. The salt concentrations vary inversely with the moisture content, thus when rainfall is infrequent, the average salt concentration is higher.
- Aeration. There is a negative interaction between waterlogging and salinity on plant growth (Refer to Section 3.4).

PRIMARY SALINITY

Primary salinity is generally confined to the Zone of Ancient Drainage and areas where the annual rainfall is <350 mm. The origin of the salt was predominantly cyclic (Hingston 1958) or some may be saline sediments and wind blown material from salt lakes and the Australian desert regions (G. Scholz personal communication). Soils derived from 'parna' (i.e. aeolian material from playas) often contain high concentrations of salt. These soils can be found in any position in the landscape (Grealish and Wagnon 1995). Many of the finer-textured soils had high salt concentrations in the subsoil before clearing.

Primary salinity is intrinsically linked to certain soils, because although the salt was mainly cyclic, soil salinity is related to the degree of leaching. The infiltration rate and the hydraulic conductivity of each soil horizon determine water movement through the profile and consequently the degree of leaching. In highly permeable sands, the salts have been leached from the soil profile but in fine-textured soils of moderate permeability, the salts are leached to the depth of wetting. Many soils that have primary salinity are also sodic and calcareous. Sodicity influences water movement through the profile and consequently the degree of leaching.

Many soil surveys dating from the early 1900s (Paterson 1917) reported extensive areas with high concentrations of salt in the soils. Teakle *et al.* (1940) surveyed 66,700 ha of non-sandplain country in the Lake King area and reported that 46% was saline before clearing. Similar levels of salinity were found in the Lake Brown (28%), Cleary (34%) and Salmon Gums (48%) areas (Burvill 1947). High salt concentrations were also found in the Mt Beaumont and Cascades districts (Scholz and Smolinski 1996). Many of these surveys were carried out in areas around salt lakes, so a high incidence of saline soils would be expected and these surveys do not indicate the extent of primary salinity over the whole region.

Following the clearing of native vegetation, the water balance changes and leaching may increase. Teakle and Burvill (1938) found that sodium chloride is leached below 60 cm in soils with a sandy texture. The soils containing high salt concentrations following clearing were the calcareous, 'fluffy' morrell soils, loams and fine-textured soils. Salt is leached completely from the soil profile (>1.5 m) only in highly permeable soils (e.g. deep sands, loamy sands). In soils with a moderate permeability, the salts will accumulate at the depth of wetting, which is a useful indicator of the maximum effective rooting depth (rainfall <350 mm).

Many of the lakes and drainage lines in the low rainfall areas of WA were saline before clearing. For example, the main drainage lines in the Zone of Ancient Drainage were saline before clearing, as are the drainage lines in the main valleys to the east of the land presently released for agriculture. The waterbodies which were 'fresh' have either been degraded after clearing or are under threat from secondary salinisation.

Table 5.3.1 Salt tolerance of selected crop and pasture species (adapted from Maas and Hoffman 1977).

Crop	Threshold ECe (mS/m)	% yield decrease per 100 mS/m increase in salinity	Salt tolerance
Barley (*Hordeum vulgare*)	800*	5.0	Tolerant
Wheat (*Triticum aestivum*)	600*	7.1	Moderately tolerant
Lucerne (*Medicago sativa*)	200	7.3	Moderately susceptible
Clover (*Trifolium* spp.)	150	12	Moderately susceptible
Perennial ryegrass (*Lolium perenne*)	560	7.6	Moderately tolerant
Tall wheatgrass (*Agropyron elongatum*)	750	4.2	Tolerant

* *Less tolerant during emergence and seedling stage, when ECe should not exceed 400-500 mS/m.*

Assessment of primary salinity

Primary salinity only needs to be assessed where the average annual rainfall is less than 350 mm. Salinity levels may vary between cleared and uncleared areas, therefore when assessing agricultural potential, always sample both cleared and uncleared areas for comparison. The important factors to consider when assessing primary salinity are summarised in the box below.

Salts other than sodium chloride may be present but may not be leached as readily as NaCl. Sodium carbonate (Na_2CO_3) is extremely toxic to plants even at low concentrations and cannot be leached without chemical reclamation. If the pH_w >8.5 then the alkalinity should be assessed (refer to Section 5.2).

Atriplex spp. (saltbush) are common on the valley floors and lower slopes in the Zone of Ancient Drainage in; (a) native vegetation and (b) roadside vegetation. They may indicate high salt concentrations in the subsoil due to limited leaching (G. Moore).

Management options for primary salinity

If the primary salinity is sufficient to restrict or stop the growth of crops and/or pasture, then the area is not suitable for agriculture and should not be cleared. If cleared, primary salinity will be an inherent feature of the soils and management options will be restricted, unless the surface soil has low permeability in which case improving infiltration may increase leaching of the salt (refer to *Soil structure decline*, Section 3.2).

SECONDARY SALINITY IN WESTERN AUSTRALIA

Secondary salinity developed in WA as a result of a change in the water balance following the clearing of perennial native species and their replacement with annual pastures or crops.

> *"Does clearing increase salt in the ground?*
>
> *A correspondent at Northam writes that wells and soaks in the district that were at one time fresh have become quite brackish since the timber was cleared. If this is so, can it be remedied by replanting? The matter was referred to Mr Mann, the Government Analyst, who replied: It has been pretty conclusively proved that the removal of trees affects the water supply materially".*
>
> *Source: Journal of Agriculture, 1907.*

Changing from perennial to annual vegetation has increased recharge and mobilised salt stored in the deeply weathered profiles, causing groundwater to rise and widespread salinisation.

The causes and effects of salinity in WA have been widely documented (e.g. Wood 1924; Teakle 1937b; Burvill 1947, 1950; Smith 1962; Bettenay *et al.* 1964; Peck 1977, 1978; Malcolm 1983; McFarlane 1991;

WHERE AND HOW TO ASSESS PRIMARY SALINITY

Where to assess

- Zone of Ancient Drainage and the Salmon Gums Mallee Zone; rainfall <350 mm.
- Cleared site (include an uncleared site for comparison).
- No groundwater present within 3 m of the surface. (If groundwater <3 m then assess secondary salinity.).
- Clay content >10% (texture is sandy loam or finer).

How to assess

- Sample each horizon (or every 20 cm) down the profile. (If possible take >4 samples from each horizon.).
- Measure ECe on samples (or EC(1:5) and convert to ECe).
- For a general interpretation of the results use Table 5.3.4. For the specific tolerance of the individual crop and pasture species, refer to Table 5.3.1 and Chapters 8 and 9.
- Compare measurements with those from a nearby uncleared site.

George *et al.* 1997). Wood (1924) identified three key elements causing salinisation in WA and these have subsequently been substantiated by research:

(i) The salt in the soils is predominantly air-borne (refer to *Origin of salt* below).

(ii) Deep drainage below the root zone increases after the native vegetation is cleared.

The deliberate clearing of about 50% of 'Lemon' catchment near Collie to study salinisation processes resulted in groundwater reaching the soil surface within 11 years (George *et al.* 1997). Saline land has been reclaimed by the extensive reforestation of a partially cleared catchment (Bari and Schofield 1992).

(iii) Water flows from surface soils to the saturated zone down preferred pathways like old root channels.

Water movement through preferred pathways to the deep groundwater has been demonstrated using fluorescent dyes (Johnston *et al.* 1983). The low hydraulic conductivity of the deeply weathered soil materials (e.g. 10^{-2} m/day; Peck *et al.* 1980) and the high concentrations of soluble salts in the unsaturated zone overlying groundwaters of lower salinity (Dimmock *et al.* 1974) support the premise that most recharge occurs via preferred pathway flow in these areas.

Origin of salt

There are many possible sources of salts in soil including: deposition with sedimentary parent materials in a marine environment, weathering of mineral particles, deposition in rainfall and dry fallout, addition in fertilisers and other chemicals including herbicides, insecticides and fungicides (Peck 1977).

In WA, most of the salt is brought in with the rainfall and accumulates in the soil over a long time. Rainwater on the west coast contains considerable concentrations of dissolved salts (13-27 mg/L NaCl), but the concentration decreases with distance inland. For example, the amount of salt deposited each year in rainfall at Geraldton, Salmon Gums and Merredin is 195, 30 and 18 kg/ha/year respectively (e.g. Teakle 1937a; Hingston 1958; Hingston and Gailitis 1976). Some of the salt has also come from the weathering of rocks. For secondary salinity to occur only a small fraction of the total salt stored in the soil needs to be mobilised.

Table 5.3.2 A summary of salt stores in different soil-landscapes and rainfall zones.

Location (Reference) average annual rainfall	**Position in landscape (parent material)**	**Total salt store above bedrock** (t/ha)	**Number of profiles sampled**
Darling Ranges (Dimmock *et al.* 1974) 1,100 mm	Various positions in the landscape (laterite on granite)	170±140* (20-510)	10
800-1,000 mm		290±170 (110-450)	4
800 mm		810±530 (50-1,920)	29+
Baker's Hill area 590 mm		950±540 (230-1,920)	19
Eastern wheatbelt (McFarlane and George 1992) 330 mm	Upper and mid-slopes (granite weathered *in situ*)	247 (7-657)	7
	Uplands (laterite on granite)	289 (139-422)	5
	Clayey hillside (colluvium over truncated laterite)	802 (43-1,798)	10
	Sandy hillside (colluvium over truncated laterite)	1,056 (109-2,231)	12
	Broad valley floors (alluvium)	2,571 (44-6,206)	14
	Soils adjacent to playas (alluvium, parna)	13,533 (5,752-21,314)	2

* *Mean ± standard deviation (with range below).*

Salt store

The salt stored in soils is a major factor contributing towards the extensive development of secondary salinity. Many soils in WA contain very large amounts of salt in the profile, especially the deeply weathered pallid zone clays of the laterite profile (Section 2.2). The salt store varies with distance from the coast, annual rainfall (salt accession, leaching) and soil properties (parent material, permeability; Table 5.3.2). For instance, in the Darling Ranges (rainfall >1,100 mm) the salt store is about 170 t/ha, but where the rainfall is <800 mm the average salt store increases to 810 t/ha (Dimmock *et al.* 1974; Malcolm *et al.* 1978).

The amount of salt stored in the profile may also indicate the mechanism of recharge. Areas with a high salt store indicate recharge probably occurs through preferred pathway flow, e.g. permeability contrast soils, while areas with low salt stores indicate recharge by matrix flow.

Water balance

Before clearing the water balance (equation 3.3.4) is near equilibrium; recharge is small (typically less than 1-2 mm per year in the central wheatbelt) and the change in soil water storage is negligible from year to year. Following clearing, when the deep-rooted native species are replaced with comparatively shallow rooted annual crops and pastures there is a small (usually <10%), but very significant, change in the water balance. Runoff and recharge increase, while evapotranspiration decreases. The results of some water balance studies from WA are summarised in Table 3.3.3.

The amount of water stored in the soil profile fluctuates with seasonal conditions, but is generally higher after land is cleared. This gives the soil less ability to 'buffer' against high winter rainfall or heavy unseasonal rainfall. For instance, a moist soil under a crop or pasture is able to store less additional water than a comparatively 'dry' soil under native vegetation. This increases the likelihood of waterlogging and recharge from saturated flow (Hall 1996).

Estimates of recharge following clearing vary with rainfall, catchment factors and vegetation. Recharge is not uniform through the landscape and is variable between years because it is very dependent on the rainfall (Peck and Hurle 1973; McFarlane 1991). Large increases in the height of watertables and spread of saline land are commonly reported following abnormally wet years or flooding of valley floors (George *et al.* 1991). On the other hand, dry years can be observed on a hydrograph as a lowering of the watertable or as a lower rate of increase. However, this does not mean there is a significant decrease in the area affected by salinity, because it takes many years to leach the salt down the profile again (Peck 1978).

In essence, recharge is now thought to be a whole catchment process. Discharge areas may act as recharge areas for parts of the year (George *et al.* 1991). Figure 5.3.1 shows a cross-section of a stylised wheatbelt catchment with the areas of recharge and discharge.

Soils and recharge

Recharge is generally a catchment-wide problem as stated previously, but the amount and the mechanisms of recharge are not the same on all soils. An understanding of these processes (Table 5.3.3) assists with identifying appropriate management options to minimise recharge.

Groundwater systems

In the Zones of Rejuvenated Drainage and Ancient Drainage the overall groundwater contours generally follow similar contours to the land surface. In these areas George *et al.* (1991) identified four groundwater systems:

(i) Perched unconfined aquifers in coarse-textured soils overlying pallid zone clays. There are two types: perched watertables (<1 m) in permeability contrast soils, and perched unconfined aquifers beneath deep (up to 10 m) sand sheets. There is connected flow from perched aquifers to the regional aquifer.

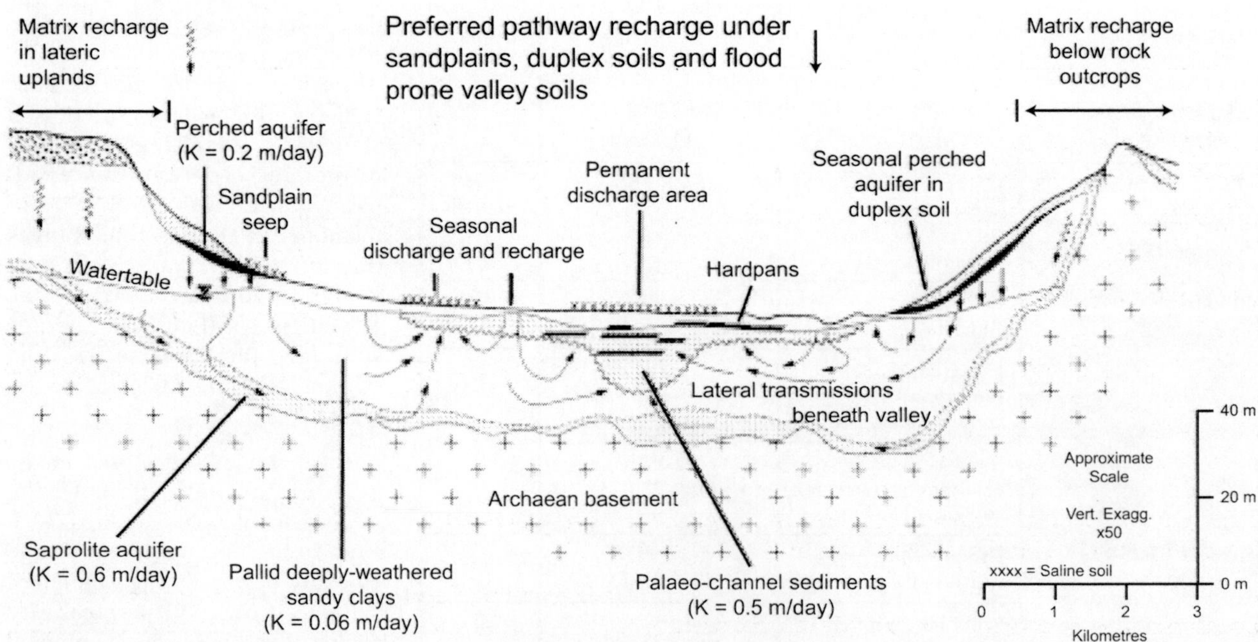

Figure 5.3.1 *Diagrammatic cross-section of a wheatbelt catchment showing where and how recharge is thought to take place (from Figure 4; George* et al. *1991).*

Table 5.3.3 Recharge processes in relation to the four profile hydrology groups identified in Section 1.2.

Profile hydrology group	Main recharge processes
Uniform coarse-textured soils	These soils are well to rapidly drained and the amount and timing of rainfall are probably the critical factors affecting recharge. They have a low capacity to store water (30-80 mm/m) and the water continues to move through the soils at the pressure used to identify the upper storage limit. Unless there are actively growing plants, soil water can rapidly drain below the root zone. High rates of recharge are expected on deep sands (Tennant and Asseng 1997), especially those with a very low water storage and/or where physical or chemical properties limit root growth. For instance, deep drainage (below 1.5 m) of 214 mm was measured on a yellow deep sand under crops at Moora in 1995 and 141 mm in 1996 (Anderson *et al.* 1997). The rainfall for 1995 (750 mm), could be expected in 20% of seasons, while the probability of receiving the 1996 rainfall was 50%. Removing impediments to root growth may be important in reducing recharge. For instance, traffic pans which restrict root elongation but not water movement, will reduce crop water use and may increase recharge. Amelioration using deep ripping is likely to increase crop yields by an average of 500 kg/ha (Jarvis and Porritt 1985) and crop water use by 25-62 mm (Nulsen 1991).
Permeability contrast soils	Permeability contrast soils are imperfectly to poorly drained. The hydraulic conductivity of the clayey subsoil is between 0.0001 and 0.2 m/day (e.g. Seow *et al.* 1988; Tennant *et al.* 1992), so matrix flow is generally very restricted. Recharge is predominantly by saturated flow down preferred pathways (i.e. water flow via large pores, cracks, old root channels). Water will only flow in these 'large' channels from a saturated layer above, e.g. a perched watertable. The June to August period, when rainfall is high and evapotranspiration is low due to slow growth (i.e. low leaf area, low temperatures), is a period of high recharge hazard on permeability contrast soils (Hall 1996). Recharge is closely linked to the incidence of waterlogging, so measures that reduce waterlogging (e.g. agronomic measures, reverse interceptor banks; Section 3.4) will reduce recharge.
Cracking clays	Cracking clays have a high water storage, but infiltration is highly variable depending on the surface condition. Recharge is likely to be minimal unless the soils have chemical properties which restrict root growth in the subsoil and therefore water use.
Medium to fine-textured soils	These soils generally have good physical and chemical properties, have a moderate to high soil water storage (90-150 mm/m) and few restrictions to root growth. They have been the subject of numerous water balance studies, mainly to study the water use efficiency of crops in low rainfall areas (e.g. Rickert *et al.* 1987; Siddique *et al.* 1990). Deep drainage was not usually measured directly, but in most years a clear wetting front was observed which made the assumption of zero deep drainage quite reasonable (Gregory *et al.* 1992). However, in years like 1984 when the annual rainfall was 310 mm (the long-term average), almost half the rain fell before the growing season commenced and deep drainage almost certainly occurred (Hamblin and Tennant 1987; Gregory *et al.* 1992). Recharge is infrequent on these soils and may be associated with flooding on valley floors, heavy unseasonal rains or a very high rainfall during the growing season.

(ii) Semi-confined aquifers above the bedrock in the less weathered saprolite (Smith 1962; Bettenay *et al.* 1964; Nulsen and Henschke 1981).

(iii) Unconfined aquifers in alluvial deposits within major valleys.

(iv) Aquifers within the bedrock.

Dolerite dykes and the weathered pallid zone clays derived from dolerite are strongly implicated in controlling groundwater movement and the occurrence of saline seeps (Engel *et al.* 1987). Mafic rocks like dolerite weather to form soils with a low quartz and a high clay content, which impedes groundwater flow. There are also few large channels or preferred pathways in pallid zone clays derived from dolerite (Dell *et al.* 1983; Johnston *et al.* 1983).

Local variation in terms of dolerite dykes, basement highs, quartz dykes, and faults often result in a highly complex hydrogeology at the subcatchment scale. In many areas, groundwater is effectively controlled by mafic dykes into 'cells' (10-500 m) and again at a regional scale by structural faults into groundwater cells 1,000-5,000 m in extent (Clarke *et al.* 1997; George *et al.* 1997). This cell-like behaviour of aquifers makes extrapolation of results between catchments difficult.

For more information on groundwater systems in south-western Australia refer to: Nulsen and Henschke (1981), Lewis (1991), George (1992a, b) and Salama *et al.* (1994).

ASSESSMENT OF SECONDARY SALINITY

The salinity status of a soil can be assessed from indicator plants, measuring the salt concentration in soil samples or with electromagnetic-induction instruments, or by measuring the depth to a saline watertable. All of these methods are satisfactory, although using indicator plants is generally the easiest. Electromagnetic-induction meters are useful for mapping the distribution and severity of salinity and for detecting salt deeper in the profile.

Indicator plants

Moderate to highly saline sites can be readily identified by indicator species (Table 5.3.4). Equally important are those species which are absent because the salt concentration is too high. On moderately to highly saline sites, reliable indicator plants include sea barley grass (*Hordeum marinum*), annual beard grass (*Polypogon monspeliensis*), spike rush (*Juncus acutus*), ice plant and curly ryegrass.

*Salt affected land with a characteristic cover of sea barley grass (*Hordeum marinum*).*

Measuring soil salinity

Soil salinity is extremely heterogenous within a short distance, both horizontally and vertically, so adequate sampling is essential. There are also seasonal variations due to leaching. Soil salinity can be measured with electromagnetic-induction meters, or from soil samples either in the field with a pocket EC meter, or in the laboratory.

Electromagnetic-induction instruments

Electromagnetic-induction instruments measure soil salinity directly and do not require direct contact with the soil, so they can be connected to a data logger and towed behind a vehicle. The EM38 (McNeill 1980) is an example of a typical instrument used in agriculture as it measures salinity at depths to 0.60 m in the horizontal mode and to 1.2 m in the vertical mode. The EM38 must be calibrated for temperature, soil moisture, texture and clay mineralogy and the presence of magnetic minerals (Bennett *et al.* 1995; McKenzie *et al.* 1997). It has been used successfully to map salinity and to predict the effect of soil salinity on plant growth (e.g. *Eucalyptus globulus*, Bennett and George 1995; pastures McKenzie *et al.* 1997).

Other electromagnetic instruments include the EM31 with an effective depth range of 3.0-6.0 m and the EM34 with an effective depth of 15-60 m (McKenzie *et al.* 1997). They are suitable for identifying high concentrations of salt in the subsoil or in deeply weathered profiles.

Soil testing

A large number of samples must be taken because salt concentrations can vary significantly within a short distance. Teakle and Burvill (1938) found salt contents ranging from less than 0.05% to greater than 2% NaC1 within a few metres. Similar variations can occur over depths of less than 0.3 m (Figure 5.3.2). The time of sampling can also affect the salt measured. Jackson *et al.* (1956) found that samples collected in summer were highly saline, while samples in winter from the same site were non-saline to slightly saline.

Table 5.3.4 Indicator plant species and the approximate soil salinity.

Soil salinity*	Indicator plant species
Non-saline	Agricultural plants not affected.
Slightly saline	*Crops:* Very salt sensitive crops such as lupins are affected. *Pasture:* Fewer salt sensitive species such as yellow serradella, strand, medic, rose and cupped clovers are present.
Moderately saline	*Crops:* Wheat is affected; barley is the preferred alternative. *Pasture:* Fewer clovers, medics and non-salt tolerant grasses are present.
Highly saline	*Crops:* Cereals only return a satisfactory yield when seasonal conditions are favourable. *Pasture:* Patchy grass and bare ground. Barley grass dominates and clovers, medics are usually absent. Balansa and Persian clover could be present.
Extremely saline	*Crops:* Salinity too high for any crops. *Pasture:* Barley grass and other salt tolerant species may be present, however samphire and/or bare ground become dominant as the salinity increases.

** Soil salinity of the surface horizon (0-30 cm). For the equivalent ECe readings refer to Table 5.3.5. If the crop is waterlogged then growth will be adversely affected at lower salt concentrations (see Section 3.4).*

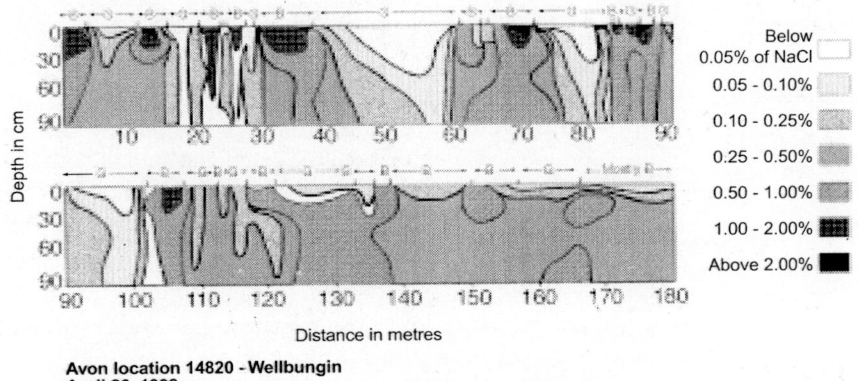

Figure 5.3.2 *The vertical and horizontal distribution of salt in saline soil; B = bare ground, G = grassy or good area (from Figure 1A; Teakle and Burvill 1938).*

Both affected and unaffected areas should be sampled for comparison. If sampling in summer, samples can be selected from the surface crust to see if salt is concentrating near the surface. Other samples should be selected down the profile (e.g. 0-1.5, 2-8, 8-15, 15-30 cm) and in the subsoil. If sampling in winter, the surface salt may have been leached down the profile, especially in permeable soils. In these cases always sample the subsoil.

In the laboratory the replicate soil samples can be bulked and subsampled for each depth interval or for good and poor growth areas.

There are two standard methods for assessing the salinity of a soil sample in the laboratory. Both methods measure electrical conductivity. MilliSiemens per metre (mS/m) are the standard units in WA (see box on Conversion of units, below).

(i) Measurement of the electrical conductivity of the saturation extract (ECe).

The ECe is preferred because it approximates the field water content (soil solution), resulting in a measurement which equates more closely to the plant's response. The method involves extracting the solution from a water saturation paste of the sample. Beatty and Loveday (1974) describe one method of preparing a saturation extract.

The soil salinity categories for ECe (Table 5.3.5) developed by the United States Salinity Laboratory staff are universally applicable (USSL 1954).

(ii) Measurement of the electrical conductivity of a 1:5 soil:water suspension.

When it is not practical to measure the ECe, the EC of a 1:5 soil:water suspension can be measured and converted to ECe using equation 5.3.1 below. George and Wren (1985) developed a relationship between the ECe and the EC_w using a wide range of WA soils.

$$ECe = (364 \times EC_w)/SP \quad mS/m \qquad (5.3.1)$$

where,

EC_w is the electrical conductivity of a 1:5 soil:water suspension,

SP is the saturation percentage of the soil and can be estimated from a relationship with texture.

Soil texture	**Saturation percentage*** (% w/w)
Sand to clayey sand	25
	↓
Sandy loam to sandy clay loam	32
	↓
Sandy clay to clay	45

* *From George and Wren 1985.*

Table 5.3.5 The assessment of saline land with its soil salt content (from USSL 1954).

Soil salinity (ECe mS/m)	**Rating**	**Effect on plants**
0-200	Non-saline	Salinity effects mostly negligible
200-400	Slightly saline	Yield of sensitive crops reduced
400-800	Moderately saline	Yield of many crops reduced
800-1600	Highly saline	Only tolerant crops yield satisfactorily
>1600	Extremely saline	Only very tolerant plants yield satisfactorily

Refer to Chapter 10 for more information on electrical conductivity.

CONVERSIONS OF UNITS ARE:

1 mS/m = 10 µS/cm
= 10 µmhos/cm
= 0.01 dS/m
= 0.01 mS/cm
= 0.01 mmhos/cm
= 5.5 mg/L (or ppm)
= 0.385 gr/gl (grains/gallon)

For *salt estimation*, assuming sodium chloride is the only salt present:

1 mS/m = 5.5 mg/L (= 0.385 grains per gallon)

Depth to saline watertable

Depth to the watertable is an alternative method of assessing salinity. The *critical depth* is the depth of the watertable from which water can reach the soil surface by capillary rise in sufficient quantities to cause a salt problem. The critical depth will vary with the concentration and composition of solutes, the frequency and amount of rainfall, soil physical properties and the crop.

Although the critical depth depends on a number of factors, an estimate can be made which is sufficient for most purposes. Nulsen (1981) obtained a good correlation between the depth to a saline watertable and the type and amount of ground cover. For instance, the critical depth for barley (*Hordeum vulgare*) was 1.5 m. Critical depths are summarised in Table 5.3.6.

Soil type has a large influence on the critical depth, but this has not been studied with regard to saline watertables under dryland conditions. Although a precise relationship between the soil and the critical depth to a saline watertable is not available, capillary rise is greater in a heavy clay than a sand. Generally capillary rise decreases in the order: medium-textured soils > uniform clays > duplex soils (sandy A horizon) > uniform sands.

Depth to the watertable refers to the water level approximately one week after a hole has been drilled. The time elapsed allows the water level to equilibrate. If the watertable is more than 2 m deep, it is unlikely that salinity will affect current crop yields, although it is useful to monitor the site.

MANAGEMENT OPTIONS FOR SECONDARY SALINITY

The effective management of secondary salinity includes managing both the catchment and the saltland. A management plan should be developed for the overall catchment to contain, and if possible reduce, the area of salinity. This involves managing the water balance of the catchment to reduce recharge.

The water use of trees, crops and pastures, especially the comparative water use is important when developing strategies to reduce recharge. Certain trees or crops have been shown to be 'high water users', however water use (especially tree water use) is difficult to measure. A number of techniques have been used, however they are often indirect or the measurements are only taken for short periods. As a result most methods have some limitations, and the results should be carefully considered. For example, the ventilated chamber technique tends to over-estimate transpiration from trees (see review by Raper 1997).

The broad management options for controlling secondary salinity are listed here and then discussed below:

- improving annual crop and pasture agronomy,
- using perennial plants,
- managing shallow water,
- managing the groundwater,
- protecting and managing the remnant vegetation,
- managing soils with major chemical and/or physical limitations,
- managing groundwater discharge sites.

Table 5.3.6 Critical depth to a saline watertable (from Nulsen 1981).

Depth to saline watertable (m)	**Effect of salinity on ground cover**
>2	Negligible effect
<1.8	Wheat yield decreased
<1.5	Barley yield decreased
<1.2	Ryegrass growth affected, replaced with salt-tolerant species

These options can be used on their own but will be most effective if combined as part of an overall salinity management system. Management options should be linked to recharge processes (matrix flow, preferred pathway flow) as outlined in Table 5.3.3.

Improving annual crop and pasture agronomy

Broadscale annual crops and pastures are the major land uses within most catchments in south-western Australia and will be for the foreseeable future, so their water use has a large effect on the catchment water balance.

There is a linear relationship between plant water use (transpiration) and potential production. Current crop yields and pasture production in WA are well below the biological potential, which corresponds to 20 kg/ha/mm for cereal crops (French and Schulz 1984) and possibly 40 kg/ha/mm of dry matter for pastures. Therefore, small increases in transpiration can result in large economic benefits. However, the effect on recharge is often minimal, because transpiration is usually increased at the expense of evaporation from the soil surface, not recharge.

There is scope for increasing transpiration (George *et al.* 1997) by:

- improving agronomy (species and variety selection, fertiliser, weed control, timing of operations),
- farming to the land's capability (i.e. match crop or pasture requirements with site conditions),
- removing impediments to root growth (traffic pans, aluminium toxicity),
- changing the rotation or length of the rotation, because the water use of annual pastures, especially pastures regenerating after a lengthy cropping phase is often lower than annual crops.
- using deep-rooted annual pastures like yellow serradella,
- improving grazing management.

Using perennial plants

> There is universal agreement that annual crops and pastures cannot control or minimise salinisation on their own, no matter how well they are managed. Annual plants do not use rain that falls in intense or prolonged events, or during the extended 'summer' drought from November through to the break of the season. A perennial component, whether it be trees (native, exotic), fodder shrubs (tagasaste, *Acacia* spp.) or perennial pastures (lucerne, perennial grasses) is essential to increase water use and reduce recharge.

Perennial pastures. Lucerne and perennial grasses are likely to use more water early in the growing season (when annual pastures are still establishing) and they can also respond to unseasonal rainfall. The water use of a range of perennial grasses was 12-70 mm higher than subterranean clover-based pastures in trials on the Esperance sandplain (Hall 1996). Lucerne can use more water than an annual pasture (Scott and Sudmeyer 1993; Latta and Blacklow 1997) and when grazed, transpired 433 mm over a 12 month period compared with 231 mm transpired by wheat (Nulsen and Baxter 1982). In practice, most perennial pastures have a significant component of annual volunteer species.

Fodder shrubs. There have been many instances of tagasaste drying out perched aquifers in deep sands. However, there are limited measurements on the water use of tagasaste. On a site with no groundwater within the root zone, the water use of tagasaste varied with the rainfall, being slightly in excess of the annual rainfall in each of the three years (Eastham *et al.* 1994).

Trees. In general, trees in plantations or agroforestry surrounded by agricultural land use more water than native vegetation. This is probably because the planted trees can access more water over a longer period, but could also be an 'oasis effect' (i.e. increased atmospheric demand for water, because the air crossing bare ground surrounding the trees is drier than air over natural vegetation).

Tree water use is highly dependent on climatic conditions, especially rainfall and access to groundwater, however there are also significant differences between tree species (e.g. Greenwood *et al.* 1985a; Raper 1997). Most measurements are from high rainfall districts and may not be applicable to other areas. In the Zone of Ancient Drainage, the water use of a mixed eucalypt plantation grown above a saline seep was 350 mm, slightly more than the annual rainfall (George 1990).

The main commercial tree crops are blue gums (*Eucalyptus globulus*) and maritime pine (*Pinus pinaster*). The water use of blue gums varied from 0.22 to 1.50 times the pan evaporation (Hookey *et al.* 1987; Greenwood *et al.* 1985a). Water use of established maritime pine plantations on the Swan Coastal Plain was between 96% and 105% of the rainfall (Carbon *et al.* 1982).

Agroforestry; an effective option to reduce deep drainage.

Protecting and managing the remnant vegetation

Remnant vegetation in good condition will have a similar water use to the native vegetation before it was cleared, i.e. recharge will be minimal, except from naturally high recharge areas like large granite rock outcrops (Bettenay *et al.* 1964). However, catchments with significant areas of remnant vegetation can still be affected by salinity, although it takes longer for the salinity to develop.

The condition of the remnant vegetation and the presence or absence of an under-storey also affect water use. In some vegetation communities, the under-storey plants use an appreciable amount of the rainfall. For instance, from limited measurements in a native Jarrah forest near Dwellingup, the understorey used 32-36% of the rainfall (Greenwood *et al.* 1985b). Even in drier areas, like at Durrocoppin Nature Reserve north of Kellerberrin, the heath vegetation used 590 mm but the shrubland used only 390 mm (Farrington *et al.* 1992).

Managing shallow water

Episodic recharge from flooding, inundation and waterlogging are major recharge processes in many areas (George *et al.* 1991). Consequently, collecting surface and near-surface water to reduce these recharge processes is an important strategy. Measures to reduce waterlogging, like interceptor drains on sloping sites (>1.5%) and relief drains on areas with a low slope, will reduce waterlogging, increase crop and pasture production and reduce recharge (McFarlane and Cox 1992; Section 3.4). The water can be re-used or safely disposed of depending on the salt content.

Managing the groundwater

Pumping groundwater or installing deep drainage systems have been used widely in irrigation schemes to control salinity. The idea is to control the height of the watertable so that evaporation from the soil surface is minimal. Similar methods have been used successfully to reclaim saline land under dryland conditions (George and McFarlane 1993). However, the cost is usually prohibitive in dryland agriculture unless the land is very valuable (e.g. land for towns, infrastructure, or of environmental value). The low permeability of many subsoil clays in WA limits lateral movement of the water, so a closely spaced drainage network is needed. There may also be adverse environmental effects from the disposal of saline water (George and McFarlane 1993).

Managing soils with major chemical and/or physical limitations

Soils with chemical and/or physical properties that severely limit production are likely to be major recharge and/or erosion sites.

Controlled grazing (left) on saline land encourages colonisation by pioneering species (G. Moore).

Acid yellow sandy earths. The acid yellow sands and sandy earths are deep (>2 m) yellowish brown sands to sandy loams, characterised by being highly acid (pH_{Ca} 3.5-4.2), especially in the subsurface soil (i.e. 10-30 cm). Aluminium toxicity is the cause of low yields on these soils (Section 5.1; Carr *et al.* 1991).

Poor plant growth and low yields translate to low water use and a high recharge hazard. In farm based trials in the eastern wheatbelt, significant recharge of between 8 and 25 mm was measured under volunteer pasture and annual crops (Russell 1996).

Applying lime or molybdenum may increase yields (Porter and Wilson 1984). Yellow serradella with its deep root system, high tolerance of aluminium and good nutritive value for stock would appear to be the best pasture option available presently. Deep-rooted crops like narrow-leafed lupins are likely to use more water than wheat. Another option is to allow these areas to revegetate passively by abandoning unproductive paddocks (Abensperg-Traun 1996).

Pale deep sands. The pale deep sands include white and pale yellow sands and are usually >2 m deep. Pale deep sands are most commonly found as 'spillway sands' or deep sandy hollows within areas of sandplain. These sands are highly leached, contain negligible clay and have a very low capacity to store water (20-50 mm/m). Crop yields are low; for example, yields average 0.4-0.8 t/ha in the 400-600 mm rainfall areas (Lantzke 1993). Low crop yields and poor pasture growth result in low water use and a high recharge hazard.

A perennial shrub like tagasaste with its deep root system could probably use all the available rainfall on these soils. In higher rainfall areas, lucerne or revegetation with maritime pine are useful options. The best annual pasture is probably blue lupins.

Shallow gravelly soils over duricrust. On the uplands in the Zones of Ancient Drainage and Rejuvenated Drainage there are shallow (<30-50 cm) gravelly or sandy soils over cemented ironstone gravel, which limit root growth resulting in low crop yields (Sullivan 1991). Poor plant growth results in low water use and a high recharge hazard. Water balance studies on slightly deeper gravelly sands in the eastern wheatbelt have shown significant recharge from these soils (Russell 1996).

Granite outcrops and high bedrock. The granite rock outcrops and areas of high bedrock were identified as major recharge areas in the Belka Valley (Bettenay *et al.* 1964). Recharge almost certainly occurred before clearing and will have increased if they have been cleared or the remnant vegetation is in a degraded condition. Bedrock highs have been identified as high recharge areas in the Esperance district as well (Short and Skinner 1996).

Managing groundwater discharge sites

There are various types of discharge areas. The management of two contrasting discharge sites, (valley floor salinity and sandplain seeps) is described here.

Management of valley floor salinity is mainly concerned with slowing the rate of salinisation or obtaining limited production from the saline land (Table 5.3.7). Valley floors often represent the discharge for a regional aquifer system so reclamation is usually not possible.

Sandplain seeps result from the discharge of perennial perched aquifers which are common below deep sand sheets in the Zones of Rejuvenated Drainage and Ancient Drainage. The groundwater has a low EC (<100 mS/m), but salt is concentrated by evaporation from the capillary fringe or where seepage water intersects the surface. Discharge is caused by the convergence of shallow groundwater, low permeability of underlying layers and geomorphology (George 1992b).

Sandplain seeps are usually small (1-10 ha) and have been successfully reclaimed with trees. Alternatively, the groundwater can be intercepted before it reaches the surface and removed for irrigation or water supply purposes (George 1990, 1991).

Table 5.3.7 A summary of management options for valley floor salinity.

Salinity 0-30 cm (ECe, mS/m)	Indicator plants	Depth to saline watertable (m)	Management options
Non-saline or slightly saline (<400)	Agricultural plants not affected	>3.0	• Monitor watertable depth on a regular basis. • If rising groundwater; consider tree planting when the watertable is within 5 m of the surface, because in this situation trees have the potential to transpire large amounts of water as they mature.
		2.0-3.0	• Salinity is likely to develop. • Increase water use on-site and off-site. • Drain surface if subject to flooding and/or waterlogging. • Use alley farming.
		1.5-2.0	• Salinity will increase without further increases in the height of the watertable. • Surface drainage if area subject to flooding, inundation and/or waterlogging. • Use alley farming to decrease the likelihood or rate of salinisation.
Moderately saline (400-800)	Barley grass (*Hordeum marinum*) dominant pasture	1.0-1.5	• Fence affected area. • Surface drainage if area subject to flooding, inundation and/or waterlogging. • Sow tolerant crops like barley if cropping. • Plant salt-tolerant forage (depends on surface waterlogging and flooding), sow a mixture of species (refer to Section 9.4 *Halophytes*).
Highly saline (800-1,600)	Patchy barley grass or bare ground	<1.0	• Fence affected area. • Surface drainage if area subject to flooding, inundation and/or waterlogging. • Plant salt-tolerant forage* (depends on surface waterlogging and flooding), sow a mixture of species (refer to Section 9.4 *Halophytes*).
Extremely saline (>1,600)	Bare ground, some samphire	<1.0	• Fence affected area. • Allow natural regeneration if mainly bare ground. • Maintain ground cover through careful grazing of samphire. • Saltbush* may be a marginal option.

** The water use of halophytes (e.g. saltbush) is low in comparison with other species, but it is probably more useful to compare their water use with the evaporation from bare ground, often the alternative 'land use'. Results from the eastern wheatbelt show water use by saltbush of 1-2 mm/day, compared with evaporation from bare ground of 0.4 mm/day. Total water use for the year was 0.9 to 1.5 times annual rainfall depending on the planting density (Greenwood and Beresford 1980). Water use is likely to be negatively correlated with soil salinity. Halophytes can reduce salinity at the soil surface (Smith 1962), allowing the establishment of more palatable salt-tolerant grasses (Section 9.4).*

Further reading

George, R., McFarlane, D. and Nulsen, R. (1997). Salinity threatens the viability of agriculture and ecosystems in Western Australia. *Hydrogeology Journal* **5:** 6-21.

Malcolm, C.V. (1983). Wheatbelt salinity, A review of the salt land problem in south-western Australia. Western Australian Department of Agriculture Technical Bulletin No. 52.

McFarlane, D. (1991). A review of secondary salinity in agricultural areas of Western Australia. *Land and Water Research News* **11:** 7-16.

McFarlane, D.J. and George, R.J. (1992). Factors affecting dryland salinity in two wheatbelt catchments in Western Australia. *Australian Journal of Soil Research* **30:** 85-100.

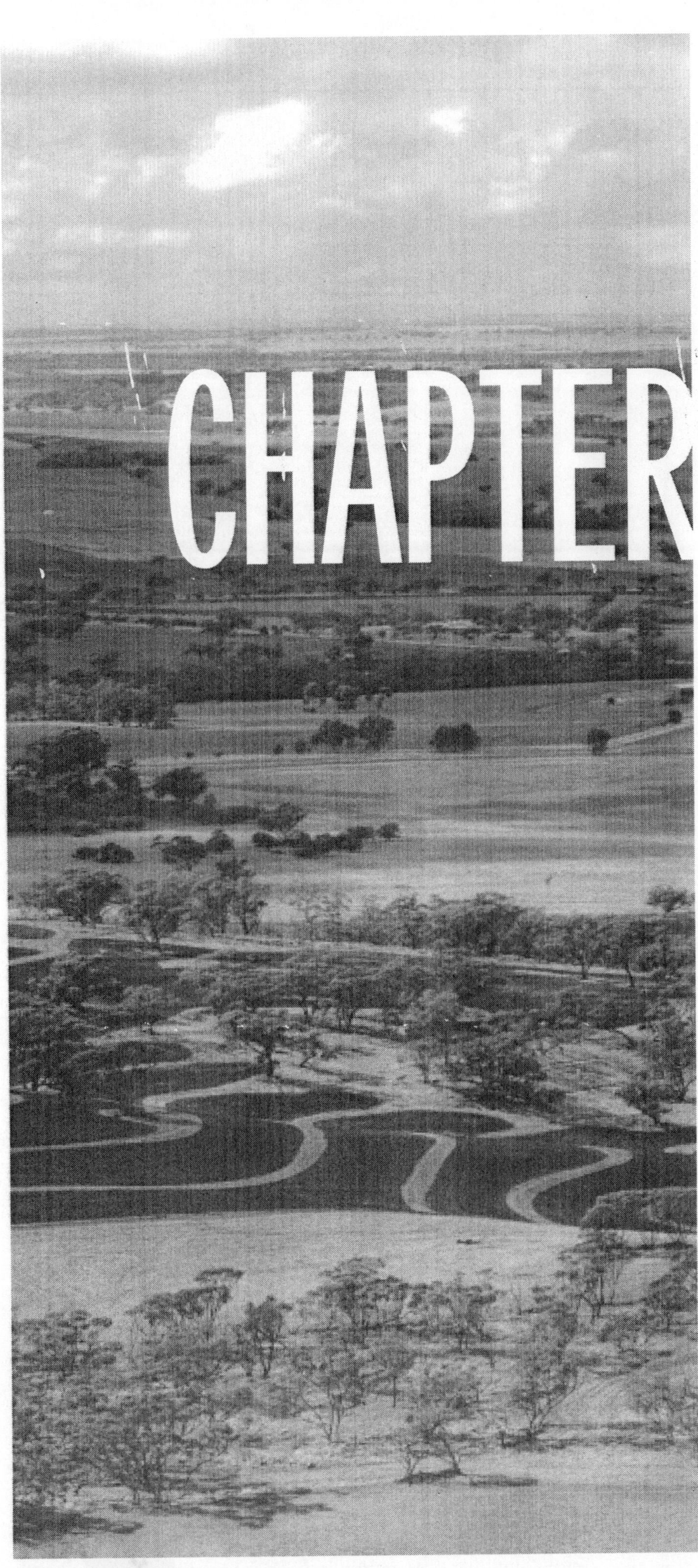

CHAPTER 6

PLANT NUTRITION

6.1 INTRODUCTION

*Mike Bolland**

For profitable production, plants require the following inorganic elements: the macro-nutrients carbon, hydrogen, oxygen, nitrogen, phosphorus, sulphur, potassium, calcium, magnesium and the micro-nutrients iron, manganese, copper, zinc, molybdenum, boron and chloride. In addition, sodium, silica, cobalt and nickel are essential for some species. Other elements may yet be shown to be essential in minor amounts. The division between macro-nutrient and micro-nutrients can be arbitrary because in some plants there is little difference between the concentrations.

The nutrient elements are involved in various metabolic processes and in the structure of plant tissue. Sources are the atmosphere and the soil. Plants take up the nutrient elements as gases from the atmosphere (e.g. oxygen, carbon dioxide), as water or ions from soil solution (e.g. water, nitrate, ammonium, phosphate, sulphate, calcium, iron, copper and zinc), or as chelates from soil solution (iron, copper and zinc). Leaves have evolved to exchange carbon dioxide and oxygen efficiently with the atmosphere. Plant roots have evolved to absorb water and nutrient elements from the soil.

Some nutrient elements can accumulate in plants without being essential at the high levels actually present e.g. manganese in *Lupinus albus* (Jarvis and Bolland 1991). Other elements accumulate at high levels and cause toxicity problems e.g. phosphorus (Loneragan *et al.* 1966), boron (Khan *et al.* 1985; Riley 1987). Plants may also contain high concentrations of non-essential nutrient elements, some of which may be toxic (e.g. aluminium, Mengel and Kirkby 1987).

If the soil does not provide sufficient nutrient elements, it may be profitable to apply fertiliser. The foliage (leaves and stems) of plants can only take up small amounts of nutrient elements. However, it can be profitable to supply some micro-nutrients by spraying solutions onto foliage e.g. zinc (Brennan 1991). When applied to the soil, fertiliser is either applied to the surface, drilled with the seed, or placed (banded) below or to the side of seed. Plant roots can only absorb nutrient elements from moist soil, so if the fertiliser is applied to a dry soil surface, the roots cannot absorb any of the nutrient elements present at the surface (positional unavailability).

NUTRIENT DEFICIENCIES OF SOILS FROM SOUTH-WESTERN AUSTRALIA

Most soils in Western Australia are ancient and highly weathered. When newly-cleared for agriculture, they are often acutely deficient in phosphorus, nitrogen, copper, zinc and sometimes molybdenum (Robson and Gilkes 1980). Profitable production has only been achieved by applying fertilisers.

* *Agriculture Western Australia, Bunbury.*

Nitrogen

Nitrogen levels are improved by growing legume-based pastures and crops, or by applying fertiliser, principally urea, ammonium nitrate, ammonium phosphate and ammonium sulphate. Fertiliser nitrogen generally has a poor residual value because it is removed in agricultural products, leached, and volatilised into the atmosphere.

Phosphorus

Phosphorus has a good residual value (Barrow 1966, 1980). Regular applications of fertiliser have improved the phosphorus status of most WA soils. Consequently, only maintenance dressings are now required for most soils.

Potassium and sulphur

For profitable production, potassium and sulphur fertilisers need to be applied regularly to pastures in the high rainfall (annual average >700 mm) areas of south-western Australia (Yeates 1985a, 1986). Most soils in the lower rainfall areas have either shown no plant yield response or unprofitable responses. The widespread use of single superphosphate containing 10.5% sulphur has maintained soil sulphur levels. Increasingly, farmers are using concentrated fertilisers, such as triple superphosphate and DAP (diammonium phosphate) which contain negligible (0 to 1.5%) sulphur. The continual removal of potassium and sulphur in agricultural produce and the use of no-potassium and low-sulphur fertiliser will result in a future need to apply these elements for profitable production.

Copper and zinc

Copper and zinc fertilisers are applied when the soil is first cleared and developed (Brennan and Riley 1988). Copper has been shown to have a good residual value, with the original levels applied lasting for about 25 to 30 years (Gartrell 1980a; Brennan *et al.* 1986). Therefore fresh copper fertiliser needs only be applied at the recommended level every 25 to 30 years. The residual value of zinc may be up to 30 years. Zinc is a contaminant of rock phosphate from Christmas and Nauru Islands which until the late 1980s was used to make single superphosphate in WA. There is approximately 400 g/g of zinc in single superphosphate which was applied regularly, usually each year, to crops and pastures and helped maintain soil levels (Ozanne *et al.* 1965; Gartrell 1984). In recent years, imported phosphatic fertilisers and rock phosphate used to make fertilisers in WA have been low in zinc. It is now added to WA-made fertilisers to maintain the levels added to the soil.

Molybdenum

Deficiencies occur on a range of acid soils affecting subterranean clover (Gartrell 1980c) and wheat (Riley *et al.* 1983). As the soil becomes more acidic so molybdenum is more strongly adsorbed and can become less available to plants. Only very low amounts are required and its residual value is good except in acid soils (Gartrell 1980c; Riley *et al.* 1983) where yearly applications may be needed. These are best provided by coating wheat seed with small amounts of molybdenum compounds or using seed grown on molybdenum-adequate soil (M.M. Riley unpublished data).

Manganese

Manganese is deficient on some WA soils, however it has a good residual value and usually only one or two fertiliser applications are necessary (Gartrell 1980b, 1984). The exception is alkaline soils close to the coast near Boxwood Hills, where it needs to be applied annually.

Iron

Deficiency occurs on some acid organic sands near the south-western coast (Yeates 1985b). It has not been shown to be economic to apply fertiliser to pastures and crops. Ferrous sulphate is usually applied when growing subterranean clover (*Trifolium subterraneum*) for seed on these soils (J.S. Yeates, personal communication).

ASSESSING NUTRIENT STATUS

Soil and tissue testing, plant symptoms and test strips are used to help identify nutrient deficiencies and subsequently determine fertiliser requirements.

Plant symptoms

Deficiency of some nutrient elements can result in plants developing specific symptoms. Many publications provide photographs and descriptions of the symptoms for different species (Bergmann 1983; Snowball and Robson 1983, 1986, 1991). However, the symptoms only provide an indication and usually need to be confirmed by soil and tissue tests and field experiments in which different levels of fertilisers are applied to the soil. Symptoms indicate fertiliser requirements for the next year, except for copper and zinc, which may need to be applied to this year's crop. Most fertiliser needs to be applied early for a profitable return.

Tissue testing

Tissue testing provides an indication of the status of nutrient elements at the time of sampling. Tissue testing usually affects fertiliser decisions in the next year. The exception is copper and zinc which affect grain yields later that year. Copper or zinc fertiliser can be sprayed on the crop before grain formation. Tissue testing cannot be used alone to provide fertiliser advice because it is affected by many non-nutritional factors and interactions (Mengel and Kirkby 1987). Also, concentrations of most nutrient elements decline as annual plants age and mature (Mengel and Kirkby 1987). For elements that are immobile in the plant, such as copper and zinc, samples of the youngest tissue provide better, more reliable, values than whole tops (zinc: Reuter *et al.* 1982; copper: Robson *et al.* 1984).

Soil testing

Soil testing for the macro-nutrients is only reliable for phosphorus and potassium. The values need to be calibrated with plant yield, and are used to provide an estimate of the phosphorus or potassium present. It is then possible to determine the likelihood of obtaining a profitable plant yield response to applications of fertiliser to the next crop or pasture. However, soil testing is not precise and can only provide a rough guide (Gartrell and Bolland 1987; Bolland *et al.* 1989).

Soil tests for micro-nutrients have been developed to a limited extent in WA. Tests for zinc have been used for both sub. clover and wheat (Brennan 1992a, b).

Test strips

Strip testing can be used by farmers. While applying fertiliser, a strip is left to which no fertiliser is applied (as a control), then subsequent strips receive half, double etc. the amount of fertiliser applied to most of the paddock. Replications of the strips can be done if desired. For pastures, it is necessary to fence an area to prevent grazing so that growth responses to the different fertiliser levels can be seen. Stock preferentially graze plots treated with high levels of fertiliser, so negative yield responses occur. By using test strips, the farmer can determine if fertiliser was warranted and profitable for individual paddocks.

FERTILISER REQUIREMENTS

NP-Decide

Most production and revenue from agriculture in WA comes from areas receiving less than 700 but more than 300 mm average annual rainfall. Nitrogen and phosphorus are the major nutrient elements applied and farmers need to estimate the most profitable amounts required. The data required to provide this advice have been generated in numerous field experiments by Agriculture Western Australia that have measured the yield response to levels of nitrogen and/or phosphorus in the year of application (freshly-applied fertiliser). In the years after application the effectiveness of fertiliser residues relative to freshly-applied fertiliser has also been measured. The data have been incorporated into a model called NP-Decide (Burgess 1988; Bowden 1989) which estimates the most profitable level of fertiliser to apply to crops or pastures in individual paddocks.

PHOSUL-K model

The PHOSUL-K model (Yeates *et al.* 1991) has been developed to provide advice for the most profitable level of phosphorus, sulphur and potassium fertiliser required for different pasture-based enterprises in high rainfall areas. Different soil types and environments can be considered.

Nutrient balance

For paddocks with a long fertiliser history and soils that can retain applied fertilisers (e.g. limited leaching of nutrients) a nutrient balance can be used to help determine fertiliser requirements.

The balance includes estimates of soil sorption, nutrient losses (leaching, erosion, volatilisation) and the amount removed in agricultural produce. The amount of macro and micro-nutrients removed per tonne of grain for some broadscale crops is listed in Table 6.1.1. The exact amount removed will vary, depending on factors such as the variety, time of sowing and root diseases.

The remainder of this chapter deals with the individual nutrient elements that are important for agriculture in Western Australia.

Further reading

Mengel, K. and Kirkby, E.A. (1987). 'Principles of Plant Nutrition, 4th edition'. International Potash Institute, Berne, Switzerland.

Table 6.1.1 Nutrient removal per tonne of grain or straw for the main broadscale crops grown in Western Australia.

Crop (Reference)		Macro-nutrients (kg/t)						Micro-nutrients (g/t)		
		N	P	K	S	Mg	Ca	Cu	Zn	Mn
Wheat (a)	Grain	16-26	2.0-3.5	3-7	2.0-3.0	1-1.5	0.2-0.4	2-4	15-30	10-50
	Straw	2-10	0.2-1.5	6-16	0.4-1.5	0.5-1	0.6-2	1-3	10-30	10-60
Narrow-leafed lupins (b)	Grain	50	4	8	3	2	3	5	40	25
Field peas (c)	Grain	40	4	10	2-3	-	-	2	40	-
	Straw	60	6	35	5	-	-	-	-	-
Faba beans (d)	Grain	40	4	10	1.5	1.2	1.3	10	30	30
Chickpeas (e)	Grain	30	3	9	2	1.4	2	7	30	30
Canola (f)	Grain	40	6-7	9	7-10	3.8	4.0	6-10	30-40	25-30
	Straw	10-15	1-2	20-30	3-4	2-3	7.0	6-8	12-15	15-18

(a) Bolland et al. *(1991)*
(b) P. Nelson personal communication
(c) Pritchard (1993)
(d) K.H.M. Siddique personal communication
(e) K.H.M. Siddique personal communication
(f) R. Brennan personal communication

6.2 NITROGEN

*Mel Mason**

Nitrogen is a very important element for plant nutrition and growth. Nitrogen deficiency is the most common nutrient deficiency in non-legume crops in Western Australia, particularly on the coarse-textured soils in higher rainfall areas. Nitrogen deficiency can result in large reductions in yield and lowers the protein levels in seeds. The atmosphere consists of 80% gaseous nitrogen, but this plentiful source is not directly available to non-leguminous plants. It is, however, used by leguminous plants through fixation as a result of a symbiotic relationship between the plant and nitrogen fixing *Rhizobium* bacteria. The use of nitrogenous fertilisers on non-legume crops in Western Australia has increased steadily since the early 1960s.

NITROGEN IN PLANTS

Nitrogen (N) is an essential part of many plant compounds such as proteins, amino acids etc. It also has an essential role in the production of chlorophyll, the compound which results in the green colour of the plant and is vital in photosynthesis.

Leguminous crops and pastures obtain much of their N requirements from the atmosphere by symbiotic fixation in nodules. The remainder of the plant's requirements is obtained from the soil or sometimes from fertiliser. Responses to N are generally in the form of greener plants due to increased chlorophyll production, and increased vegetative growth, photosynthetic area and seed yield. Increased yield of cereals and grasses is largely due to increased tillering (Halse *et al.* 1969; Feyter and Cossens 1977), but nitrogen can also increase seed numbers per ear (Darwinkel 1983). In addition, a good nitrogen supply can increase the concentration of grain protein (Rowland *et al.* 1994; Mason and Madin 1996), but in crops such as canola this can be at the expense of oil concentration in the seed (Mason 1995).

On very deficient sites, the plant will use N to increase tillering and vegetative growth to set up yield potential through the production of more ears and seeds per ear. This N also increases the photosynthetic leaf area and this increases the potential for carbohydrate production to fill the seeds. With lower levels of supply, most of the N is used for growth and relatively little is available for seed protein. However, as N supply is increased and the requirement for growth is satisfied, more N is available for transfer to the seed for protein production (Mason 1987, 1992b).

Nitrogen for the seed protein comes both from late uptake by the plant and remobilisation from the senescing leaves. Therefore, the more N in the leaf tissue the greater the potential supply for seed protein production.

Soil organic nitrogen or N in legume residues can be more effective in raising the protein concentration of seeds than fertiliser N applied early in the growing season (Rowland *et al.* 1994). This is because the fertiliser N has to be applied early to set up potential yield, and unless excessive amounts are applied, little is left in the soil for uptake later in the season. On the other hand, organic sources of N continually break down throughout the season resulting in a continual supply of available N. Later in the season the N supplied will have little effect on yield, but will be available for transfer to the seed for protein formation. This could also be accomplished by late applications of fertiliser, but is rarely economic unless it results in a crop such as wheat being improved so that it can be delivered into a segregation with a higher price.

Nitrogen is required by crops and pastures in relatively large amounts and can easily become deficient. Because it is required early for setting up yield potential in many crops (e.g. tillering in wheat), the availability at various times can be critical. For a crop such as wheat, if there is only a low supply of N early in the season, yield potential will be low because tillering and hence ear numbers will be reduced. Even if the N supply is increased later in the season, this will only have a limited effect on final seed yield because the plants cannot produce new (effective) tillers or new ears late in the season.

NITROGEN IN THE SOIL

In all of our soils, most of the N (>98%) is organic. However, plants can only use inorganic forms such as ammonium-N and nitrate-N. Organic matter is mineralised in the soil and inorganic N is produced for crop growth.

Soil and environmental factors have a marked effect on various processes in the N cycle and therefore affect the availability of the forms of N absorbed by plants. In addition, these factors may directly affect the plants' ability to take up N. For example, under very dry conditions the root zone may be too dry for the plant to take up N. A representation of the N cycle is shown in Figure 6.2.1.

Mineralisation of soil organic matter depends on many factors. It requires adequate soil moisture and increases with increasing temperature, within normal growing season ranges. Crops will often green-up in spring due to a flush of mineralisation and increased root growth. The rate of mineralisation slows as soil pH falls (Dancer *et al.* 1973) but is increased by aeration. Cultivation will usually stimulate mineralisation (Powlson 1980).

* *Formerly Agriculture Western Australia, now Mel Mason and Associates, Perth.*

The rate also depends on the organic matter present (Mason and Rowland 1992). Organic matter in soils ranges from lignified compounds (which decompose and release mineral N very slowly) to recently incorporated high-N residues of crop or pasture legumes, which break down readily. This inorganic N is produced throughout the growing season, but the amount produced is often not enough to satisfy the requirements of high-yielding non-legume crops, except on very fertile soils.

The initial product of mineralisation is ammonium nitrogen (NH_4^+). This form is not lost readily from the soil, except possibly as ammonia gas (NH_3) if it volatilised in highly calcareous soils (Fenn and Kissel 1976). There is an equilibrium or balance between ammonium (NH_4^+) and ammonia in the soil. In acid conditions this equilibrium is overwhelmingly in favour of ammonium, but under alkaline conditions it swings in favour of ammonia, which can result in gaseous losses.

Ammonium nitrogen is not leached readily from most soils (Broadbent *et al.* 1958; Gasser 1964). Because of its positive charge, it is held by the abundant negative charges on the exchange complexes. However, some downward movement can occur in coarse-textured soils with a very low cation exchange capacity in high rainfall areas.

If the ammonium is not taken up quickly by plants it is converted to nitrate by the process of nitrification. Nitrate is absorbed readily by plants, but is also easily lost from the root zone by leaching (Shaw 1962). Because it is a negatively charged ion there is little to prevent its downward movement with water. Nitrate can also be lost by denitrification (Stefanson 1972). This occurs in anaerobic conditions, such as those associated with waterlogging, and results in the production of gases such as nitrogen (N_2), nitrous oxide (N_2O), nitric oxide (NO) and nitrogen dioxide (NO_2) which are lost to the atmosphere.

The rate of nitrification varies considerably and is affected by pH, temperature (with the rate increasing as temperature rises), moisture and other factors. Nitrification is very slow in highly acid soils and the ammonium-N from fertiliser may not completely nitrify during the growing season, but in alkaline soils the process may be complete within two weeks (Mason 1992a). The process also depends on having a readily available source of energy for the denitrifying bacteria.

During the conversion of ammonium to nitrate, either from soil or fertiliser sources, hydrogen ions are released into the soil contributing to acidity (Pierre 1928; Helyar 1976; Mason *et al.* 1994). The actual change in soil pH will depend on the extent of leaching and the pH buffering capacity of the soil. Thus, increasing acidity can result from using ammonium-based fertilisers, or by introducing legume-based pastures and legume crops into the rotation (Section 5.1).

Nitrogen can also become temporarily unavailable through fixation. Some is fixed by micro-organisms during the decomposition of plant material with a high carbon/nitrogen ratio e.g. cereal straw (Black 1973; Powlson *et al.* 1985).

Soil N fertility can be raised by increasing organic matter and nitrogen using legume-based pastures, or for a shorter term benefit, by including legumes such as lupins, field peas, faba beans and chickpeas in the rotation. For current crops the supply of N can be increased by applying fertilisers.

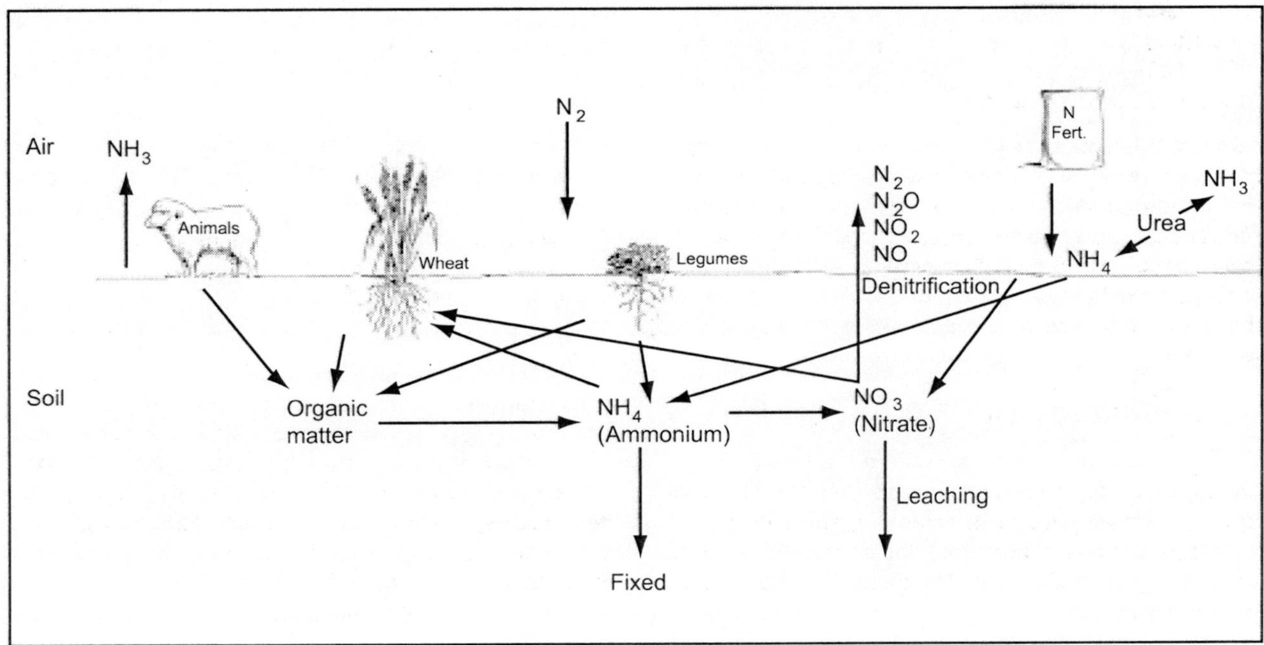

Figure 6.2.1 *The nitrogen cycle.*

DISTRIBUTION OF NITROGEN DEFICIENCY

In WA all soils are potentially N-deficient. The light, or coarse-textured soils are very deficient. The heavy, or fine-textured soils generally contain higher levels of N before they are cleared, but levels can be depleted by intensive cropping with non-leguminous plants.

The occurrence and degree of deficiency in crops and pastures depend on many factors other than soil type. Deficiencies are more likely in higher rainfall areas because of the greater chance of leaching, especially in uniform coarse-textured soils, or because of losses through denitrification due to waterlogging, especially in fine-textured or permeability contrast soils. In drier areas, leaching and denitrification are less likely.

In higher rainfall areas, potential yields are higher, increasing the demand for N and increasing the possibility that deficiency will limit yields. On the other hand, there is generally a greater input of N into the soil due to more productive legume pastures and longer pasture phases in the rotations.

In many paddocks, N deficiency will occur in patches because of differences in soil type or local conditions (e.g. as a result of differences in leaching, waterlogging or patchiness in previous pasture legume stands).

TREATMENT OF DEFICIENCY

Many different fertilisers are available for overcoming N deficiency. Some are used solely as a nitrogen source e.g. urea (46% N), ammonium sulphate (21% N) and calcium ammonium nitrate (27% N), but many compound fertilisers are applied as a source of both nitrogen and phosphorus. In some cases other elements are also important e.g. the sulphur in ammonium sulphate.

Rates of application vary according to many factors such as expected rainfall, soil type, paddock history, economics (determined by cost of the fertiliser and price received for the product) and time of sowing. Optimum rates of application vary from nil in highly fertile situations e.g. the first non-legume crop following a period of medic pasture on fine-textured soils in the eastern wheatbelt, to more than 70 kg N/ha in high rainfall areas with coarse-textured soils and two or more non-legume crops in succession.

The optimum time to apply nitrogenous fertiliser depends on achieving a balance between the need for an adequate supply early enough in the season to set up a good crop yield potential, and avoiding N losses by processes such as leaching. If there is a low probability of leaching then all the N could be applied around sowing and still be available later. However, if severe leaching is likely (e.g. in deep coarse-textured soils in high rainfall areas) then application should be delayed for three to four weeks until the crop has an effective rooting system. The N-P compound fertilisers are designed to be drilled with the seed to obtain maximum benefit from the phosphorus. For pastures, while a spring application may result in a larger response, the reason to apply N may be to increase early growth to fill a 'feed gap' and so it will be applied early (autumn to early winter).

Nitrogen fertilisers can be applied in several ways. They can be topdressed before sowing, drilled with the seed, deep or side-banded at sowing, topdressed on the growing crop, or applied as sprays. The selection of the appropriate method depends on many factors. Some were discussed previously (i.e. the possible need to have phosphorus near the seed and the need to avoid leaching losses).

Urea is the most common fertiliser used if N alone is required. It can be readily leached while it remains in the urea form (Mason 1991a), but it usually breaks down (producing ammonium) within two to three days or at the most two weeks. However, the initial products of its breakdown result in a local alkaline reaction around each granule. This alkalinity results in the production of free ammonia, which has several implications. Free ammonia is toxic to germinating seedlings. Therefore urea should not be drilled at more than 30 kg/ha when in contact with the seed of crops such as wheat, and not at all with more sensitive crops such as canola. The second implication is for urea topdressed on the surface. If it remains on the surface, ammonia gas can be volatilised and lost to the atmosphere. These losses can be largely avoided if the urea is covered (e.g. topdressed in front of conventional planting machines) or if application is followed by rain to wash it into the soil. The worst situation is when urea is applied to moist soil, so that it begins to break down, followed by a warm dry period.

Other nitrogenous fertilisers can cause germination problems under some circumstances, but the mechanism is different from that involving urea and the damage is caused by a salt effect. If high rates of soluble fertiliser are placed in contact with seed in moist soil and this is followed by a dry period, the concentration of salt around the seed is high and can delay or reduce germination.

There is little or no advantage of deep or side-banding the fertiliser in terms of availability of N to the plant, because N is very mobile in the soil. However, banding helps to avoid fertiliser toxicity or volatilisation and may make the N less available to weeds germinating on the surface.

There is little advantage in using N as a spray rather than as a solid fertiliser, except perhaps for increasing grain protein when the spray is applied late (e.g. around the 'boot' or anthesis stages for wheat), but results obtained with this practice have been variable (Mason and Rowland 1988). For applications early in the season, high volumes of spray must be applied to get the required amount of N but most of the spray falls on the soil. As the season progresses, the advantage of foliar spraying increase, but it is generally ineffective.

RESIDUAL EFFECTIVENESS

At the application rates normally used by farmers there will be little residual N at the end of a season, except perhaps following a drought. However, there is considerable residual N in the organic matter from legume pastures (Mason and Rowland 1990). Soils that are developed from the virgin state using legume pastures in the rotation reach a permanently higher yield plane because of higher levels of organic matter, which also affects water and nutrient-holding capacity. The residual effect of legume crops (Rowland *et al.* 1988, 1994) appears to be confined to one year, with perhaps some small benefit carrying into a second year. This appears to be related to the use of the legume residues with little or no increase in soil organic matter.

PADDOCK ASSESSMENT

Plant symptoms

The characteristic symptom of N deficient plants is yellowing of the older leaves and perhaps some red or orange tinges (Mason 1986; Mason and Howieson 1988; Mason 1996). In general, N deficient crops appear pale green to yellow depending on the degree of deficiency. Nitrogen is a very mobile nutrient and can easily be remobilised in the plant. Therefore, when it is in short supply N is translocated from old tissue (or leaves) to younger, actively growing tissue. These symptoms help distinguish N deficiency from those of less mobile nutrients (e.g. sulphur) where the younger leaves become yellow. In grasses and cereals, N deficiency results in reduced tillering, and in crops such as canola, the number of pods is reduced.

Tissue tests

Tissue testing can be used for diagnostic or prognostic purposes. For diagnosis, the concentration of N in the plant or plant part can be compared with established critical levels. If it falls below the critical range it is deemed to be deficient *at that time*. The critical level will vary with season and conditions, and will change with the stage of growth. Critical ranges have to be determined experimentally for different growth stages of different crops. In grape vines, a deficiency can be identified by analysing leaf petioles.

Diagnosing N deficiency is not the same as *predicting* that there could be a deficiency at some later stage or *predicting* a response to applying N. These predictions are called prognosis. Prognostic tissue testing for annual crops has not been very useful (Mason 1991b). This work commenced in 1979 and used whole tops, youngest fully expanded leaves, oldest leaves and sap measurements. Tests included total N, nitrate nitrogen in tissues and sap nitrate measurements using rapid field tests. The problem is not with the chemical measurement, but with how to relate this measurement to plant responses.

Soil tests

A soil test is generally of little use in determining the soils N status accurately, or the plant's requirement for fertiliser. Within a broad range, measurements of total soil nitrogen or total organic carbon can give a general idea of the N fertility, but not an accurate assessment of the N available to a crop. Incubation tests have also proved unsuccessful predictors of N requirement. Measuring of inorganic N in the soil is also generally unsuccessful in WA except as a comparison between paddocks, because inorganic N levels are transitory and do not necessarily indicate how much N will be available during the growing season.

While these soil measurements cannot be directly related to subsequent crop performance or response to N fertiliser, they can often be used by experienced people with other information (soil type, paddock history, crop yield potential) to determine a recommendation for N fertiliser.

Further reading

Mason, M.G. (1991). Nitrogen. *In* 'The Wheat Book - A technical manual for wheat producers' (eds. M. Perry and B. Hillman). Department of Agriculture Western Australia. Bulletin No. 4196 pp. 62-65.

Mason, M.G. and Rowland, I.C. (1993). Effect of some agronomic practices and rotations on grain protein in wheat. Western Australian Department of Agriculture, Division of Plant Industries Technical Report No. 55.

Mason, M.G. (1994). Nitrogen fertilisers for cereal production. Western Australian Department of Agriculture Bulletin No. 4283.

6.3 PHOSPHORUS

*Mike Bolland**

Most soils in Western Australia are ancient and highly weathered with very low levels of natural (indigenous or native) phosphorus (P). Profitable production of crops and pastures has only been possible by applying phosphatic fertilisers, which with nitrogenous fertilisers are the most important crop fertilisers. Phosphatic and sulphur fertilisers are widely used for pastures in high rainfall (>800 mm) areas and P fertilisers are used for pastures in the lower rainfall areas. In south-western Australia, about 100,000 t of elemental P were applied annually to pastures and crops in the early 1990s.

PHOSPHORUS IN PLANTS

Phosphorus is an essential component of cell membranes, genetic material and energy storage and transfer systems for the chemical reactions in plants. It is particularly required during the earliest stages of plant growth when there is rapid cell division and expansion. Deficiency during the critical early stage can greatly reduce yield potential. It has been found to be uneconomic to apply P fertiliser to wheat crops more than 10 days after sowing (Smith 1967).

During the early growth of annual plants, the initial source of P is the seed itself. Concentrations in excess of 0.3% P in the seed of cereal crops, lupins (*Lupinus angustifolius*) and pasture legume species (*Trifolium subterraneum, T. balansae, Medicago polymorpha, Ornithopus* spp.) are usually adequate for early growth (Bolland and Baker 1988; Bolland and Paynter 1990; Bolland *et al.* 1989b, 1990). Concentrations of less than 0.2% P in the seed can reduce yields by up to 70% during early growth (Bolland and Baker 1988; Bolland and Paynter 1990; Bolland *et al.* 1990). The seedlings rely on the seed's P only until they develop a root system capable of absorbing significant amounts of P from the soil. In mature plants, most of the total P has been derived from the soil.

Role of mycorrhizal fungi

Vesicular arbuscular mycorrhizal fungi form a symbiotic association with plant roots. The fungi derive photosynthates from the plant and increase uptake of nutrient elements by the plant (Abbott and Robson 1982). The fungi extends the root system of the plant enabling it to more intensively explore a greater volume of soil, and this increases uptake of P by the plant in deficient soils (Abbott and Robson 1982; Bolan *et al.* 1983). This should enable the plant to use both fertiliser residues and freshly-applied fertiliser more effectively, thus reducing requirements (Abbott and Robson 1982, 1991; Bolan *et al.* 1983). Different species of the fungi have different effects in increasing uptake. Glasshouse experiments have shown that when the roots of subterranean clover are infected with appropriate mycorrhizal species, fertiliser requirements can be halved in soils with low to deficient phosphorus levels (A.D. Robson personal communication).

A range of fungal species occurs in WA soils, including indigenous species (Abbott and Robson 1991). However, when the more desirable mycorrhizal species have been inoculated into a field soil they fail to overcome competition from the established species and so do not establish in significant numbers in the roots of agricultural plants (Abbott and Robson 1982). Consequently it has not been possible to exploit mycorrhizal fungi to increase the effectiveness with which agricultural plants absorb P.

PHOSPHORUS IN THE SOIL

In the year of application, less than 5% of the total P applied as superphosphate or ammonium phosphate fertiliser is absorbed by plants. This is because the water-soluble P that moves from the fertiliser into the wet soil reacts rapidly with other elements, principally calcium, iron and aluminium in the soil solution (precipitation) and on the surfaces of soil constituents (adsorption), to form a range of insoluble compounds (Barrow 1980). Both precipitation and adsorption occur concurrently in the soil and are collectively called sorption. The reactions continue, even in dry soil, to form more stable compounds that are less soluble and therefore less available to plants (Barrow 1980).

Phosphorus is adsorbed by iron and aluminium atoms exposed at the edges of crystals of iron and aluminium oxides in the soil or on the edges of clay mineral crystals (Bowden *et al.* 1980). Phosphorus is generally more strongly adsorbed as soil pH decreases.

When soils are newly cleared, P is initially adsorbed onto the most reactive sites which bind it strongly. As more P is added, the most reactive sites for adsorption become saturated, so P is adsorbed onto less reactive sites where it is more available to plants. However, cultivating the soil (ploughing, scarifying, or sowing crops with tined equipment) incorporates the P and exposes it to soil whose adsorption sites are less saturated. Consequently, cultivation usually increases P adsorption and reduces its availability for plant growth (Williams and Simpson 1965; Bolland and Gilkes 1990b). The top 10 cm of the soil is normally

* *Agriculture Western Australia, Bunbury.*

cultivated, so as fertiliser is applied over the years the adsorption sites in this layer become saturated, subsequent applications of P are less strongly sorbed and are more available to plants.

Phosphorus is strongly adsorbed by the soil particles so there is very little present in the soil solution. Therefore very little P is moved over large distances in the soil solution, either by diffusion or as water is drawn towards plant roots because of the lower potential induced by transpiration. Consequently absorption of P by plant roots is very dependent on interception by the roots as they grow. That is why fertilisers, such as superphosphate or ammonium phosphate, are drilled with the seed when sowing crops. Following rain the water soluble P in the fertiliser dissolves rapidly and almost completely and moves into the soil solution. The seedling roots grow into soil with plenty of water soluble P and the plants absorb most of their P at this time. Similarly for pasture, fertilisers are applied to the soil surface, close to the start of the growing season, so that plant roots have a better chance of taking up the water-soluble P dissolved from the fertiliser granules.

PHOSPHORUS DEFICIENCY AND TOXICITY

Deficiency

Acute deficiency occurs in newly cleared soils. However, most soils used for agriculture are 'old land' soils and have received fertiliser applied over many years. Fertiliser P has a good residual value, so that these days P is not a major deficiency.

The 'old land' soils most likely to develop deficiency if fertiliser is not applied regularly are the high P sorbing soils that are managed intensively for agricultural production, e.g. for dairy, or continuous cropping in higher rainfall areas where large cereal yields (6 t/ha) or lupin yields (3 t/ha) are obtained. If fertiliser P is not applied for 5 to 10 years, deficiency is likely if most of the P is removed in produce, by leaching or by sorption reactions which continue even in the dry state. Sorption reactions continually produce compounds that are more and more stable and they release less P for plant absorption.

Toxicity

Drilling 150 to 200 kg/ha single superphosphate (i.e. 15 to 20 kg P/ha) with wheat seed on very sandy soils in the Quairading area results in toxicity, particularly during the early stages of growth (Loneragan *et al.* 1966). This is because the sandy soils are able to sorb only a small proportion of the water-soluble P and the proportionally higher amounts of P in soil solution are toxic to the plants.

Narrow-leafed lupins (*Lupinus angustifolius*), which are grown widely on these sandy soils, are more susceptible to P toxicity than wheat. With lupins, 100 kg/ha single superphosphate drilled with the seed can cause toxicity, resulting in lower plant densities and grain yield losses of up to 30% (Jarvis and Bolland 1991). Placing superphosphate on the soil surface before sowing lupins avoids toxicity but the effectiveness of the fertiliser is often reduced (Jarvis and Bolland 1991).

Placing superphosphate 5 to 9 cm below the seed sown at 5 cm not only reduces toxicity problems but the effectiveness of the fertiliser is generally improved (Jarvis and Bolland 1990, 1991). Subsequent experiments on very sandy soils (Phosphorus Retention Index <2 mL/g, see later) in the >350 mm rainfall areas have shown that banding the superphosphate below the lupin seed is more effective than applying the fertiliser to the surface (Bolland and Jarvis 1996). Phosphorus leached in these soils, and in time enough fertiliser was leached below 10 cm so there was no yield advantage from banding below the seed. Applying P fertiliser to the soil surface was then as effective as banding.

When growing narrow-leafed lupins on sandy soils, superphosphate should be banded below the seed if the total fertiliser applied in previous years is <250 kg P/ha (~2.5 t single superphosphate/ha) and the Colwell soil test P (see later) at 10 to 20 cm is <16μg P per/g soil. Surface application is as effective as banding on soils with a fertiliser history of >250 kg P/ha and a Colwell soil test of >16μg P/g soil.

TREATMENT OF DEFICIENCY

Effectiveness of phosphatic fertilisers

Plant roots take up P from the soil solution as water-soluble orthophosphate (PO_4^{3-}). Consequently fertilisers in which most (>80%) of the P is water-soluble are the most effective for plant production (Barrow 1980). These fertilisers are made by adding strong acid, such as sulphuric acid or phosphoric acid, to rock phosphate to produce single, double and triple superphosphate, or by adding ammonia to phosphoric acid to produce the ammonium phosphate fertilisers such as mono-ammonium phosphate (MAP) or di-ammonium phosphate (DAP). In WA, local ammonium phosphate fertilisers are made by adding ammonium sulphate to phosphoric acid so these fertilisers contain sulphur as well as N and P (e.g. Agras No. 1, Table 6.3.1).

Single superphosphate has been the most widely used fertiliser. However, the more concentrated double and triple superphosphates and ammonium phosphate fertilisers that supply both N and P are being used more widely now. Concentrations of P and other nutrient elements are listed for these fertilisers in Table 6.3.1. Double and triple superphosphate contain about twice as much P as single superphosphate, so they are cheaper to transport and spread. The various superphosphate and ammonium phosphate fertilisers are about equally effective per unit of P.

Rock phosphate fertilisers

The superphosphate and ammonium phosphate fertilisers are much more effective than rock phosphates in most WA soils, therefore rock phosphate fertilisers are not used widely. The rock phosphates contain no plant available (i.e. water-soluble) P and to produce water-soluble PO_4^{3-} special conditions are required:

- The soil needs to provide a large supply of hydrogen ions to dissolve the rock phosphates. The hydrogen ions are present on soil constituents and in the soil solution (pH) and are measured by titrating the soil with alkali (Gilkes and Bolland 1990). Extensive research in south-western Australia has shown that most soils do not possess enough hydrogen ions to dissolve the rock phosphate (Bolland and Gilkes 1990a).
- Moist soil throughout the growing season, because dry soil disrupts rock phosphate dissolution (Bolland and Gilkes 1990a). Most soils in areas receiving <800 mm average annual rainfall dry out during the growing season between rainfall events, either because of the low rainfall and/or the low water storage capacity.
- Rock phosphate dissolution is enhanced if the products, P and Ca, are removed from solution. Rock phosphates dissolve to appreciable extents in soils with very large capacities to sorb both P and Ca (Khasawneh and Doll 1978).

In summary, rock phosphates are effective fertilisers in soils containing large numbers of hydrogen ions and large capacities to adsorb P and Ca and which remain wet throughout the growing season. The organic sands in south-western Australia, where the average annual rainfall is >800 mm, meet these requirements, though P and Ca are removed by leaching rather than sorption.

Laboratory studies have shown that the organic sands will extensively dissolve about 70% of the reactive apatite rock phosphates (Hughes and Gilkes 1994). Pot experiments using these soils with no leaching have shown that reactive apatite rock phosphate from North Carolina is about 80% as effective as superphosphate (Bolland 1994). Field experiments have likewise confirmed that rock phosphates are equally or more effective than superphosphate (Fitzpatrick 1961; Yeates *et al.* 1984, 1986; Yeates and Clarke 1993; Bolland *et al.* 1995a). This is because the rock phosphate is extensively dissolved in the organic sands and the P and Ca that dissolve from the rock phosphate are leached rapidly. Another reason is that the water-soluble P from superphosphate is leached rapidly (Bolland *et al.* 1995a). Therefore, rock phosphates are relatively more effective than superphosphate because most of the P from superphosphate is leached, whereas rock phosphates dissolve in the soil and maintain the P near the plant roots.

Models to determine fertiliser rates

Many P fertiliser experiments have been undertaken by Agriculture WA. Some trials are unpublished, but most of the data have been incorporated into the 'NP-Decide' model (Burgess 1988; Bowden 1989) which is used to help provide fertiliser advice. NP-Decide considers soil properties, environmental factors, management systems, plant species and economic factors. Using the model, farmers and advisers can see the effect on profitability when one or more factors is altered (e.g. cost of fertiliser, price for grain, assuming different residual values of fertilisers etc.).

Pastures in high rainfall (>700 mm) areas generally require regular applications of P, S and K. The PHOSUL-K model (Yeates *et al.* 1991) has been developed to provide fertiliser advice for these pastures using data generated from a large number of field experiments. The P component is the same as that developed for the original 'Decide' model and later, for the NP-Decide model used for the lower rainfall areas.

An alternative approach is the 'P budget', but it can only be used for soils with a long history of P fertilisation. Fertiliser P is only applied to maintain the current status of the soil; to replace P removed in agricultural produce, or lost through other means; P strongly sorbed by the soil or incorporated into very stable organic compounds, or lost by

Table 6.3.1 Nutrient element contents of some phosphate fertilisers used in Western Australia.

Fertiliser	Phosphorus %	Sulphur %	Calcium %	Nitrogen %
Single superphosphate	9.1	11.5	20.0	0
Double superphosphate	17.5	3.5	16.0	0
Triple superphosphate	20.0	0-1.5	15.0	0
Di-ammonium phosphate (DAP)	20.0	0-1.0	Nil	17.5
Mono-ammonium phosphate (MAP)	22.6	0-1.0	Nil	12.0
Agras No. 1	7.6	17.0	Nil	17.5
Summit Topyield	20.0	1.7	Nil	18.0
Summit Pasture	18.6	10.0	14.0	0

leaching or erosion. However, all losses from the system need to be estimated, requiring accurate records of produce exported from the farm and their P content, and estimates of other losses, many of which are difficult to quantify for the range of soils and environments encountered. The budget approach does not consider economic factors, consequently is not widely used compared with models such as NP-Decide which consider economic and biological factors in a flexible and interactive way.

Applying phosphorus

In most soils, P is not mobile and stays close to its point of application. The P is only moved through the soil by cultivation and seeding operations. The most effective method of applying P fertilisers to crops is to place the fertiliser in a band with the seed (drilling the fertiliser) so that the seedling roots are likely to absorb more P and proportionally less is adsorbed by the soil.

Phosphorus fertilisers are usually applied to the surface of pastures (i.e. topdressed or broadcast), because these are usually not cultivated. Applying fertiliser to the surface is a straightforward, but inefficient, practice because surface-applied P is positionally unavailable to plant roots and the surface soil is most prone to drying out. Phosphorus drilled with the seed is more likely to be in moist soil for longer and is positionally available to plant roots, making it more effective.

RESIDUAL EFFECTIVENESS

Fertiliser P applied to the soil is not only effective for plant production in the year of application, but the residues also provide water-soluble PO_4^{3-} for plant uptake (Barrow 1980). The residues include P left in old fertiliser granules, P sorbed by the soil, P incorporated into plant material, grazing animals and soil flora and fauna and the P which is returned to the soil as dead plant and animal remains, urine and faeces.

Phosphorus is a major deficiency for newly-cleared soils in WA, but after regular fertiliser application over many years, the residues eventually provide enough P for plants and it is then no longer a major deficiency (Barrow 1980). Subsequently, only enough fertiliser needs to be applied to maintain the status of the soil, which is called the maintenance dressing or maintenance application. Maintenance dressings are required to replace P removed in agricultural products, P which is strongly sorbed by the soil or incorporated into stable organic compounds and is therefore no longer available to plants, and P lost by leaching or soil erosion.

Leaching and eutrophication

Sands with very low clay (<5%), iron and aluminium oxide contents have very low capacities to adsorb P. In WA, applied P has been found to leach from these soils in the >450 mm rainfall areas (Ozanne *et al.* 1961). In the high rainfall (>800 mm) coastal areas, most agricultural production is from subterranean clover-based pastures (dairy, meat, wool, hides, hay). Phosphorus from these pastures is known to enter the many natural and man-made drains and contribute to eutrophication of the water bodies (Hodgkin and Hamilton 1993; Section 7.3).

Other soils in these high rainfall areas include very shallow sandy soils (<10 cm) over slowly permeable subsoils (clay or lateritic ironstone gravel), clayey and loamy soils. These soils are usually waterlogged for most of the growing season, and P, either in solution or sorbed onto soil constituents, is carried into the drains and water bodies (Weaver *et al.* 1994). Despite the losses, fertiliser P has a good residual value (Barrow 1966; Bolland *et al.* 1995a). Therefore, although most of the soils were very deficient when newly cleared, the long history of superphosphate applications has resulted in an increase in the P status and they are no longer deficient.

To reduce eutrophication while maintaining profitable pasture production, farmers are encouraged to apply only enough fertiliser to maintain the P status of the soil. The PHOSUL-K model (Yeates *et al.* 1991) is used to help determine the maintenance level required. Alternative fertilisers containing low amounts of total and water-soluble P are being developed for these soils (e.g. Yeates *et al.* 1984; Bolland *et al.* 1995a). However, reduced applications are unlikely to greatly reduce the amount being lost to waterways. This is because the residual P from fertilisers applied previously provides most of the P that contributes to eutrophication. Transfer of residual P from the pastures into the waterways needs to be reduced. One way is to grow a buffer of trees and shrubs between the waterways and the pasture, and fence this buffer zone to prevent grazing. These buffer zones greatly reduce the amount of P from fertilised soils entering waterways (e.g. Weaver *et al.* 1994; Section 7.3).

PADDOCK ASSESSMENT

The P requirement of a paddock can be monitored using plant symptoms, tissue testing, soil testing or can be inferred from the fertiliser history. Soil testing is the best method currently available. The capacity of the soil to adsorb P also needs to be considered, so routine procedures to estimate P sorption are discussed.

Plant symptoms

Phosphorus deficiency affects early growth and so deficient plants remain very small as the season progresses. Therefore, deficiency symptoms become more obvious as the season progresses. Phosphorus is a mobile nutrient in the plant, consequently symptoms become obvious in the older leaves as P moves from old to new tissue. The deficient plants have few, small leaves and the older leaves die and wither away. Acutely deficient plants produce little or no seed or grain.

Deficient subterranean clover plants remain very small and dark green and often develop purple-red stems and leaf margins. Severely deficient lupin plants remain very small and darker green than healthy plants. The older leaves dry, and starting at the leaf tip, the leaves twist and eventually drop off, so deficient plants have very few, young small dark green leaves. Wheat plants also remain very small, with reddening of stems and dead lower, older leaves. The tips of the older leaves become yellow, followed by progressive death of the leaf. Deficient plants are small with few, small young leaves with yellow tips.

Tissue tests

Tissue testing is used in two ways:

- Diagnostically, to provide an assessment of the current P status of plants (i.e. at the time of sampling);
- Prognostically, to predict future yields based on the concentration of P in the tissue at the time of sampling. *Tissue testing for P has generally not been a successful predictive (prognostic) tool in south-western Australia* (Gartrell and Bolland 1987; Bolland 1991).

In WA, commercial laboratories use dried, whole plant tops for testing, so these have been used in research studies (Bolland 1991; Bolland and Paynter 1994; Bolland *et al.* 1995b).

At any sampling time (diagnostic), there is usually a good relation between yield and P concentration. The calibration is similar for different fertiliser types (e.g. freshly-applied fertiliser and residues of superphosphate or rock phosphate). It is also independent of the method of applying the fertiliser (drilled with the seed or applied to the soil surface), but it usually varies for different plant species. The major problem with tissue testing is that the concentration of P in dried tops usually decreases with increasing maturity of annual plants and the rate of this decrease varies from year to year depending on seasonal conditions and other factors.

The concentration of nutrient elements in plant tissue cannot be used alone to determine how much fertiliser to apply to agricultural or horticultural crops, or how to accurately predict plant responses to fertiliser applications. For both diagnostic and predictive purposes, the tissue-test calibration is affected by many non-nutritional factors and interactions (Mengel and Kirkby 1987; Walworth and Sumner 1988). For prognostic purposes, the calibration varies for different fertilisers, sampling times within and between years and plant species.

Tissue testing can indicate whether there is a gross deficiency or luxury supply of P, which indicates the need at the start of the next growing season. It is usually too late to get an economic yield response to P fertiliser applications by the time tissue testing indicates a deficiency, because economic responses occur early in the growth of annual plants. Another problem is that advice is sought for a single or a small number of test values from a crop or pasture in a particular paddock. The difficulty is in deciding the appropriate calibration, relating yield to tissue test values, considering the soil, environment, cultivar and plant species and stage of development. It is therefore difficult to interpret the tissue test value correctly. Current research may provide better tests, involving the selection of specific leaves for analysis.

Soil tests

In order to calculate the amount of P fertiliser to apply to the next crop or pasture, there are three steps:

1. Define the relationship between yield and the level of P applied (*yield response curve*; Figure 6.3.1) for a defined soil type, environment and enterprise. The yield response curve differs for different soils and is largely affected by the capacity to adsorb P. For example, in Figure 6.3.1 soil 1 has a lower capacity to adsorb P than soil 2 (see *Phosphorus sorption*, later). Relative yields are used rather than absolute yields to reduce the effect of seasonal variation.
2. Determine (for each paddock) the yield with no additional P, or the *current P status*. Soil testing, though not perfect, is the simplest and most convenient way of estimating the current P status (Thomas and Peaslee 1973; Fixen and Grove 1991). To do this it is necessary to define the relationship between yield and the concentration of P in the soil (soil test P), which is called the *soil test calibration* (Figure 6.3.2, Fixen and Grove 1991). The calibrations need to be determined for particular soil types, environments and enterprises,

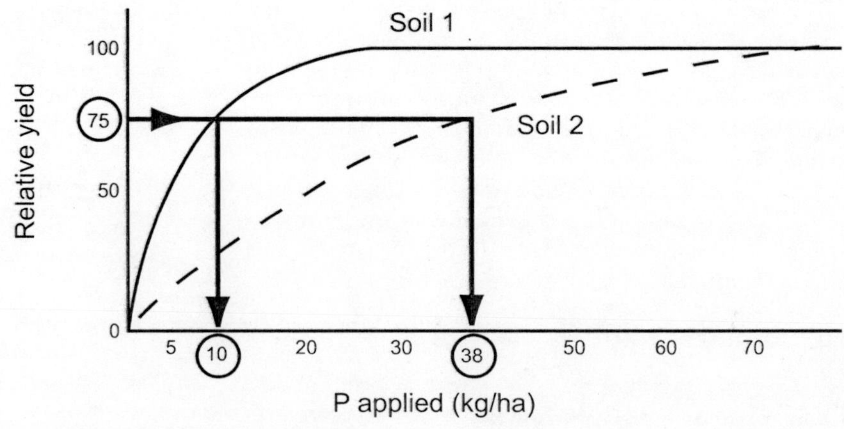

Figure 6.3.1 *The yield response curves to applied phosphate of two soils with different adsorption capacities.*

as for the yield response curves. Soil samples are collected in summer (November to March, but usually January to February), and the concentration of P in the soil is related to plant yields measured later that year.

Both the yield response and soil test calibration curves are determined using simple field experiments lasting one year. They are then repeated on many sites and in different years. In this way both curves provide robust yield responses and soil test calibration curves, covering different seasons and soil types.

3. The soil test calibration (Figure 6.3.2) is used to estimate the percentage of the maximum (relative) yield produced by the amount of P alreadypresent in the soil. This yield is then transferred to the appropriate yield response curve (Figure 6.3.1). It is then possible to assess the likelihood of obtaining a yield response to freshly-applied fertiliser. If the test suggests that without additional P fertiliser, the yield will be only 10% of the potential, then there is a very high likelihood of a yield response. On the other hand, if the soil test value suggests that the soil P will produce yields approaching the maximum, then there is a very low chance of obtaining a yield response to extra fertiliser.

The Colwell (1963) sodium bicarbonate soil test procedure is used in WA. There are numerous other procedures, but WA researchers have found that all of the test reagents studied predict yields equally well (Bolland *et al.* 1988, Kumar *et al.* 1994). Each reagent produces different values and has a different soil test calibration curve, but they predict similar yields for the P already present. Consequently there is no need to change the testing procedures (Bolland and Wilson 1994b).

Research has shown that when a single cultivar is grown at one site, the soil test calibrations differ between years and different fertiliser types (Bolland *et al.* 1989a; Bolland and Gilkes 1992). Calibrations also differ between plant species (Bolland *et al.* 1994b) because:

- Soil testing reagents extract all the P in the soil solution together with a proportion of that present in the fertiliser residues in the soil. The amount extracted from the residues varies depending on seasonal conditions (e.g. summer rainfall, temperature before collecting the soil samples).
- Plant production and how effectively the plants use the P from the residues, vary with seasonal conditions.
- Soil test P, measured on samples collected in January to February, may not exactly match the amount actually released from the residues and used by the plants later in the year.
- Different fertiliser types leave different residues in the soil. These differ in their effectiveness for plant growth and in their solubility in soil test reagents.

For these reasons, plus some other unknown factors, soil test calibrations for P only provide a crude estimate of the current P status of soils (Gartrell and Bolland 1987; Bolland *et al.* 1989a). However, there is no suitable alternative to soil testing, because alternative methods require more complex experiments which are costly, and as a result can only be replicated at a few sites. The results of all field experiments, whether simple (i.e. soil testing and yield response curves) or complex (i.e. measuring the residual value of fertilisers), are management and site specific, and the results are confined to the seasonal conditions experienced during the experiments.

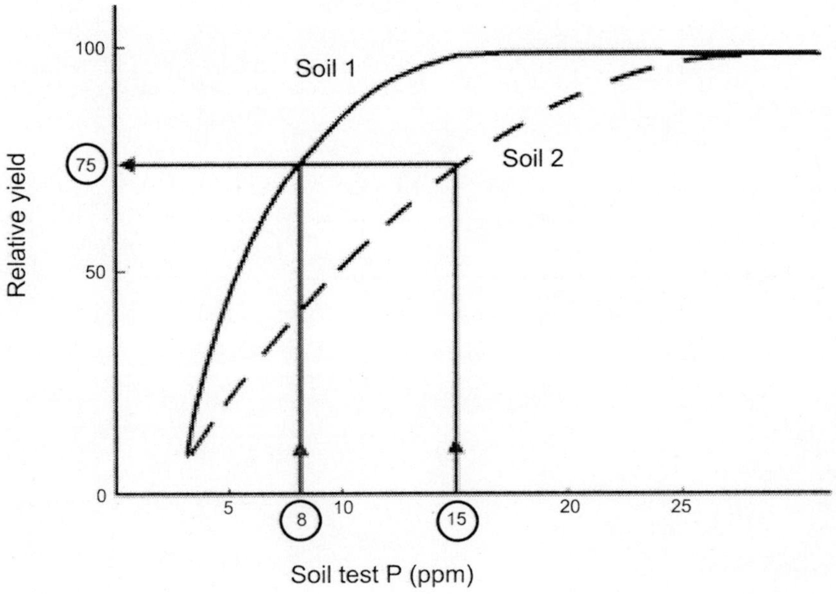

Figure 6.3.2 *The soil test phosphate calibration curves of two soils with differing abilities to adsorb phosphate.*

Phosphorus sorption

The capacity of the soil to sorb P greatly influences both the yield response and soil test calibration curves. For the yield response curve, as the capacity of the soil to sorb P increases, so more fertiliser needs to be applied to produce the same yield. For the soil test calibration, as the capacity of the soil to sorb P increases, a higher concentration of P in the soil (soil test P) is required to produce the same yield.

The Ozanne and Shaw (1967) procedure (see below) is used widely in WA to study P sorption by soils. The amount sorbed is the difference between the initial amount added and the concentration remaining in the equilibrium solution. The relationship between the amount of P adsorbed by the soil and the concentration in the equilibrium solution is then plotted (*phosphorus sorption* isotherm; Figure 6.3.3), from which the following two values are determined:

(a) When the equilibrium solution contains 0.3 μg P/mL, the amount of P sorbed by the soil is called the P adsorption (Figure 6.3.3). A concentration of 0.3 μg P/mL was selected from research done by Asher and Loneragan (1967), because this concentration of P in solution is believed to be non-limiting for pasture production in WA (Ozanne and Shaw 1967).

(b) The amounts of P sorbed by the soil when the concentration of P in the equilibrium solution is raised from 0.25 to 0.35 μg P/mL. The difference between these amounts is called the P buffer capacity (PBC) of the soil (Ozanne and Shaw 1967).

> *P sorption* (Ozanne and Shaw 1967). In this method, 2.5 g of soil is shaken for 16 hours with 50 mL 0.01M $CaCl_2$ to which one of at least five levels of P, ranging from 0 to 25 μg P/mL, is added so that the concentration after 16 hours in the solution (called the equilibrium solution) is in the range of 0 to 1 μg P/mL. The soil and solution are separated by centrifuge, and the P concentration in the equilibrium solution is measured colormetrically by the Murphy and Riley (1962) procedure.

Phosphorus sorption isotherms are affected by fertiliser history. As the amount of applied fertiliser increases, so the isotherm moves to the right and downwards. This changes the value of P adsorption, but the PBC is unaffected (Figure 6.3.3). Consequently, PBC can be regarded as an intrinsic soil property and has been used in research studies to measure the capacity of soils to sorb P. However, PBC is not a straightforward measure, so two other methods, reactive iron and Phosphorus Retention Index (PRI), are used.

Reactive iron is the amount of iron extracted from the soil by ammonium oxalate (Schwertmann 1964; McKeague and Day 1966). It is usually reasonably well correlated with the P sorption capacity of the soil (Singh and Gilkes 1991; Brennan *et al.* 1994). The correlation is positive so that the trend is for P sorption to increase as the amount of iron that is soluble in ammonium oxalate increases. Although WA soils contain less aluminium that is soluble in ammonium oxalate than iron, it seems that the aluminium plays a more important role in determining P sorption (Singh and Gilkes 1991; Brennan *et al.* 1994).

PRI (Allen and Jeffery 1990) is an estimate of the slope of the P sorption isotherm and is similar to indices of P sorption suggested by Bache and Williams (1971). The PRI has been found to be reasonably well correlated with PBC (Allen and Jeffery 1990; Brennan *et al.* 1994; Bolland *et al.* 1994c).

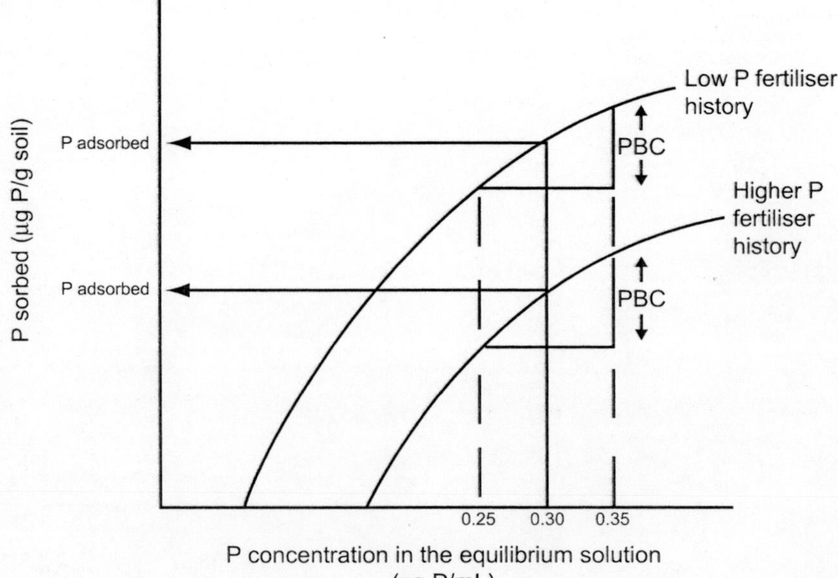

Figure 6.3.3 *Phosphorus sorption isotherm, showing how PBC (P buffer capacity) and P adsorption are derived.*

FACTORS TO CONSIDER WHEN COLLECTING SOIL SAMPLES FOR ANALYSIS

Temporal variation. Both field and controlled experiments have shown that between March and November there is no appreciable variation in soil test results due to sampling time (Bolland 1995).

Spatial variability. Soil tests vary considerably due to spatial variation even where the soils appear relatively uniform (Bolland and Wilson 1994a). This is due to the natural variations in physical, chemical and mineralogical properties and uneven distribution of applied fertiliser. Adequate replication is essential to obtain meaningful results, and a bulked sample consisting of at least 30 individual soil samples is recommended for each paddock.

Sampling depth. The standard depth for soil sampling is 10 cm. Failing to sample to this depth will result in higher results, especially in soils with a high P sorption. The P will tend to be concentrated near the surface in pasture paddocks and paddocks that are cropped using direct-drill or no-till methods (Bolland 1992).

Mechanical soil testing. A 'rotary blade soil sampler' was developed by M.J. Baker to sample hardsetting soils on which it is difficult to sample to 10 cm using a conventional 'pogo' sampler. The samples from these soils are frequently 'shallow' (<10 cm) with few replicates, resulting in an unrepresentative sample. The mechanical sampler uses a rotating blade to cut a 2 cm wide slot in the soil with many samples taken per paddock and then bulked for analysis. The soil is sub-sampled for testing and is more reliable than conventional methods on hardsetting soils (Bolland *et al.* 1994a).

Reactive iron. The soil is shaken with 0.3M ammonium oxalate at pH 3.25 for two hours at 23°C using a soil to solution ratio of 1:33.3. After centrifuging, the concentration of iron in the equilibrium solution is measured by atomic absorption spectrophotometry.

PRI (Allen and Jeffery 1990) is measured by shaking 5 g of soil with 100 mL of 0.02M KCl containing 10 µg P/mL. Thus only one level of P is added to the suspension. The PRI (mL P/g soil) is the amount sorbed (µg P/g soil) divided by the concentration of P in the equilibrium solution (µg P/mL).

Further reading

General

Khasawneh, F.E., Sample, E.C. and Kampratt, E.J. (eds.) (1980). 'The role of phosphorous in agriculture'. American Society of Agronomy, Madison.

Rock phosphate

Khasawneh, F.E. and Doll, E.C. (1978). The use of phosphate rock for direct application to soils. *Advances in Agronomy* **30:** 159-206.

Soil and tissue testing

Westermann, R.L. (ed.) (1991). 'Soil testing and plant analysis'. Book Series No. 3, Soil Science Society of America, Madison.

6.4 POTASSIUM

Noeleen Edwards[*]

Potassium (K) is the seventh most abundant element in the earth's crust and is found in a number of primary minerals, especially feldspars and micas, as well as soil clay minerals. It is regarded as the third major plant nutrient after nitrogen and phosphorus.

In Western Australia, many of the sandy-surfaced soils have low levels of extractable K levels in the virgin state and these levels have generally declined since clearing due to management practices. Potassium fertilisers were first required for horticultural crops and for pastures on dairy farms. As cropping has intensified on the coarse-textured soils of the wheatbelt, deficiency has become more common in crops and pastures. Sandplain soils on the west and south coasts and the coarse-textured duplex soils are most likely to develop deficiency.

POTASSIUM IN PLANTS

Potassium was first identified as important for plant growth by Justus von Liebig in the 1840s. Potassium uptake is highly selective and closely coupled to metabolic activity. It is required in many of the plant's physiological processes, but does not form part of the plant's structure, so relatively large amounts are needed for plant growth compared with the amount exported from the farm in produce. It is used in photosynthesis, transport of sugars, enzyme activation, maintenance of plant turgor and regulation of stomata. Potassium is the most abundant cation in the cytoplasm and is very mobile within plants at all levels - within individual cells, tissues and in long-distance transport. Plants deficient in K cannot use other nutrients and water efficiently. They are less tolerant of stresses such as drought and waterlogging, and also less resistant to pests and diseases.

Subterranean clover is relatively more sensitive to deficiency than the deeper rooted legumes (i.e. narrow-leafed lupins, serradella, lucerne, sandplain lupins), but cereals and grasses are generally less sensitive (Table 6.4.1), apart from barley and veldt grass (Asher 1964). In solution culture, the K concentration needed for subterranean clover to attain 50% of maximum growth was about twice that needed by annual ryegrass, silver grass and ripgut brome (Asher 1964).

This sensitivity is partly due to the internal requirement and partly to the ability of the root system to obtain K from the subsoil. Legumes with the ability to develop a deep root system (e.g. lupins) are able to grow satisfactorily on deep sands, but subterranean clover on the same soils is severely deficient (Toms and Fitzpatrick 1961). Pasture species grown on deep sand took up most of their K requirements from the surface soil, but when K was supplied at depth, those species with deeper roots obtained a greater proportion from depth than the more shallow-rooted species (Ozanne *et al.* 1965). In soils with a large subsoil supply, spring wheat obtained 34% of its total K requirements from the subsoil (Kuhlmann 1990).

Table 6.4.1 Relative susceptibility of common crop and pasture species to low potassium.

High susceptibility	Moderate susceptibility	Low susceptibility
Medic species	Serradella	Wheat
Subterranean clover	Narrow-leafed lupins	Oats
	Canola	Annual ryegrass
	Barley	Silver grass
		Capeweed
		Erodium
		Brome grass

POTASSIUM IN THE SOIL

The different forms of K in the soil have varying degrees of availability to plants. Potassium occurs in primary minerals, clay and fine minerals, soluble forms, and crop and microbial residues. Water soluble and exchangeable K are the forms usually extracted in soil testing. The respective components are:

Weathering of primary minerals ⇒ K non-exchangeable ⇔ K exchangeable ⇔K in soil solution ⇔ K in plant

A potassium cycle is illustrated in Figure 6.4.1.

Most of the total K in soils is in the mineral form and is assumed to be only slowly available to plants. It occurs as a structural component of primary minerals such as micas (biotite, muscovite) and K-bearing feldspars ($KAlSi_3O_8$ orthoclase, microcline). Mineral K becomes available as these minerals weather or decompose. Feldspars weather more readily than micas so they are an important source of K. Feldspars are abundant in acid igneous rocks such as granite which underlies much of south-western Australia, but the soils are highly weathered (e.g. by laterisation) and so in most soils the only unweathered primary mineral left is quartz, because it is highly resistant to weathering.

[*] *Agriculture Western Australia, South Perth.*

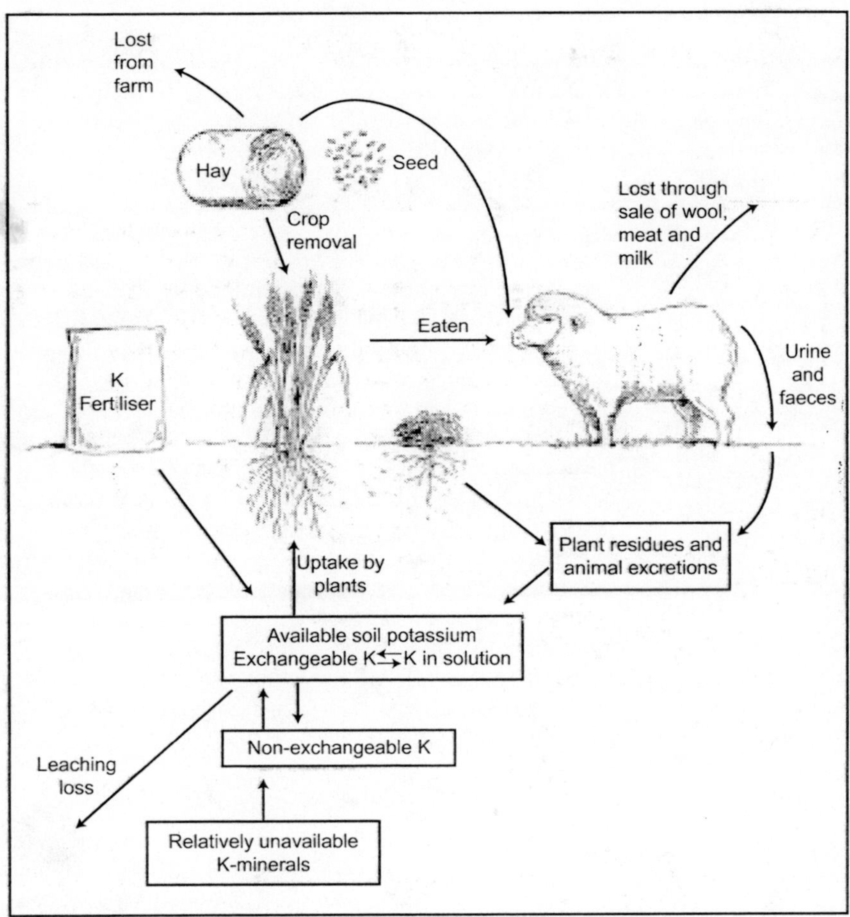

Figure 6.4.1 *Potassium cycle under normal farming conditions (Moreschi 1991).*

Soils formed on rock weathered *in situ* (i.e. acid igneous, metamorphic and basic rocks) may have adequate reserves of K.

Non-exchangeable K is unlike mineral K in that it is not bonded covalently within the crystal structures of the soil mineral particles. It is held between adjacent tetrahedral layers of micas, vermiculites and intergrade clay minerals. Non-exchangeable K is released to the exchangeable form when levels of exchangeable and soil solution K are decreased by crop removal and leaching. Soils that contain significant amounts of illite usually have satisfactory K levels as it is the dominant cation between layers. As soils age under leaching, illite disappears and kaolinite dominates, as in the highly weathered soils. Relatively few pure kaolins exist in nature, with K-containing impurities such as micaceous and smectitic layers being present (Sparks 1987). Singh and Gilkes (1990) have shown that the clay fraction of WA soils is very finely divided and poorly crystalline. This produces clay with a relatively large surface area and high cation exchange capacity.

Soil solution K is the form absorbed directly by plants and is subject to leaching. The concentration in the soil solution fluctuates greatly, but is usually low. Soil solution K peaks in late autumn in the surface horizons because it is released from organic matter and microbial tissue (Roberts 1968). Levels of soil solution K are determined by the equilibria and kinetic reactions between the other forms of K, soil moisture, the divalent ion content in solution and the exchange phase (Sparks and Huang 1985).

Exchangeable K is held by the negative charges of organic matter and clay minerals. It is easily exchanged with other cations and readily available to plants. As it is absorbed, it is replenished from non-exchangeable K held in clay particles.

In coarse-textured soils, exchangeable K is mainly associated with organic matter, therefore the K content decreases with depth. Most of these soils are deficient in K or have the potential for deficiency to develop, but this is not universally true because some sandy soils contain unweathered primary minerals in the sand and silt fractions (e.g. Busselton sand which contains 7% feldspars, Drover 1961). Most deep sands in WA are low in total K (<0.02%) as they are highly weathered and the only primary mineral they contain is quartz.

In medium and fine-textured soils, the exchangeable and non-exchangeable K is mainly associated with the clay fraction. Potassium content tends to increase with depth, because in most soils the clay content increases down the profile. There are detailed reviews of K in soils by Sparks and Huang (1985) and Sparks (1987).

DISTRIBUTION OF POTASSIUM DEFICIENCY

Potassium deficiency has been widespread in annual legume pastures in high rainfall areas of south-western Australia since the 1950s (Fitzpatrick and Dunne 1956; Toms and Fitzpatrick 1961; Cox 1973, 1980). It is only more recently that deficiencies in wheatbelt cereal and lupin crops have been identified (Colwell and Grove 1976; Cox 1978).

The incidence of deficiency is increasing as native supplies of K are exhausted. Leach and Easton (1991) reported that about 30% of soil samples from the Great Southern region (400 to 600 mm) were marginally to highly deficient and that at least 20% of sites sampled would respond to K fertiliser. The development of deficiency is related to soil properties (i.e. parent material, degree of weathering, clay minerals, texture), land use and rainfall.

Soil properties. Soils containing K-bearing primary minerals (e.g. orthoclase, microcline, biotite, muscovite), illite or with a high clay content have good reserves of K. This does not imply that all other soils will be deficient, but they are likely to have lower native K reserves which could be depleted over time, so they have the potential to become deficient. The size of the exchangeable and non-exchangeable K pools will have a large influence on the time it takes for deficiencies to develop. These reserves are mainly related to the soil's parent material, soil texture and the degree of weathering (Table 6.4.2). Uniform coarse-textured soils (or deep sands) are particularly susceptible to deficiency because of their low clay content and because they are mainly formed from highly weathered materials. The development of deficiencies on permeability contrast soils depends on the depth to the clayey subsoil in relation to plant root depth and the supply of K in the subsoil.

Rainfall. The higher the annual rainfall, the more K crops and pastures are likely to need. Therefore deficiencies are observed more frequently in higher rainfall areas. The potential for leaching is also increased in high rainfall areas, especially in highly permeable, uniform coarse-textured soils with a poor ability (low clay and organic matter) to retain K. However, because improved crop management has increased potential yields in the lower rainfall areas of WA, K deficiency has also become a limiting factor on soils with marginal levels of K.

Land use. The amount of K removed in produce will determine how fast the reserves are depleted. Hay removes a large amount, e.g. 45 kg of K in a 5 t oaten hay crop and while much is returned in urine and faeces, this is often not in the paddock where the hay was cut. Therefore, frequent hay removal from a paddock may induce deficiency unless fertiliser is applied or the soil has large K reserves. For crops, most of the K remains in the vegetative parts of the mature plant and K contents in grain are low e.g. 4 kg/t in wheat and 8 kg/t in lupins. Very little K is removed in animal products e.g. <1 kg of K/ha in wool and meat, and <6 kg of K/ha in milk. Most of the ingested K is recycled through urine and faeces. This recycling can lead to transfer of K to stock camps and is a particular problem on dairy farms.

Table 6.4.2 Soil parent materials, exchangeable and total potassium reserves and incidence of deficiency for south-western Australia (McArthur 1991; Toms and Fitzpatrick 1961).

Parent material	Main soils	Typical K levels		Incidence of K deficiency in field
		Total K (%)	Exchangeable K (ppm)	
Acid igneous and metamorphic rocks weathered *in situ* (e.g. granite, gneiss)	Red and brown shallow and deep loamy duplex soils Red loamy earths	0.3-0.6	150-300	Rare
Basic rocks weathered *in situ* (e.g. dolerite, greenstone, gabbro)	Alkaline red shallow and deep loamy duplex soils	0.4-0.6	>300	Rare
Laterite on granite	Sandy gravels Loamy gravels Yellow deep sands	0.02-0.1	40-150	Pale deep sands
Laterite on sedimentary rocks	Sandy gravels Sandy duplex soils Yellow deeps sands Pale deep sands	0.01-0.06	20-120	Pale deep sands and coarse-textured permeability contrast soils
Shallow colluvium derived from laterite overlying truncated lateritic profile (i.e. mottled or pallid zone)	Grey shallow and deep sandy duplex soils	0.02-0.0.8	50-130	Occasional
Alluvium	Various	Reflects properties of source material		
Aeolian	Pale deep sands Yellow deep sands	0.01	15-40	Common
Parna (Aeolian deposits from playas and arid interior)	Calcareous loams	0.5 - 0.8	>300	Nil
Organic soils	Organic (peaty) sands	0.01	30-50	Common

SOIL AND ENVIRONMENTAL FACTORS AFFECTING UPTAKE

The availability of soil K to plants is influenced by many factors, in addition to clay mineralogy. These include particle size, water content, acidity, aeration, organic matter and the presence of other ions (Hosking 1986). Soil moisture content affects the rate of root extension and therefore the ability of roots to reach nutrients, the speed of diffusion of K to the root surface and the total amount in the soil solution. Potassium uptake is an active process, therefore low concentrations of oxygen in waterlogged soils will reduce uptake.

More K becomes available as the soil becomes more acidic and liming reverses this causing some fixation. However, liming can improve root growth and the supply of other nutrients so the net effect is unpredictable (Hosking 1986). The effect of other ions on uptake is not clear. Large quantities of calcium at the root surface are thought to impede K uptake. Soil organic matter makes an important contribution to the CEC of soils and on sandy soils, organic matter can hold most of the exchangeable K.

TREATMENT OF DEFICIENCY

The most commonly used potassium fertiliser in WA is muriate of potash (KCl). Sulphate of potash (K_2SO_4) is more expensive and generally only used on horticultural crops that are sensitive to chloride. The high solubility of these fertilisers can be a disadvantage because they are leached easily and because plants could absorb a luxury supply of K. Pasture plants containing a luxury level of K may present problems to grazing animals. Coating KCl with sulphur has been used experimentally to reduce the rate of dissolution (Saunders *et al.* 1988), but slow release fertilisers have not been successful commercially. The highest requirement for K is in the vegetative growth stage. For crops and pastures, one application of fertiliser is recommended within four to six weeks after seeding or germination to maximise the response obtained. The fertiliser should be topdressed because drilling KCl with the seed can inhibit germination. Split applications may be of value for pastures in the very high rainfall zone where a spring flush of growth can be expected (Cox 1973).

Mobility of applied K

Potassium is a mobile nutrient and leaches in sandy soils (Drover 1961; Roberts 1968). The rate depends on rainfall and the CEC of the soil. Measurements of K in a sandplain soil at Badgingarra have shown that applied K moves slowly down the profile over a number of seasons and that most of the K applied over a four-year period remained within 1 m of the surface (Edwards 1993). While shallow-rooted species would be unable to obtain K that has leached downwards, plants with deeper roots are still able to use it. These species will also recycle some K from deeper in the profile to the surface, where it will be available in the following season after being leached out of the plant residues.

RESIDUAL EFFECTIVENESS

Residual effectiveness of fertiliser is affected by rainfall, soil type and farming system. Total rainfall and its distribution during the year affect leaching. Losses will be highest in the high rainfall regions on sandy soils. The most common K fertiliser used in WA, muriate of potash, is very soluble and susceptible to leaching.

Potassium leached from loams and clays is minimal because K is attached to the clay and organic matter. In soils containing weathered 2:1 clays (e.g. degraded illite or vermiculite), K can be absorbed back into the interlayer spaces. This fixation can occur when soluble K increases in the soil solution and the equilibrium is shifted towards non-exchangeable K. On these soils, large amounts of fertiliser must be applied before soil test values increase. On sands with very low contents of clay and organic matter, K fertiliser can move out of the root zone in one season. Although organic matter can hold relatively large amounts of exchangeable K, this K is easily removed.

The farming system also has a large effect on the residual value of K fertiliser. Recycling will be highest in a grazing system, but the K excreted in urine and dung is returned unevenly within the paddock and is concentrated in stock camps. The high concentration of K in urine patches will make it vulnerable to leaching. Large quantities will be removed from a paddock that is cut for hay or silage. Although the amount of K removed in grain is much smaller than in hay, continuously cropping soils that have marginal levels of K will cause a significant decline in reserves. If the legume phase of a rotation is responsive to K, nitrogen supply may be increased for a following cereal crop. To minimise leaching, recommendations for K fertiliser are at levels to supply only one or two crops. Regular applications are needed on soils with low K levels.

PADDOCK ASSESSMENT

Plant symptoms

Potassium is mobile in plant tissue. In deficient plants it is translocated to new growth, so symptoms first appear in the oldest leaves. Deficiencies are most likely to be seen during periods of maximum growth. Symptoms vary with species and are often not clearly defined. In cereals, yellowing and death of the tips of oldest leaves progresses down the margin to give an arrow of normal tissue. With severe deficiency the whole leaf dies. In narrow-leafed lupins, the oldest leaflets yellow and are shed rapidly, so that the plant has a bunched appearance with leaves at the top of the plant. In subterranean clover, leaf margins become scorched with some fine spotting on the surface. Plants have small leaves and turn a dull yellow. In serradella, the oldest leaves turn yellow, with the leaflets closest to the tip affected first, before being shed (Cox 1980). In grasses, symptoms include necrotic tipping (Asher 1964).

An early indication of deficiency in pastures is a change in composition, with grasses replacing clovers which are then restricted to urine patches. In highly deficient pastures, there is an invasion of undesirable species such as silver grass (Fitzpatrick and Dunne 1956).

Tissue tests

Deficiency symptoms in a crop or pasture can be confirmed by plant analysis. A sample of whole tops is used. Unless the sampling is done very early in the season, it is usually too late to correct a deficiency in that season. Critical levels differ between species, and decline with increasing plant maturity. Prediction of future yield from a tissue test is difficult, because yield is affected by many other factors as well as K concentrations. Tissue testing can also be used to monitor luxury consumption of K when K fertiliser is used. Very high levels of K can cause an imbalance between K, Ca and Mg in the tissue, which in pasture plants may lead to health problems for grazing animals.

Critical tissue levels are:

Subterranean clover - 1.5% K in plants up to 8 weeks and 1.0% in plants older than 10 to 12 weeks (Cox 1973).

Cereals - 2.0% in whole tops before flowering (Yeates 1986).

Narrow-leafed lupins - 3.0% at the 10-leaf stage, 1.5% at first flowering, 1.0% post-flowering (Cox 1978).

Soil tests

The most common test used in WA measures bicarbonate-extractable K. It is a variation of the Colwell phosphorus test, where 1 g of soil is shaken for 16 hours at 23°C with 100 mL of 0.5M $NaHCO_3$ at pH 8.5 (Colwell 1963). Potassium is measured in this extract using atomic absorption spectroscopy with in-line neutralisation of the bicarbonate extract (Jeffery 1982). The soil test is based on a sample taken from the top 10 cm of the soil profile and is useful for identifying deficient areas. Calibration data for predicting yield responses of wheatbelt crops and pastures are limited and calibrations vary between seasons.

On sandplain soils with a medium to high yield potential, economic responses for lupins and cereals are likely if the soil test is less than 30 ppm. On duplex soils, where the depth to clay is more than 30 to 40 cm, responses are likely if the soil test is less than 40 ppm. This apparent difference in the responsiveness of soil types is probably related to the ability of roots to penetrate and explore the subsoil. Responses in productive subterranean clover pastures are likely if the value is less than 60 ppm. Pastures in the high rainfall zone may be responsive at levels above 60 ppm, depending on the level of productivity.

Where plants can obtain substantial amounts of K from the subsoil, the soil test could give an inaccurate estimate of available K. However, it cannot be assumed that a clay subsoil will contain high levels of K. Recent sampling of duplex subsoils in the Great Southern has shown that many have little K at depth, possibly because it has been removed over a long period of cropping and hay cutting. Measuring the K level in the subsoil on texture contrast soils may be necessary to determine the likely fertiliser response.

Further reading

Haynes, R.J. and Williams, P.H. (1993). Nutrient cycling and soil fertility in the grazed pasture ecosystem. *Advances in Agronomy* **49:** 119-199.

Sparks, D.L. (1987). Potassium dynamics in soils. *Advances in Soil Science* **6:** 1-63.

Sparks, D.L. and Huang, P.M. (1985). Physical Chemistry of Soil Potassium. *In* 'Potassium in Agriculture (ed. R.D. Munson). American Society of Agronomy, Crop Science of America, and Soil Science Society of America, Madison, WI, pp. 201-276.

6.5 SULPHUR

*Mel Mason**

Sulphur (S) is an essential element for plants and an important constituent of protein. It is also important for the function of several plant enzyme systems. Plants obtain most of their sulphur from the soil, though in areas close to the sea and industry there can be significant input from the atmosphere.

Sulphur deficiency was very rare in crops and pastures in most agricultural areas of Western Australia. This was due to the almost universal use of single superphosphate as a source of phosphorus for crops and pastures. The 11.5% sulphur in single superphosphate generally satisfied plant requirements and masked potential deficiency. The exception was pastures grown in high rainfall (>600 mm) areas where deficiency tended to show in late winter-spring even after application of single superphosphate in autumn (Barrow 1966, 1967; Glencross and Cox 1969). Sulphur deficiencies are more likely in legume pastures and crops such as canola, which have a high requirement for sulphur, than in pasture grasses and cereal crops.

SULPHUR IN PLANTS

Sulphur is an essential part of certain amino acids such as methionine, cystine and cysteine, which are essential components of most proteins. Consequently, deficiency affects protein production. Sulphur also has a role in producing good quality protein in grain. If S is deficient, the production of sulphur-containing amino acids is reduced, which in turn reduces the quality of the protein, which has a flow-on effect in reducing the quality of flour for bread making (Moss *et al.* 1981; Randall *et al.* 1981; MacRitchie and Gupta 1993).

Sulphur is relatively immobile in the plant and early uptake from a limited soil supply may only be sufficient for the needs of early leaf production, leading to deficiency for new growth. A continuing supply then depends on the rate of mineralisation of soil organic matter and fertiliser additions. Applying S fertilisers results in greener plants, increased growth and photosynthetic area, increased seed yield and better quality seed due to enhanced protein production.

Relative susceptibility to low supply

Sulphur is one of the major plant nutrients and so it is required by crops and pastures in relatively large amounts. It can easily become deficient without a continual supply in an available form. Different plants have different susceptibilities to S deficiency (Table 6.5.1). Legume pastures have a higher requirement than grasses and cereal crops. Canola also has a high requirement for S (Grant 1991). Disturbing or cultivating the soil results in increased mineralisation of soil organic matter (Powlson 1980), with a consequent increase in available sulphate sulphur. This is an extra reason for the greater incidence of S deficiency in undisturbed pastures than in crops.

Sulphur deficiency is becoming more common in WA because of the increased use of low-sulphur fertilisers and increasing area of canola.

Table 6.5.1 Plant differences in susceptibility to a low supply of sulphur.

High susceptibility	Moderate susceptibility	Low susceptibility
Canola	Clovers	Wheat
Lucerne	Grasses	Oats
	Lupins	Barley

SULPHUR IN THE SOIL

Most soil S is contained in organic matter (Evans 1975). The ratio of N:S in soil organic matter is usually about 8:1 or a C:N:S ratio of 140:10:1.3. The plants take up S in the form of inorganic or sulphate sulphur. There is often a limited supply of sulphate in the soil, but this is replenished throughout the season by mineralisation of organic matter. Whether the plants develop deficiency depends on the initial supply of sulphate sulphur and the ability of the soil to replenish that supply through mineralisation, to keep up with the plants' needs. Mineralisation is affected by factors such as moisture and temperature (Swift 1985) and is stimulated by cultivation. Symptoms of S deficiency that sometimes are noticeable early in the season, can often disappear in spring when the rising temperatures result in a flush of mineralisation and an increased supply of sulphate sulphur. Warmer conditions can also cause a rapid increase in root growth, which can often result in roots accessing supplies of sulphate, that were beyond the previous rooting depth.

Sulphate sulphur can occur in the soil as adsorbed sulphate (Williams 1975) or soluble sulphate. Both forms are available to plants, but sulphate that is not adsorbed can be leached readily. The susceptibility of soils to S deficiency will therefore depend on their ability to build up organic matter, adsorb sulphate and resist leaching. Increased soil organic matter under

* *Formerly Agriculture Western Australia, now Mel Mason and Associates, Perth.*

pasture increases S in the soil. Soil sulphate levels are generally high in autumn, low in winter and rise again toward the end of spring (Barrow 1969b). The S cycle is illustrated in Figure 6.5.1.

In WA, soils derived from granite, gneiss, or epidiorite that are in rainfall zones of 625 to 750 mm/annum have little ability to adsorb sulphate. Within this rainfall zone there is a positive correlation between rainfall and the ability to adsorb S (Barrow *et al.* 1969). However, the increased leaching with increasing rainfall, is greater than the extra S received in rainfall accessions, resulting in a negative correlation between rainfall and S accumulation in the soil (Barrow 1969a).

DISTRIBUTION OF SULPHUR DEFICIENCY

Deficiency is most common on soils with a low capacity to adsorb S. This means that most deficiencies occur in high rainfall areas on coarse-textured soils susceptible to leaching, because they have a low clay content and poor water and nutrient-holding capacity (Hingston 1959). Sulphur deficiency is least likely in low rainfall areas on fine-textured soils, where losses are generally confined to that removed in produce.

Many soils are potentially deficient, but until recent years this was masked by the regular use of single superphosphate as a source of phosphorus. This fertiliser contains 11.5% S, which was usually sufficient to prevent S deficiency, especially in the wheatbelt. In recent years, with the use of low S fertilisers such as di-ammonium phosphate (DAP) and triple superphosphate (TSP), S deficiencies are becoming more common. On some soils, such as the loamy earths (Kandosols) and deep or shallow duplex soils (Chromosols and Sodosols) in the Avon and Chapman Valleys, S deficiency readily shows after a few years unless there have been substantial additions from fertiliser (Mason *et al.* 1975). On many duplex soils, although the coarse-textured surface soil may have a low adsorption capacity, the clayey subsoil is often able to adsorb and then supply enough sulphate for the plants' needs depending on the depth of the surface layer.

Sulphur accessions from rainfall vary with location and though very high near highly industrial areas such as in northern Europe and parts of the USA (Evans 1975), contributions are low in most agricultural areas in WA. Accessions are greater close to the sea. Recent figures (M. Bolland unpublished data) indicate levels of 7 to 9 kg/ha/year near Busselton. Inland, S accession is much lower and values of 0.5 to 4 kg sulphate/ha/year were measured by Hingston and Gailitis (1976) from several locations in the wheatbelt.

TREATMENT OF DEFICIENCY

Where deficiency is known, S will generally be applied at sowing in the form of a compound fertiliser, which also supplies the requirements for phosphorus and/or nitrogen. In situations prone to heavy leaching (e.g. coarse-textured soils in high rainfall areas) the soluble sulphate applied early in the season can be leached beyond the root zone with the consequent appearance of S deficiency. Application should be delayed for at least three to four weeks after sowing or a second application made at this time. In pastures in very high rainfall areas, even though S is applied with phosphatic fertilisers at the break of the season, a spring S application should be routine.

If S deficiency is diagnosed early enough in the growing season it can be treated by applying fertilisers, without having a major effect on final yield and the quality of crops and pastures. Soluble sulphate is present in many fertilisers with %S ranging from 10 to 17% (Table 6.3.1). For most S deficient situations the sulphate form in gypsum or ammonium sulphate is used to obtain a quick response, particularly where deficiency has been diagnosed earlier in the growing season.

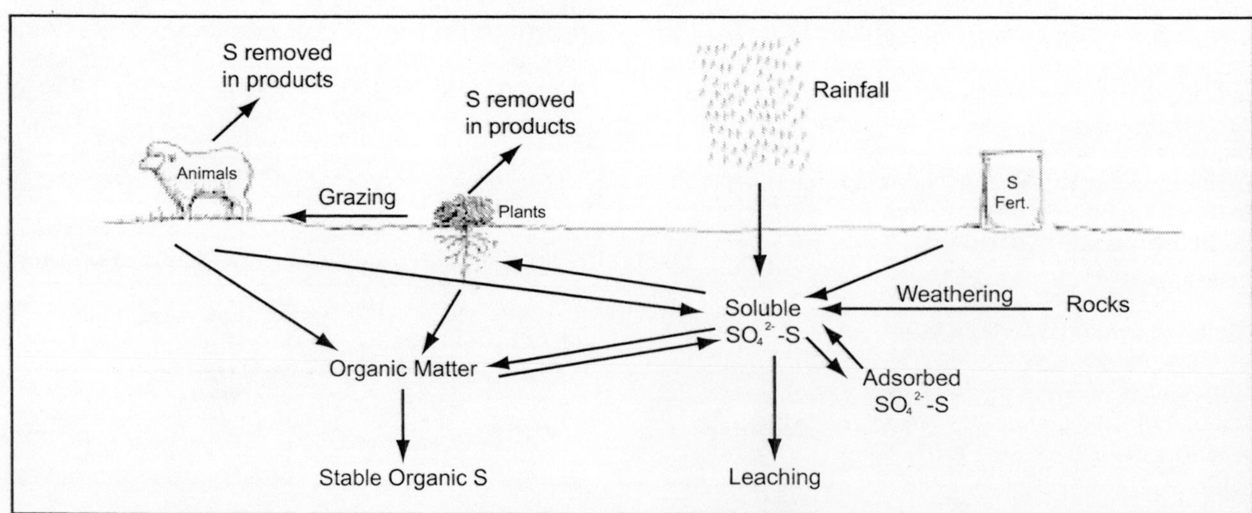

Figure 6.5.1 *The sulphur cycle.*

In addition, elemental sulphur can be used for a long-term effect. Elemental sulphur is insoluble but is broken down slowly by *Thiobacillus* bacteria to more soluble forms (Boswell and Friesen 1993). Rate of breakdown depends on several factors including the relative abundance of the bacteria and temperature. The finer the S particles, the quicker the oxidation (Barrow 1971) but finely ground material can cause handling difficulties and poses a risk of explosion. Elemental sulphur will not oxidise fast enough under normal circumstances to meet the demands of deficient crops or pastures. Elemental sulphur is completely immobile in the soil.

There are no special placement requirements for S fertilisers. Crops such as canola have a much higher requirement for S than cereals and therefore deficiency is seen more often when low S rates are applied. A 2 tonne crop of canola will remove 14 to 20 kg of S compared with about 4 kg for an equal weight of wheat and 5 to 8 kg for lupins. Therefore, fertilisers containing S should be used as an insurance on canola grown where there is any risk of S deficiency.

Residual effectiveness

Residual effectiveness of S fertilisers depends on the source, soil type, rainfall and cropping activity. As outlined previously, elemental sulphur has a greater residual effect than sulphate, due to its low solubility. The rate of dissolution depends on particle size. Sulphate sulphur can be readily leached, especially in coarse-textured soils with a low water and nutrient-holding capacity, and low S adsorption capacity. Residual effectiveness on leaching soils will be less in the high rainfall zone than in the drier areas. Thus S deficiency can develop quickly in high rainfall areas on uniform coarse-textured soils or deep duplex soils with a coarse-textured surface soil. Because most of the soil S is stored in organic matter and cultivation stimulates its breakdown, the S supply will be depleted more readily under crops than under permanent pasture.

PADDOCK ASSESSMENT

Plant symptoms

Sulphur deficiency is generally characterised by plants with pale or yellow young tissue, while older leaves can often remain quite green (Yeates 1981, 1987). Symptoms vary with the degree of deficiency, stage of growth and plant species. With canola, the deficiency shows as a rolling of younger leaves with interveinal mottling and a pink colour on the undersurface. The leaf petioles become turgid and brittle (Mason and Walton 1993). Sulphur deficiency and its symptoms are more marked with high levels of available nitrogen.

Tissue tests

Tissue testing appears more reliable than soil testing. In canola, levels of S less than 0.2% in young or recently mature leaves usually indicate deficiency. But perhaps the N:S ratio is more important. At high levels of nitrogen there is a greater demand for S for manufacturing protein. Sulphur deficiency symptoms are often seen with concentrations of S in the tissue considerably more than 0.2%, but with an N:S ratio greater than 19:1 (Yeates 1987). This ratio has been also used as a critical level for cereals, but both critical S concentrations and the N:S ratio in young wheat plants change sharply with plant age, making them unsuitable as a diagnostic aid unless the maturity of the plant is known (Robson *et al.* 1995). However, for lupins this study also showed that S concentrations in the young leaves (0.28%), stems (0.07%) and whole shoots (0.15%) and the critical N:S ratio in young leaves (22:1) are likely to be good diagnostic indices for deficiency because they did not appear to be affected by plant maturity.

Total S levels above 0.15% in subterranean clover tops at early flowering usually indicate an adequate S supply (Yeates 1981). The higher external requirement of grasses compared with subterranean clover appears to be due to the poorer ability of grasses to obtain S from the soil rather than differences in the internal requirements of the two plants (Gilbert and Robson 1984).

Soil tests

Until recently there has been no readily interpretable soil test for S. Reliance had been placed on a measure of sulphate adsorption capacity (Barrow 1975), which is a measure of a soil's ability to hold sulphate against leaching rather than a measure of the supply of available S. Sulphate adsorption capacity is lowest in the uniform coarse-textured soils that have low clay and organic matter contents, and also on deep duplex soils with a coarse-textured surface soil. Medium and fine-textured soils have a much higher ability to adsorb sulphate.

A research group at the University of New England (Armidale, NSW) have done considerable work to find a reliable soil test for S and are optimistic about measuring ester sulphur in the soil. In this study, unfertilised soils had soil test values of <3 mg ester S/g, while fertilised soils had values >7 mg ester S/g (Anderson *et al.* 1992; Blair *et al.* 1993). This test is now being used commercially.

Further reading

McLachlan, K.D. (1974). Handbook on sulphur in Australian Agriculture (ed. K.D. McLachlan - various authors). CSIRO, Melbourne.

McLachlan, K.D. (1975). Sulphur in Australasian Agriculture (ed. K.D. McLachlan - various authors). Sydney University Press.

6.6 COPPER

*Ross Brennan**

In Western Australia, copper (Cu) deficiency is widespread and an estimated 8 million ha, or one-third of south-western Australia has received applications of Cu (Gartrell and Glencross 1968). Being able to identify and correct multiple trace element deficiencies, for example copper (Cu), zinc (Zn) and sometimes molybdenum (Mo) has had a large impact on agriculture. The use of fertilisers containing these elements allowed large tracts of the 'light land' in the wheatbelt to be cleared for agricultural production during the 1960s. These soils have been fertilised with 0.7 to 2.5 kg Cu/ha at least once. Where soils are deficient, farmers require accurate information about the length of time that Cu applications will remain effective in supplying the needs of crops, pastures and livestock.

COPPER IN PLANTS

Copper is an essential component of many plant enzymes; including ascorbic acid oxidase, plastocyanin, superoxidase dismutase and amine oxidases. Copper is also involved in providing the plant with structural strength through lignification. The appearance of 'wilting' on copper deficient plants is caused by structural weaknesses as a result of decreased lignification (Bussler 1981).

Copper deficiency also interferes with reproductive processes resulting in sterile pollen in wheat plants (Graham 1975). Observations suggest that it is also specifically required in symbiotic nitrogen fixation (Snowball *et al.* 1980). There is variation in the relative susceptibility of plants to low Cu supplies (Table 6.6.1). Cultivars vary in their response to external supply, but not in their internal requirement (Robson and Reuter 1981).

Copper has limited mobility in the plant, but under certain conditions, it can be translocated from older to younger leaves. Mobility varies with the copper status, so that well supplied plants are able to translocate Cu, but Cu-deficient plants cannot (Loneragan 1981).

Table 6.6.1 Susceptibility of some common crop and pasture plants to low copper (adapted from Table III, Gartrell 1981).

High susceptibility	Moderate susceptibility	Low susceptibility
Wheat	Barley	Peas
Lucerne	Oats	Cereal rye
	Clover	Pasture grasses
		Lupins
		Canola

Copper and take-all. There are reports of beneficial effects of Cu fertilisers in reducing the severity of take-all even where Cu is not limiting plant growth (Graham 1983). In these situations it has been suggested that Cu fertiliser may be acting as a fungicide. Brennan (1991) reported the effect on the incidence and severity of take-all of wheat grown in five field experiments naturally infested with *Gaeumannomyces graminis* var. *tritici.* Copper-deficient wheat plants were more susceptible to take-all than adequately supplied plants. The addition of fertiliser beyond that required for maximum plant growth and grain yield had no effect on the incidence and severity of take-all. This work strongly suggests that there is no beneficial effect of extra Cu fertiliser in controlling take-all disease of wheat (Brennan 1991).

COPPER IN THE SOIL

The adsorption and binding of Cu by colloidal particles is sufficiently strong to make it important in controlling the availability of Cu to plants.

Adsorption onto clays. In addition to cation exchange mechanisms, Cu and other micro-nutrients enter specific adsorption processes through covalent bonding to certain functional groups on clay surfaces. The amounts adsorbed increase with increasing pH (James and Barrow 1981), hence it would be expected that availability to plants would decrease with increasing soil pH.

Adsorption onto oxide surfaces. Copper is strongly adsorbed onto iron and aluminium oxides and oxyhydroxides (Grimme 1968). The number of surface OH^- groups could determine the maximum level of adsorption, but McBride (1981) suggested that surface OH^- groups co-ordinated to two Al^{3+} or Fe^{3+} ions would probably not be active in adsorption. Therefore, adsorption on crystalline minerals (gibbsite) is likely to occur at crystal edges where the OH^- group is co-ordinated to one Al^{3+} ion.

Manganese oxides have also been found to specifically adsorb Cu, with adsorption increasing as a function of pH (Murray 1975). McKenzie (1980) found the affinity of synthetic Mn oxides for Cu^{2+} ions to be even greater than that of Fe and Al oxides.

Adsorption onto organic matter. The ability of organic matter to form stable complexes with micro-nutrients has been well established (Hodgson 1963).

* Agriculture Western Australia, Albany.

Organic constituents in the soil form both soluble and insoluble complexes with Cu. The insoluble copper-organic matter complexes are those associated with the humic acid fraction while soluble complexes are mainly associated with the fulvic acid fraction and individual organic molecules. Several factors influence the quantity of Cu bound by organic matter, including soil pH, ionic strength, molecular weight and functional group content. The strength of binding by humic acids decreases with an increase in the amount of Cu applied, which explains why small amounts of Cu in highly organic soils are so tightly complexed that they are unavailable for plant uptake.

DISTRIBUTION OF COPPER DEFICIENCY

Soils deficient in Cu for agricultural plants occur on all continents of the world. Copper deficiency is often related to soils derived from specific parent materials i.e. organic soils and peats and coarse-textured soils derived from acid igneous rocks, limestone or sandstone (Donald and Prescott 1975; Gartrell 1981; Jarvis 1981). Copper deficiency has not been reported in soils derived from basic igneous rocks (e.g. dolerite) or red-brown earths (Donald and Prescott 1975).

In WA, Cu deficiency is widespread and has been identified on: calcareous aeolian sands formed on coastal dunes (Donald and Prescott 1975); organic soils (Gartrell 1981); a range of soils formed on sandstones in the Gingin and Dandaragan districts (Teakle and co-workers 1938-45); calcareous loamy sands and siliceous sands formed from marine limestone (Dunne and Throssel 1948; Brennan *et al.* 1984); soils with coarse-textured A horizons (<15% clay, 0-10 cm) formed from acid igneous or sedimentary materials such as laterite, dissected laterite, granite, gneiss (Teakle and co-workers 1938-1945; Dunne 1948; Toms 1958; Gartrell 1969); and occasionally fine-textured soils formed from acid igneous or sedimentary materials (Gartrell 1981).

More information on the distribution of deficient soils in WA can be found in Gartrell and Glencross (1968) and Brennan and Riley (1988).

FACTORS AFFECTING UPTAKE

Interaction with other nutrients. Copper deficiency can be induced if zinc fertilisers are applied (Gartrell 1969) and appears to involve competition in uptake between Cu and Zn by plant roots (Chaudhry and Loneragan 1970). This interaction is only of significance if the Cu supply is marginal.

Cu and liming on acid soils. The amount of Cu adsorbed by soil surfaces increases with pH (James and Barrow 1981; Jarvis 1981) and so it would be expected that availability of Cu to plants would decrease with higher pH. This has not always been observed. Soil pH was found to have little effect on the availability of Cu as measured by plant uptake (Piper and Beckwith 1949; Williams 1977). However, other work has shown that liming (which increased pH) decreased available copper to plants (Jarvis 1981).

Copper and nitrogen. An increase in soil nitrogen either from the addition of fertilisers or by legume pastures can induce Cu deficiency (Gartrell 1981). Increased N increases the yield potential, thus there is a greater Cu requirement, which can induce deficiency in soils with marginal supply. This is an indirect effect which is common on soils containing a marginal supply of Cu.

Chaudhry and Loneragan (1970) observed that nitrogen severely depressed the Cu concentration in plant shoots in the presence or absence of added Cu. The main effect can be explained by plant growth as nitrogen fertiliser has no overall effect on the absorption of Cu. Nitrogen enhanced the rate of Cu absorption per gram of dry root during early growth, but was found to promote the growth of plant tops more than roots so that it depressed the Cu concentrations in the shoots.

Loneragan (1975) showed that Cu can move from old to new leaves but its movement depends on the N status of the plants. In Cu-deficient plants, N levels remain higher for longer than in Cu-treated plants. The Cu movement was found to parallel the movement of N from old to young leaves.

Copper and organic matter. Many reactions can occur with organic matter that lead to the formation of stable organic-copper complexes (Stevenson and Fitch 1981). The more stable these complexes, the less available the Cu becomes for plant uptake (Ennis 1961; Goodman and Cheshire 1976; Petruzzelli and Guidi 1976). Weakly held copper can be absorbed by plants.

The effect of adding organic matter on Cu availability has often been measured chemically (with various extractants) which never really indicate the level that a plant can absorb (Beckwith 1963). However, incorporating unharvested brassica root crops in the UK increased the severity of Cu deficiency in the following crop (Davies *et al.* 1971) and severe Cu deficiency has appeared in crops planted in haystack residues in WA (Gartrell 1981).

Copper and herbicides. Glasshouse experiments have shown that chlorsulfuron and diclofop-methyl can decrease the uptake of Cu (and Zn) and therefore plant growth, when the herbicide and Cu were incorporated together in the soil (Osborne and Robson 1992). In the field, chlorsulfuron decreases the concentration of essential nutrients in wheat shoots by as much as 30%. Of the micro-nutrients, Cu and Zn were more affected than manganese. This would be important where the Cu supply is marginal.

TREATMENT OF DEFICIENCY

To correct deficiency, Cu can be applied as a fertiliser (the most common method), foliar sprays or seed treatment. Foliar sprays are often used on crops diagnosed as deficient. With seed treatments it is difficult to apply adequate amounts of Cu.

Fertiliser. An application of 3 to 9 kg of copper sulphate per ha, drilled with the seed, achieves maximum yields for most deficient soils in WA. Re-application has not been necessary for up to 30 years after the initial application, *except* where the initial application was not high enough to satisfy crops grown under an extremely high N supply (see further discussion in *Residual effectiveness*).

The effectiveness of fertilisers and the rate required to correct deficiency depend on their placement geometry, the effectiveness of the source and reactions with the soil.

Placement. Because Cu is immobile in the soil, plant roots can only take up Cu if they are in the vicinity of particles containing Cu. The effectiveness of fertiliser is therefore influenced by the number and position of Cu particles per unit volume of soil (Gilkes 1981). This is why topdressed fertilisers are usually ineffective in correcting deficiency unless subsequently mixed through the soil. Banding Cu with the seed is more effective than topdressing (Gilkes 1981) and cultivating soil that has been banded with Cu also increases the effectiveness of the fertiliser (Gartrell 1981).

Copper fertiliser is ineffective in soil that contains few roots, e.g. dry surface soils where plants are growing on subsurface moisture (Grundon 1980). Plants can be severely deficient in these situations even though fertiliser has been applied to the surface 5 to 10 cm.

Large granules of fertiliser are much less effective than fine granules, mainly because doubling the granule size reduces the number of granules per unit volume of soil by one-eighth (assuming spherical granules). Gartrell (1981) found that five times as much Cu had to be applied if the Cu-super granules were more than 3 mm in diameter, to equal the effectiveness of granules less than 1 mm in diameter.

Forms of Cu. Many compounds are effective in supplying Cu to plants when applied as fine powders (Gilkes 1981). However, copper oxide mixed dry with superphosphate was only half as effective as copper sulphate (R. Glencross personal communication). Both sources have been equally effective when granulated with superphosphate.

Slow release sources do not have any real advantage over copper sulphate in the field because copper sulphate reacts with the soil to become a slow release fertiliser. Copper chelates that have high mobility and solubility in soil solutions may appear to offer increased availability for plant uptake. However, this advantage is lost or partly lost in soils when iron (Fe^{3+}) and calcium (Ca^{2+}) ions replace the chelated Cu (Lindsay and Norvell 1969).

Table 6.6.2 Summary of the processes controlling the copper balance.

Gains	Losses
Amount of Cu fertiliser applied The amount of Cu fertiliser applied on deficient soils will partly determine the residual effectiveness (RE). Large additions are not advised because they can induce other problems (e.g. zinc deficiency). Recommended rates varied from 0.75-2.75 kg/ha (Gartrell and Glencross 1968). Some farmers have re-applied Cu fertiliser.	*Removal in crop and livestock products* Small quantities are removed in agricultural produce e.g. 8 g/ha of Cu is removed in the grain of a 2 t cereal crop, while the wool from 12 wethers contains 0.3 g, and 10 t of lucerne hay contains 80 g (Gartrell 1981).
Contaminants in fertilisers The addition of Cu as a contaminant in superphosphate and other fertilisers is a substantial addition to the soil Cu balance. The addition depends on the type and rate of fertiliser applied (Table 6.6.3). Applying 150 kg/ha of superphosphate with an average Cu content of 50 ppm would result in the incidental addition of 7.5 g Cu/ha.	*Reactions of fertiliser with the soil* Soil reactions reduce the availability of applied Cu over time. These reactions are with organic matter, clay minerals and sesquioxides to form substances which are not available to the plant. (Refer to *Copper in the soil*).
Deposition in dust and water Normally a minor factor except around major industrial areas.	*Erosion* Copper is immobile and confined to the topsoil. With both wind and water erosion, the loss of trace elements may be greater than the proportion of the Ap horizon removed, because of the sorting of particles on a size basis. For instance, in many soils, wind will sort the particles to a depth beyond that removed from the paddock, resulting in the preferential removal of clay and silt-sized particles.
Cu sprays Some fungicides (e.g. Bordeaux spray) contain Cu and when applied to certain crops (e.g. vines) are an additional source of Cu.	*Leaching* Negligible loss through leaching in all soils (refer to following section on *Leaching of applied copper*).

Table 6.6.3 Copper concentrations in some common fertilisers in Western Australia; average and range (from Table 2.2; Brennan 1986).

Fertiliser	Average Cu concentration (ppm)	Range (ppm)
Rock phosphate		
Nauru	13	7-18
Ocean Island	40	35-60
Christmas Island	120	60-240
Superphosphate	50	30-100
Triple super (19.7% P)	50	30-100
Agras No. 1 (17.5% N, 7.6% P)	120	20-250
Agras No. 2 (12.0% N, 10.4% P)	120	20-250
Agran 34 (34% N)	10	5-20
Urea	10	5-20
Muriate of potash (KCl)	10	5-25
Di-ammonium phosphate (DAP) (18.0% N, 20.0% P)	50	-

Many experiments have demonstrated the reduced effectiveness of freshly-applied Cu compared with that applied in previous years (Gartrell 1980). The limited mobility in the soil means that cultivation is necessary to disperse the Cu and increase the likelihood of interception by plant roots (Gartrell 1980, 1981).

RESIDUAL EFFECTIVENESS

Copper applications generally have a long-lasting residual effect (Toms 1958; Gartrell 1980). A single application of 3 to 9 kg of copper sulphate per ha provides enough Cu for more than 100 years of cropping with present agricultural systems in WA. No rapid decline in the availability of Cu to plants has been measured in field trials on a range of soils.

Gartrell (1981) reported no evidence of declining residual effectiveness when 1.5 to 3 kg Cu/ha had been applied to deficient soils up to 35 years previously. Brennan (1994) measured the residual value of previously-applied Cu compared with freshly-applied Cu on properties through the Lakes, Jerramungup and Esperance Districts and found it still fully effective 23 years after the initial application.

Residual effectiveness (RE) of fertilisers will vary with soil type and the controlling factors can be summarised in terms of the copper nutrient balance. The processes controlling the balance are described in Table 6.6.2. The gains are dominated by the amount of copper applied as fertiliser and incidental gains as a contaminant of other fertilisers e.g. superphosphate (Table 6.6.3).

As copper is only required in very small quantities by plants, the Cu balance is measured in g/ha.

Leaching of applied copper

The movement of copper in soils is restricted and it can be considered immobile in most soils (Jones *et al.* 1957; Jones and Belling 1967; Korte *et al.* 1976). Jones and Belling (1967) used the radioactive isotope ^{64}Cu to study leaching and found nil to minimal movement down the profile. Near Margaret River in the high rainfall area of WA, all of the applied Cu could be found in the top 10 cm of a sandy soil 15 years after initial application (Brennan and Riley 1983).

Factors that control mobility include reactions with organic matter, adsorption to clays and sesquioxides and the precipitation of insoluble copper compounds. In coarse-textured soil containing very little clay (<3%), immobility is more difficult to explain, but its conversion to insoluble copper compounds and reactions with the colloidal material may help to hold the copper near the surface.

PADDOCK ASSESSMENT

Plant symptoms

Susceptibility to Cu deficiency varies with plant species (Table 6.6.1), therefore on marginal sites deficiency is more likely to show up during the cereal phase of a rotation. The symptoms of deficiency in cereals reflect its function within the plant.

Slight deficiency

- Most ears of cereals are filled with plump grain but there are a few shrivelled grains or empty ears. The straw is weak below the ear, stems bend and snap easily. The head is prone to wind and hail damage at this stage. The straw often turns purplish-grey on the sunny side and darkening and blackening may be seen around nodes and on the glumes. *There are no symptoms on the leaves.*
- Slight deficiency is unrecognisable before maturity and is usually only noticed during harvest. Grain yield losses may be up to 20%, or higher if ears fall to the ground.

Mild deficiency

- Most ears contain shrivelled grain, some ears are empty and an occasional tiller is aborted. Straw is often purplish-grey on the sunny side, nodes are blackened and glumes darkened. Cu-deficient plants often remain green after unaffected plants have matured. Occasionally, leaf symptoms are seen but generally the deficiency is unnoticed until healthy crops are mature or during harvest. Grain yield losses are often between 20 and 60%.

Moderate deficiency

- Few normal ears. The grain is shrivelled and many white empty heads, called 'rat-tails', are formed. Aborted tillers are common.
- Straw is darkened and plants tend to remain green and continue tillering.
- The youngest fully emerged leaf is white back from the tip or has broad yellowish streaks. Often this is seen only on the flag leaf while older leaves appear normal. The emerging leaf is yellow-white, it dies and the tiller aborts; leaf tips turn pale brown, roll and twist after death.
- The deficient plants appear pale and limp and can go unnoticed unless compared with normal plants.
- Grain yield losses are often between 50 and 90%.

Severe deficiency

- Nearly all tillers abort, nearly all ears die prematurely and most fail to emerge from the boot. Plants continue to tiller to produce a 'grassy clump'. Plants are generally pale and limp (even older leaves) so there is no contrast between young and old leaves.
- The tips of youngest leaves are white to yellow-white. They die, become pale brown and often roll or twist. Roots are often poorly developed.
- Symptoms of severe deficiency can be seen from the early growth stages through to when the ears normally emerge.
- Grain yield losses can exceed 90%.

Symptoms can be confused with other effects. For example, the restricted grain filling of mildly deficient plants which results in shrivelled grain may be confused with effects of root rots, frost, drought, herbicide damage, or molybdenum deficiency. Likewise, the symptom of rat-tailed ears can also be caused by root rots, frost, herbicide, drought or molybdenum deficiency.

Tissue tests

A reliable diagnosis of deficiency from plant symptoms alone is not always possible, so much work has been done to assess deficiency from plant analysis. The objective is to separate the Cu-adequate from Cu-deficient plants. The development of plant analysis to diagnose deficiency requires an understanding of the relationship between the Cu concentration within the plant and growth.

Analysing the youngest emerged blade (YEB) of wheat at the five to eight leaf stage, but before flowering, provides the most accurate method of assessing its Cu status. Provided the sample is not a mixture of deficient and healthy plants, interpretation is relatively simple:

ppm	Deficiency state
>2	Healthy
1.6-2.0	Marginal
1.3-1.6	Mildly deficient
<1.3	Moderately to severely deficient

Analysis of grain has limited usefulness, especially if the sample is from both deficient and sufficient plants. However, in trials from WA where there was no grain yield increase after copper applications, the concentration of Cu in the wheat grain was frequently between 1.3 and 2.0 mg/kg (J. Gartrell unpublished data). The critical Cu concentration is about 1.0 mg/kg. Farmers with grain containing <1.2 mg Cu/kg would be advised to apply fertiliser.

Soil tests

A host of chemical tests have been devised for estimating the ability of soils to supply Cu for plant growth. However, these tests are empirical since there is no certainty that the Cu extracted is necessarily that which the plant roots can absorb. Furthermore, most soil tests extract much more Cu than is accumulated by the shoots and roots of annual crops (Robson and Reuter 1981), so using soil tests to diagnose copper-deficient soils can easily give erroneous figures and interpretations.

When used correctly, Cu extracted with ammonium oxalate can show the quantity of Cu the soil is capable of supplying to plants. When extracted by ammonium oxalate a Cu level of 0.3 mg/kg (ppm) or less in the top 10 cm of soil indicates almost certain deficiency for wheat, 0.3 to 0.8 mg/kg Cu may be deficient; and >0.8 mg/kg Cu is almost certainly adequate. This method can also indicate whether fertiliser has been applied to a paddock previously, which is useful if a property has changed ownership.

The DTPA procedure was specifically designed to simultaneously identify deficient levels of Cu, Zn, Mn and Fe in neutral and calcareous soils. This test has not been calibrated adequately for WA soils which are often more acidic than those on which the test was developed. Any extractant which is to be used for testing Cu requires calibration for particular soil types and the plant species grown.

Further reading

Loneragan, J.F., Robson, A.D. and Graham, R.D. (eds.) (1981). 'Copper in soil and plants'. Academic Press, Sydney.

6.7 ZINC

Ross Brennan[*]

Zinc (Zn) deficiency in agricultural plants has been widely reported in Western Australia. The need for this essential micro-nutrient was first established in the 1930s, when its importance to the growth of cereals and pastures was demonstrated. Since that time zinc application has become an important aspect of agriculture in south-western Australia.

ZINC IN PLANTS

Worldwide, Zn deficiency in agricultural plants is recognised as one of the most common micro-nutrient deficiencies and it is becoming increasingly significant in crop production. The agricultural importance was recognised in the early 1930s, but a specific role for Zn in plants was not identified until the 1960s. Since then a series of Zn-containing enzymes has been identified and considerable progress has been made in identifying the physiological effects of deficiency in plants.

As a component of proteins, zinc acts as a functional, structural and/or regulatory factor in a large number of enzymes. Therefore, many of the physiological effects resulting from Zn deficiency are associated wth the disruption of normal enzme activity. Photosynthesis, membrane leakiness, auxin metabolism and reproduction are all affected by Zn deficiency. Zinc may also play a role in RNA and DNA structures (Brown *et al.* 1993).

The susceptibility of agricultural plants to low Zn varies (Table 6.7.1) and deficiency is likely to appear first in wheat or oats.

ZINC IN THE SOIL

The incubation of Zn with warm moist soil decreases its availability for uptake by plants. This reduction in the effectiveness of applied Zn is interpreted as being evidence of reactions that convert some of the Zn into forms unavailable to plants.

The continuing reactions of applied Zn which can be assessed by the *residual effectiveness* (RE), differed among soils. Decline in the availability could not be specifically related to any one soil property. However, multiple linear regression analysis between the RE of Zn determined by dry matter production, Zn uptake and soil properties showed that pH, clay (%), organic carbon and free calcium carbonate were important properties (Brennan 1990, 1992a). Martens (1968) found that pH, organic matter and clay were important in determining the uptake of Zn by corn plants. The role of organic matter in binding Zn is well documented (Hodgson *et al.* 1966; Martens *et al.* 1966; Follet and Lindsay 1970; Stevenson and Ardakani 1972).

Zinc is adsorbed by clays (Reddy and Perkins 1974; Farrah and Pickering 1976) however the effects of adsorption on its availability are not well documented. Brennan (1990) suggests that the reactions with clay are important in determining the effectiveness of Zn incubated in the soil compared with freshly-applied Zn.

In acid soils, the availability of Zn for plants is decreased by the addition of calcium carbonate (Lucas and Knezek 1972). The effect of pH in controlling Zn availability in alkaline calcareous soils may be through precipitation of compounds of lower solubility (Clark and Graham 1968; Saeed and Fox 1977) or by the adsorption of Zn by carbonates (Udo *et al.* 1970).

DISTRIBUTION OF ZINC DEFICIENCY

Extensive Zn-deficient areas occur in Australia (Stephens and Donald 1958; Anderson 1970; Williams and Andrew 1970). In western and southern Australia, widespread deficiency occurs in several million hectares of clover/grass pastures growing on calcareous and siliceous sands and loams, as well as on calcareous or strongly acid soils of coastal areas (Riceman 1948). The most extensive Zn and Cu-deficient province, comprising 8 million ha, is in south-western Australia (Gartrell 1974; Donald and Prescott 1975).

Table 6.7.1 Susceptibility of some common crop and pasture plants to low zinc (adapted from Lucas and Knezek 1972).

High susceptibility	Moderate susceptibility	Low susceptibility
Wheat, Oats**	Clover	Lucerne
Beans	Barley	Rye
Maize	Rice	
	Potatoes	
	Tomatoes	

*** Western Australian/South Australian experience.*

[*] *Agriculture Western Australia, Albany.*

Deficiency has been reported to be associated with soils with a high pH; soils with a low pH and a low total or extractable Zn; limed acid soils; calcareous soils; cut (i.e. levelled) soils; sodic soils; soils with very low or very high concentrations of organic matter; sandy soils; wetland or poorly drained soils; and soils with high available P concentrations.

Soils deficient for cereals and subterranean clover occur in all agricultural districts of Western Australia. For land that has not previously had Zn fertiliser, the requirements of crops and pastures have been clearly defined by many field experiments and are summarised in Brennan and Riley (1988). Deficiency in plants appears where Zn has not been applied to soils that are inherently deficient or where plant-available Zn in the soil has declined to inadequate levels.

SOIL AND ENVIRONMENTAL FACTORS AFFECTING UPTAKE

Plants are more susceptible to Zn deficiency under adverse climatic conditions, and this results from effects on both the plant and soil. Deficiency is more frequently associated with flooded than dryland soils as a result of Zn reacting with free sulphides (Mikkelsen and Shiou 1977) or sesquioxides (Sajwan and Lindsay 1988). Poor aeration can cause deficiency in crops grown in orchards, camping sites and old corrals because of trampling and puddling of the soil (Lucas and Knezek 1972).

Zinc deficiency is also more prevalent in cool and wet seasons. Soil temperature effects appear to largely affect the rate of mineralisation of Zn. High light intensity and long day-length have been shown to be major factors in the development of foliar symptoms of deficiency (Ozanne 1955; Marschner and Cakmak 1989). This effect may be mediated through a number of roles that Zn has in photosynthesis, such as in the enzymes superoxide dismutase and carbonic anhydrase, and in protein synthesis.

TREATMENT OF DEFICIENCY

Initial soil applications of 1 to 2 kg of zinc oxide per ha will correct deficiency, with the rate depending on the soil type (Gartrell and Glencross 1968; Brennan and Riley 1988). Zinc sulphate (23% Zn) and carbonate are suitable alternative sources to zinc oxide (75% Zn). However, rates must be adjusted so that the same rate of Zn is supplied. Evenness of application is important.

A foliar spray of 1 kg of zinc sulphate per ha in 50 to 100 L of water applied as soon as deficiency is detected will prevent its development (Brennan 1991). However, late foliar sprays applied to crops which are already recovering naturally in drier, sunnier conditions may provide little additional benefit.

RESIDUAL EFFECTIVENESS

Zinc has a significant residual effectiveness, therefore it is not necessary to apply it every year. The residual value depends on losses from the system, incidental gains (impurities in fertiliser) and sorption reactions with the soil.

Superphosphate containing 400 to 600 ppm Zn has always been sufficient to provide adequate Zn to wheat crops decades after the initial application. Experiments show that on sandy and gravelly wheatbelt soils, an initial application of 1 kg of Zn/ha generally prevents deficiency in wheat for at least 16 years if imported DAP fertiliser with a very low Zn content is used (J. Gartrell and R. Brennan unpublished data). Farmers' experiences suggest that residual effectiveness can be 30 years or more. There is little knowledge of the residual effectiveness on the fine-textured soils which have occasionally produced Zn deficiency.

Herbicides from the sulfonyl-urea and 'FOP' groups have initiated Zn deficiencies where the soil supply has been marginal.

> *The Zn supply is maintained at adequate levels for plant growth if superphosphate (>150 kg/ha) is applied annually. The issue of residual effectiveness and maintenance applications only arises when superphosphate is not applied regularly, or at rates <150 kg/ha/annum, or when fertilisers low in Zn (e.g. Di-ammonium phosphate) are used.*

Gains to the system

Zinc is present as an impurity in ordinary superphosphate (~400 ppm) which is equivalent to the addition of 40 g of Zn/100 kg of superphosphate. Zn applied in this way has limited mobility (Gilkes 1975).

Losses from the system

The removal of Zn in agricultural products is not a major factor. Typical amounts removed vary from 15 to 45 g of Zn/ha for pasture hay, 15 to 30 g/ha for cereal grain and 3 to 9 g/ha for wool (J. Gartrell unpublished data; Brennan 1990).

Cultivation results in a redistribution of immobile trace elements like Zn through the topsoil (Ap horizon). With both wind and water erosion, the loss of trace elements may be greater than the proportion of the Ap horizon removed. Wind erosion results in the sorting of particles on a size basis. In many soils, wind will sort the particles to a depth beyond that removed from the paddock. The clay and silt particles removed in the dust (suspension), contain most of the potentially available Zn (Armour *et al.* 1990).

Leaching of zinc

The mobility of Zn in soils is generally limited, but it has been reported to vary with the soil texture. Jurinak and Thorne (1955) found that Zn applied as a chloride to the surface of a silty clay moved less than 3 cm under strongly leaching conditions. Zinc applied as either the oxide or sulphate did not move appreciably, either in heavily leached columns of sandy loam or in silt loams (Brown *et al.* 1962). Moreover, radioisotope studies have shown that most of the Zn applied was retained in the upper 3 cm of a loamy sand (Singh 1974), although there was some movement to depth (12 to 18 cm) depending on the soil texture.

In WA, fertiliser Zn applied to the surface of an acid sand (low cation exchange capacity, 4 cmol(+)/kg; low clay content, ~3.5%) remained close to the soil surface even after 1,438 mm of rain (Brennan and McGrath 1988). At levels of Zn typically applied in agriculture and forestry (0.7 kg of Zn/ha) there was no movement below 2.5 cm. Where Zn was applied at 22.5 kg/ha, 95% of the applied Zn could be accounted for in the top 5 cm. At higher application rates (68 kg of Zn/ha), only 37% of the applied Zn was recovered below 5 cm (Brennan and McGrath 1988).

This relative immobility has important implications for the placement of fertilisers. The limited movement may explain why surface applications have sometimes failed to correct deficiency. In drier districts, Zn (which is generally banded with superphosphate), probably remains in the shallow cultivated layer where most of the roots of annual plants grow. The plants can use the fertiliser Zn in this layer when the soil is moist.

PADDOCK ASSESSMENT

Plant symptoms

Symptoms of severe deficiency are characteristic and easy to identify. These symptoms are useful for both recognising Zn deficiency and for delineating responsive soils. Numerous publications have described and illustrated symptoms of acute deficiency (e.g. Asher *et al.* 1980; Snowball and Robson 1983, 1986; Grundon 1987). In more recent publications, useful systematic keys have been developed for recognising nutritional disorders in specific plant species.

The most common symptoms of acute deficiency include stunted growth, shortened internodes and petioles and small malformed leaves (little leaf) which result in the classic 'rosette' symptom in young growth of dicotyledons (e.g. Snowball and Robson 1983, 1986) and 'fan shaped stem' in monocotyledons (Grundon 1987). With careful, frequent observations, symptoms will normally be seen first in young leaves (Zn is considered immobile under conditions of deficiency). These leaves remain small, cup upwards, and develop inter-veinal chlorosis; on upper leaf surfaces, necrotic spots appear which later coalesce to form brown necrotic and brittle patches.

In monocotyledons, chlorotic stripes adjacent to the midrib become necrotic. The necrosis is often more noticeable on middle-aged leaves, which may wilt, bend or collapse. Zinc deficiency is typically patchy, even within a single field (Kubota and Allaway 1972) and symptoms develop rapidly but, depending on the degree of stress, are sometimes transient. In annuals, symptoms are more noticeable during the earlier stages of growth. Seed and fruit formation can be aborted and yields are severely reduced.

Symptoms are not exhibited by plants suffering moderate to mild levels of stress (e.g. Reuter *et al.* 1982). Indeed, in some studies, dry matter production has been reduced by 40% or more by Zn deficiency in plants which show no visible symptoms (Carroll and Loneragan 1968). Procedures for diagnosing such 'hidden hunger' are needed for many plant species.

Tissue tests

Analysing plant tissue is an alternative means of diagnosing deficiency, as soil tests are often highly soil and species specific. Tissue analysis measures the Zn status of the plant at the time of sampling. Plant analysis needs to be related to specific stages of development and specific plant organs or tissues. For example, Bates (1971) suggested sampling when symptoms first appear and sampling tissues of a similar physiological age.

There are problems with diagnosis when there are substantial growth differences among plants sampled. For example, Viets *et al.* (1954) found little difference in the Zn concentration in leaves of crops growing on Zn-deficient soils (15.4 mg Zn/ha average) and the concentration in the same crops growing on Zn-adequate soils (16.8 mg Zn/ha). Many published values for concentrations of Zn in plant species have been established by comparing deficient and adequately supplied plants.

Another approach is to use the critical concentration; that concentration just deficient for maximum plant growth (Bates 1971). For most studies, the 'critical deficiency concentration' (CDC) is the concentration which results in a 10% reduction in the yield (Ulrich and Hills 1967). CDCs are usually determined in glasshouse experiments, but need to be confirmed in the field.

Principles of plant analysis and sampling procedures have been reviewed by Smith (1986) and Jones (1972). The plant part used for analysis must be easily identifiable, easy to collect, and related to the mobility and redistribution of the nutrient within the plant. The mobility of Zn varies strongly with the adequacy of supply and leads to anomalies in the relationships between Zn supply, concentration in plants and yield.

General recommendations for tissue tests to diagnose Zn deficiency in wheat plants are:

Zn concentration in plant tops of wheat (ppm)	**Interpretation**
<12	Noticeably deficient
12-16	Some visual and or growth effects recorded
>18	Adequate supply

Alternatively, the youngest emerged blade (YEB) is useful for detecting deficiency in wheat before flowering. Plants containing less than 10 ppm in the YEB are deficient, while some visual symptoms and growth effects have been recorded with Zn concentrations between 10 and 13 ppm. Zinc concentrations in the YEB of more than 15 ppm indicate an adequate supply.

Soil tests

A reliable soil test for Zn is desirable in the following situations:

- When accurate fertiliser histories are not available, for example where properties have changed hands. The low Zn status of a soil may have been masked by the effect of impurities in plain superphosphate.
- Occasional cases of Zn deficiency on fine-textured soils mean that although deficiency in these soils is uncommon it cannot be ruled out.
- High clay contents and/or an alkaline soil are both likely to reduce the residual effectiveness of Zn, but to what extent is unknown.
- Some commercial laboratories offer Zn soil tests which appear to have been developed and calibrated outside WA, leading to recommendations that are unsuitable for local soils and conditions (usually too much Zn, too often).

DTPA extraction (Lindsay and Norvell 1978) closely mirrors Zn uptake by wheat plants on individual soils, but the relationship may vary markedly between soils. The differences can be accounted for by considering the pH, clay content and organic carbon (Brennan 1992b; Brennan and Gartrell 1990).

Equation 6.7.1 provides a reliable prediction of the critical DTPA-extractable Zn concentration for wheat (Brennan 1992b):

$$\text{Critical DTPA-extractable Zn (ug/g)} = 0.04 + 0.019\ pH_{Ca} + 0.003\ \text{clay} + 0.004\ \text{OC} \qquad (6.7.1)$$

$(n=36, R^2=0.878)$

Where,

pH_{Ca} is the soil pH measured in 0.01M calcium chloride,

clay is the % clay content,

OC is the % organic carbon.

For subterranean clover growing in acidic to neutral soils (pH_w <7.5) the equation becomes:

$$\text{Critical DTPA-extractable Zn (ug/g)} = -0.087 + 0.039\ pH_w + 0.005\ \text{clay} \qquad (6.7.2)$$

$(R^2=0.933)$.

For subterranean clover growing in alkaline soils with free calcium carbonate (pH_w >7.5):

$$\text{Critical DTPA-extractable Zn (ug/g)} = 0.005 + 0.024\ pH_w + 0.005\ \text{clay} + 0.016\ CaCO_3 \qquad (6.7.3)$$

$(R^2=0.952)$.

Where,

$CaCO_3$ is the percentage of free calcium carbonate in the soil,

pH_w is the pH measured in a 1:5 soil:water suspension.

Critical DTPA-extractable Zn levels increased by 0.018 for every 1% of $CaCO_3$ up to 20%, the limit to which this equation applies (Brennan and Gartrell 1990).

Further reading

Robson, A.D. (ed.) (1993). 'Zinc in soils and plants'. Kluwer Academic Publishers, Dordecht, The Netherlands. 208 p.

6.8 MOLYBDENUM

Michael Riley*

Western Australia has vast areas of highly weathered acid soils whose natural levels of plant available molybdenum (Mo) are so low as to render them deficient for the maximum growth of susceptible legumes and also the more tolerant cereals. Deficiency of Mo in a crop depends not only on inherent properties of the soil (such as pH and levels of molybdenum, iron and aluminium), but also on the species, seed content and form of nitrogen fertiliser added with the crop.

Correcting a Mo deficiency is relatively simple and is usually achieved by adding a small amount of Mo to the soil. A single addition may prevent deficiency in crops for a lifetime on some soils, but for only a year or two on others.

Molybdenum cannot be added to some soils with each crop, because if plant levels become too high from cumulative additions, problems may occur with grazing stock (e.g. induced copper deficiency). Moreover, future additions of lime may accentuate this problem.

MOLYBDENUM IN PLANTS

Molybdenum is required by all plants when nitrogen is absorbed in the nitrate form because it is a critical constituent of the enzyme nitrate reductase, which catalyses the reduction of nitrate to nitrite (Nicholas and Nason 1954).

It is also required in the nodules of nitrogen-fixing legumes as a constituent of the nitrogenase enzyme. Legumes require Mo when grown on either atmospheric or nitrate nitrogen, but the requirement is higher for nitrogen fixation than nitrate reduction (Anderson 1956).

In cereals, Mo is required for grain formation because impaired or incomplete grain fill is often observed in deficient plants. Oats deficient in Mo have been characterised by blue glumes and pinched grain, or completely empty husks. In wheat, deficiencies have produced 'deaf ears', and 'whiteheads' (Gartrell 1966), which are barren of grain or contain shrivelled grain.

Relative plant sensitivities

The annual crops, tomato, lettuce, spinach, beet and all of the brassicas (cauliflower, broccoli, rape) are very sensitive to restricted Mo supply (Johnson 1966).

The relative sensitivity of a species to a deficiency of any essential nutrient may be due to either a higher internal requirement or a reduced ability to absorb that nutrient from the soil. Legumes are considered less sensitive to restricted Mo supply than the crops above (Johnson *et al.* 1952), yet have an internal Mo requirement two to three times greater than non-legumes.

Large-seeded legumes are less sensitive than small-seeded legumes (Williams 1971), but this appears to depend on the Mo content in the seed (Gurley and Giddens 1969). If the concentration in large-seeded legumes is high, then the seed may supply the crop's total requirement. For small-seeded pasture legumes there appears to be no relationship between the Mo concentration in the seed and the response to added Mo (Sherrell 1984).

MOLYBDENUM IN THE SOIL

The total concentration of Mo in most agricultural soils is between 0.6 and 3.5 µg Mo/g of soil. In Australia, the concentration of Mo extracted with 0.1M HCl from soils with 13 different parent materials was between 0.3 and 3.0 µg/g, except granitic and gravelly soils which had concentrations as low as 0.05 µg/g (Williams and Moore 1952). The most important consideration however, is the plant-available Mo rather than the total concentration in the soil. Gupta and Lipsett (1981) suggested that Mo can exist in soils in five fractions as:

(i) primary crystalline material

(ii) water soluble molybdates in the soil solution

(iii) organically-complexed Mo

(iv) molybdate adsorbed on positively charged surfaces

(v) discrete, secondary compounds, either crystalline or amorphous.

If there is little Mo available for plant uptake, it may be because there is either a low total content in soil, or a low proportion in the soil solution due to a high proportion of adsorbed Mo in fraction (iv) and the formation of a high proportion of fraction (v). Several soil properties influence the distribution of Mo between these five fractions, as discussed below:

Soil pH

Soil pH appears to be the major factor affecting the availability of Mo to plants. There have been many reports of increased uptake with increasing pH by liming (e.g. Gupta and Kunelius 1980; Petrie and Jackson 1982), although it depends on the total Mo content of the soil.

* *Formerly Agriculture Western Australia, now Anderson Consulting, North Sydney.*

Molybdenum is thought to exist predominantly as the molybdate ion (MoO_4^{2-}) in soils above pH_w 4.0 (Barrow 1978; Gupta and Lipsett 1981). As such, it seems to react with soils similarly to the phosphate ion (Barrow 1970). Sorption sites include the surfaces of oxides such as Fe_2O_3 and Al_2O_3 and the weathered edges of clay particles containing metal ions which are not fully co-ordinated in the 'ideal' lattice (Barrow 1978).

Iron and aluminium

Iron and aluminium as free ions, colloidal particles, or coatings on soil constituents, seem to be a major factor in determining the availability of Mo in soils. These metal ions appear to offer the major sites for the adsorption of Mo.

Several reports have shown good correlations between the adsorption of Mo and the extractable content of free iron oxides in soils (e.g. Karimian and Cox 1978). Aluminium oxides also adsorb Mo from aqueous solutions, but do so less strongly than iron oxides (Jones 1957; Reisenauer *et al.* 1962). However, Barrow (1970) found the adsorption of Mo was better correlated with the amount of free aluminium than free iron extracted from several Australian soils.

DISTRIBUTION OF MOLYBDENUM DEFICIENCY

In Australia, deficiency has been found on podzolic soils derived from laterites, quartzites, granites, shales and slates (Williams 1971). In general, highly weathered acid soils tend to be more deficient. Gupta and Lipsett (1981) suggest that Australian soils low in Mo invariably show signs of having formed in earlier climatic periods under heavy leaching. The present low levels therefore reflect substantial earlier losses that were never replenished.

In WA, there have been isolated responses by subterranean clover to added Mo on sandy gravelly soils, and on the red-brown and brown soils (range of textures) developed from gneisses, granites and schists in the Bridgetown, Balingup, Donnybrook and Nannup districts (Dunne 1950, 1958; Fitzpatrick 1957, 1962).

Responses by cereals to applications of Mo are restricted to the yellow sandy earths and acid yellow sandy earths developed on laterised granites and gneisses in the Zone of Ancient Drainage (Doyle *et al.* 1965; Gartrell 1966). An estimated 1.5 million ha (about half) of these soils were deficient for crop and pasture growth following clearing (Gartrell 1980). The yellow sandy earths vary considerably in pH and in clay and iron and aluminium sesquioxide content.

The acid yellow sandy earths are commonly known as 'wodjil soils' because the vegetation is often dominated by *Acacia* spp., but can also include tamar (*Casuarina campestris*), mallee (*Eucalyptus burracoppinensis*), flame grevillea (*Grevillea excelsior*) and quandong (*Santalum acuminatum*). There appears to be no association between differences in native vegetation and the likelihood of a Mo response in crops and pastures after clearing (Riley 1984).

SOIL AND ENVIRONMENTAL FACTORS AFFECTING UPTAKE

Phosphate, sulphur

Plant uptake of Mo is usually enhanced by phosphorus and decreased by sulphur (Gupta and Lipsett 1981). The stimulating effect of phosphorus appears to be due to the combination of phosphate competing with (Gorlach *et al.* 1969), and releasing (Barrow 1974), Mo from adsorption sites; and the formation of a complex phospho-molybdate anion which is absorbed more readily by plants.

The addition of sulphur decreases the Mo content of crops except where it has corrected a sulphur deficiency in plants (McLachlan 1955). The inhibitory effects of SO_4^{2-} on Mo uptake occur primarily during the absorption process, with some antagonistic mechanism involved during translocation from roots to leaves (Gupta and Lipsett (1981). The acidifying effect of SO_4^{2-} may also be partially responsible for decreasing the availability of Mo to plant roots. *Regardless of the mechanism, the addition of sulphur has been found to decrease the Mo content of crops on many occasions* (e.g. Singh and Kumar 1979).

Single superphosphate supplies both phosphorus and sulphur and the net effect is generally to decrease plant-available Mo (Gupta and Cutcliffe 1968).

Nitrogen

Mo uptake is greater in the presence of nitrate (NO_3^-) than ammonium (NH_4^+) nitrogen (Ruszkowska 1968a, b; Mishra *et al.* 1970).

Application of Mo alleviated damage caused by the accumulation of NO_3^- in wheat seedlings (Lipsett and Simpson 1973). Cheng and Ouellette (1970) found that oats grown on a Mo-deficient soil were free of symptoms of deficiency when supplied with ammonium sulphate, but developed severe symptoms when supplied with potassium nitrate or urea. Results such as these have led many researchers to believe that plants deficient in Mo will generally show no symptoms of Mo or nitrogen deficiencies if nitrogen is supplied as NH_4^+.

However, a deficiency of Mo can occur in plants which have been provided with NH_4^+. In general, nitrification of NH_4^+ will occur in soils and NO_3^- accumulation may result in the plants. Results from WA show yield reductions in wheat and oats due to Mo deficiency are accentuated when ammonium sulphate is applied (Doyle *et al.* 1965; Gartrell 1966).

Organic matter

Associations between organic matter and Mo in soils are unclear. Certain soils high in organic matter have been found to be deficient in Mo. Karimian and Cox (1978) found the level of Mo adsorbed to be closely related to soil organic matter. The presence of organic matter may enhance the availability of Mo to plants if it contains readily exchangeable molybdate ions (Gupta and Lipsett 1981).

Soil moisture and drainage

Wet soils, such as those in swamps that are not leached, tend to be high in organic matter and contain large amounts of Mo (Kubota *et al.* 1961). Flooding a soil can increase the availability of Mo.

TREATMENT OF DEFICIENCY

The methods of supplying Mo to crops include application to the soil, to the seed or as a foliar spray. Molybdenum is usually applied to soils for crops and pastures in combination with superphosphate. Application as a foliar spray or as a seed treatment can be more effective than soil application, but is required with each crop.

Molybdenum can be applied to crops in several ways: sodium molybdate ($Na_2MoO_4.2H_2O$ - 39% Mo), ammonium molybdate ($(NH_4)_6Mo_7O_{24}.4H_20$ - 54% Mo), and molybdenum trioxide (MoO_3 - 66% Mo). Molybdenum trioxide is less soluble than the molybdate compounds, but appears to be as effective for plant uptake when applied to the soil (Anderson 1946), and to have a similar residual effect (Johansen *et al.* 1977). The primary mineral, molybdenite (MoS_2 - 60% Mo), has a low solubility and is ineffective for plants (Anderson 1946).

The recommended rates of Mo for soil application with superphosphate are 50 to 60 g of Mo/ha in Victoria, and 75 g of Mo/ha in NSW and WA (Gartrell 1980; Riley *et al.* 1983).

RESIDUAL EFFECTIVENESS

The availability of Mo applied to soils declines with time (Barrow and Shaw 1975). Field evidence suggests the residual value varies with soil type and rainfall (Newman 1961; Riley 1984).

In WA, the residual value on the yellow sandy earths in the eastern wheatbelt varies considerably. For example, Gartrell (1981) found an application of 92 g of molybdenum trioxide/ha to a yellow sandy earth (topsoil pH_w 5.2) was still effective for maximum wheat production 15 years later. Riley *et al.* (1983) found the effectiveness of applications declined rapidly in one to two years on some yellow sandy earths of the central and north-eastern wheatbelt. The topsoil pH_w was usually below 5.5, while at 10 to 20 cm the pH_w was less than 4.5.

On a yellow sandy earth (topsoil pH_w 5.0) at Tammin, about 120 g of Mo/ha was applied as molybdenum trioxide for three consecutive years. Freshly-applied Mo was compared with that previously applied. The effectiveness of this soil-applied Mo declined rapidly in just one year and restricted the wheat yield (Riley 1984, 1987). Molybdenum is not leached from these soils (Riley *et al.* 1987).

Barrow *et al.* (1985) showed that the relative effectiveness of Mo applied one year previously was about half that of the current application, while Mo applied two to three years previous was only about 20% as effective. He concluded that differences in the length of time that applied Mo remains effective for plant growth depends primarily on a soil's ability to retain molybdate.

Recommendations for yellow sandy earths in WA

The topsoil pH is used to determine the frequency of Mo re-applications on the yellow sandy earths.

Topsoil pH_{Ca} <4.2

If the pH_{Ca} is less than 4.2 (pH_w <5.0), Mo needs to be applied with every crop. On these soils, there is evidence that seed-dressing is more effective than soil-applied Mo. By seed-dressing relatively small amounts of Mo, the cumulative amounts of Mo added to a paddock are minimised, thereby reducing the potential harmful effects of too much Mo being made available if these acid soils are limed in the future.

Topsoil pH_{Ca} >4.2

If pH_{Ca} is more than 4.2 (pH_w >5.0), there are no firm recommendations with regard to frequency of re-application, but an annual application is not required. Assuming no symptoms of deficiency are detected, an 'insurance' re-application to the soil at the recommended rate of 75 g/ha could be made 'once a generation' i.e. about every 30 years (Figure 6.8.1).

The requirement for re-applications of Mo can be minimised by using seed taken from 'high fertility' paddocks (i.e. those with high Mo levels; avoiding seed from highly acid paddocks) and by avoiding the use of acidifying fertilisers like ammonium sulphate on soils which are already acidic.

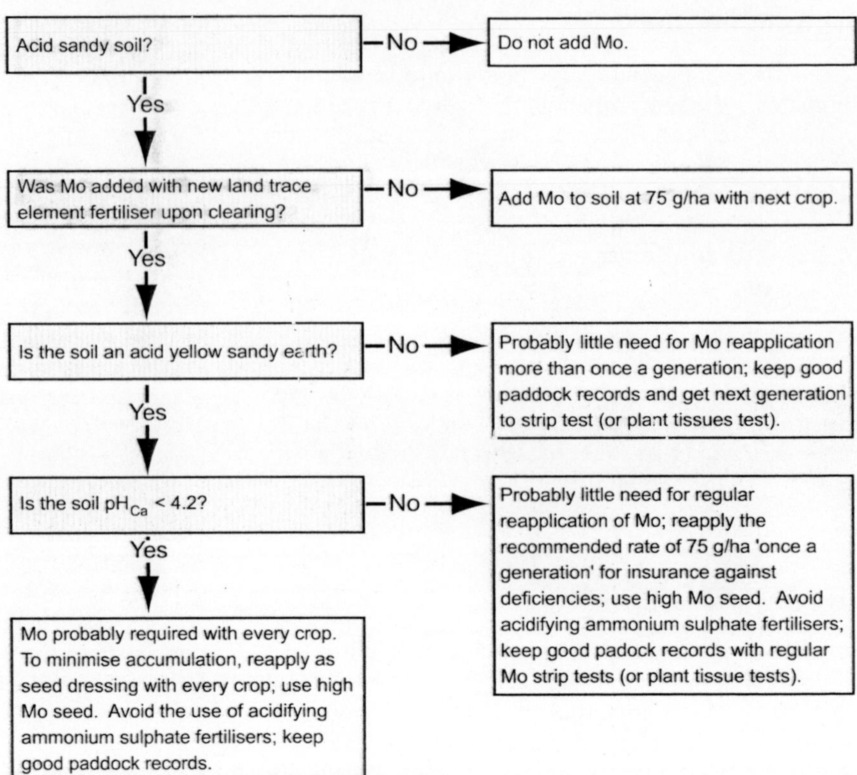

Figure 6.8.1 *Decision tree for re-application of molybdenum on WA soils.*

PADDOCK ASSESSMENT

Plant symptoms

Plant symptoms alone are not a reliable method for diagnosing deficiency as the symptoms are often non-specific and yield losses of 30% can occur before visual symptoms appear (Riley 1984).

In wheat, slight to moderate Mo deficiency has the same symptoms as nitrogen deficiency, namely reduced tillering, less foliage, shortened internodes and generally paler plants. In moderate to severe deficiency there are more distinct symptoms: delayed maturity, often with empty heads; severely affected plants may die while healthy plants grow nearby; and in the glasshouse white necrotic areas have been observed extending down from the leaf tips.

In legume pastures, Mo-deficient plants show the same symptoms as nitrogen deficiency, namely ill-thrift and paleness of the leaves, particularly the oldest leaves. Nodulation is not affected, but nitrogen fixation is, which can be detected by the distinct greenness of the haemoglobin in the nodules or by numerous small whitish nodules (Gartrell *et al.* 1994).

Tissue tests

A plant tissue test can theoretically diagnose Mo deficiency in some crops at the time of sampling. The critical concentration of Mo in the youngest emerged blades of wheat appears to lie between 30 and 50 ng/g. In legume pastures a deficiency is likely if the tops contain less than 0.1 ppm of Mo.

However, the tissue tests are expensive and few laboratories can analyse small samples of plant tissue accurately enough at the trace level needed for this diagnosis. A plant tissue test for Mo also has little prognostic value, as few inferences about future needs for the re-application of Mo can be made from current tissue levels.

Soil tests

There is currently no soil test available for identifying Mo-deficient soils.

Test strips

Test strips remain a relatively simple method for farmers to determine whether more frequent applications of Mo are required. A strip of plants sown with a superphosphate-Mo mix should be compared with a strip sown with plain superphosphate applied at the same rate and applied on the same day.

6.9 MANGANESE

*Ross Brennan**

Manganese (Mn) was the first trace element deficiency identified in broadscale agriculture in Western Australia, when it was associated with the 'grey speck' disease of wheat and oats (Carne 1927). Deficiency in cereals is widely distributed throughout south-western Australia, although it is usually confined to irregular, well defined patches rarely exceeding 20 ha. Cereals are little affected on the deep grey sands where the 'split seed' disorder, caused by Mn deficiency, has devastated narrow-leafed lupin (*Lupinus angustifolius*) crops.

Manganese toxicity is a common problem in acid soils in eastern Australia, but has not been recorded in WA.

MANGANESE IN PLANTS

Manganese has a role in many metabolic processes and chlorophyll production requires Mn. The complete role of Mn is not fully understood, however it is present in chloroplasts in a complex which oxidises water to produce oxygen. It is also involved as a cofactor of many enzymes (e.g. decarboxylases, dehydrogenases), but magnesium can be substituted as the enzyme activator.

Manganese is absorbed by the plant as Mn^{2+} and the content in plants can vary widely. In deficient plants, Mn is relatively immobile and little is translocated from older tissues to growing points. The relative susceptibility of various plant species to low Mn is summarised in Table 6.9.1. Oats appear more sensitive than wheat or barley, while cereal rye is very tolerant of low Mn.

MANGANESE IN THE SOIL

Manganese is a primary rock mineral and is found particularly in ferro-magnesian materials (Gilkes and McKenzie 1988). The Mn released from these rocks by weathering forms a number of secondary minerals, the most prominent being MnO_2 (pyrolusite).

The most important soil fractions of Mn are Mn^{2+} and the oxides, in which Mn is present in trivalent (Mn^{3+}) and tetravalent (Mn^{4+}) forms. Divalent Mn is adsorbed to clay minerals and organic matter and is the most important form in soil solution for plant nutrition. Its soil chemistry is complex and is described in reviews by Bartlett (1988) and Norvell (1988).

Manganese participates in many soils reactions, including oxidation and reduction (redox reactions), ion exchange, specific adsorption, and solubility equilibria. The distribution of Mn between solution and solid phases is related to pH, redox conditions, and the characteristics of ligands and surfaces. Mn^{2+} is the predominant form found in solution and in association with exchange sites on soil surfaces, while Mn^{3+} and Mn^{4+} are found predominantly in a variety of oxide-rich solid phases (Norvell 1988). Any attempt to understand the solubility of Mn is complicated by the potentially wide variety of solid phases involved and because of the many solubility-limiting reactions involve redox reactions.

As the Mn^{2+} level in the soil depends on oxidation-reduction reactions, all factors influencing these processes have an impact on availability. These include soil pH, organic matter content, microbial activity and soil moisture (Norvell 1988). The availability of Mn increases with increasing soil moisture and declines with increasing pH. Availability is higher in acid soils due to the higher solubility of Mn compounds at low pH. Lindsay (1972) showed that Mn^{2+} solubility decreased 100 fold for each unit increase in soil pH. Therefore, liming decreases Mn availability, whereas applying acidic fertilisers like ammonium sulphate has a beneficial effect on Mn uptake by plants (Gartrell 1980).

DISTRIBUTION OF MANGANESE DEFICIENCY

Soils that induce Mn-deficiency in cereals do not correspond with those that induce deficiency in narrow-leafed lupins. The reason for this apparent discrepancy is unknown, although a number of theories have been proposed. The two deficiencies are discussed separately.

Table 6.9.1 Relative susceptibility of some common crop plants to manganese deficiency (adapted from Lucas and Knezek 1972).

High susceptibility	Moderate susceptibility	Low susceptibility
oats, wheat	barley, rice	maize, rye
peas, beans	clover, lucerne	pasture grasses
lupins**	potatoes,	
apples, peaches	tomatoes	
grapefruit, orange		

** *Western Australian/South Australian experience.*

* *Agriculture Western Australia, Albany.*

Cereals

Manganese deficiency in cereals is widely distributed throughout the south-west of Western Australia, but it is usually confined to irregular, well-defined patches rarely exceeding 20 ha (Teakle and Wild 1940; Smith and Toms 1958). With severe deficiency, grain losses may exceed 1.3 t/ha.

The main regions of Mn deficiency include:

- The western edge of the wheatbelt from the gravelly white gum country south of Moora and New Norcia, through the powder-bark wandoo and brown mallet country of West Brookton, Wandering, Narrogin, Katanning and east to the gravelly, fluffy red morrel, blue mallet and blue mallee country of Dumbleyung, Kukerin, Kulin and Corrigin. Patches are seen every year.
- The narrow, discontinuous coastal strip of calcareous deep sands from west of Northampton to Israelite Bay. Patches occur every year.
- South Jerramungup and the Esperance Plains where deficiency is seen in drier periods only, often on broad gently sloping, alkaline shallow duplex soils.
- Minor occurrences have been seen throughout the remainder of the south-west, usually on gravelly soils on erosional surfaces, or on deeply leached, sometimes peaty, sands.

Narrow-leafed lupins

Manganese deficiency in narrow-leafed lupins is mainly a problem in slightly acidic grey sand and gravelly sand areas, which are common in the south-west where the average annual rainfall is more than 450 mm. However, as lupin growing has moved into drier areas, further patches of split seed have shown up in some seasons, usually on the coarser and deeper, more leached sands.

Deficiency causes the grain to split, delays maturity and decreases yield. Its occurrence varies with the maturity of the lupin variety, planting date, rainfall received during and toward the end of the growing season, and soil type (Perry and Gartrell 1976; Walton 1978; Radjagukguk 1981; Hannam *et al.* 1985; Hannam and Ohki 1988).

TREATMENT OF DEFICIENCY

Cereals

Manganese deficiency is controlled by:

- Mixing manganese sulphate with superphosphate and drilling it with the seed. On all except highly alkaline soils, 15 kg of manganese sulphate per ha drilled with the seed gives profitable grain increases. Some deficiency may still develop in the most deficient situations. When topdressed, manganese sulphate is only half as effective.
- Spraying leaves with manganese sulphate (4 kg/ha in 100 L of water/ha) when symptoms first appear.
- Drilling ammonium sulphate fertilisers, such as Agras, with the seed.

Narrow-leafed lupins

Manganese deficiency is controlled by:

- Early sowing and use of an early maturing variety to reduce the risk of split seed so that the seed fills and matures before subsoil moisture is exhausted.
- Soil or foliar applications of Mn (Hannam *et al.* 1984). Manganese sulphate topdressed before seeding is usually only about half as effective as when it is drilled with the seed, and is even less effective if topdressed after seeding.

Soil application

On grey sands, gravels and yellow coastal sands north of Perth, 30 kg of manganese sulphate per ha drilled with the seed either completely eliminates split seed or reduces it to <7% in most seasons. Higher rates of Mn will do little to reduce the level of split seed any further. In years when the soil containing the Mn fertiliser band is dry for extended periods during the growing season, fertiliser Mn may be only partially effective. In these situations, the roots cannot take up most of the Mn because it remains near the fertiliser band, high in the soil profile.

Along the south coast and on the yellow sands of the eastern edge of the West Midlands, 15 to 20 kg of manganese sulphate per ha drilled with the seed has usually been sufficient to give either a complete or acceptable level of control. Increases in grain yield from manganese sulphate treatments are large. Again, fertiliser Mn has been less effective in dry seasons, particularly in later sown or later maturing crops.

Commercial Mn superphosphate (5% Mn) contains about 20 kg of manganese sulphate in each 100 kg of mix.

Foliar application

Foliar sprays using 4 kg of manganese sulphate per ha in 75 to 100 L of water have been effective in most but not all cases. The development stage of the seed at the time of spraying is critical. Spraying when the pods on the main stem are 2 to 3 cm long and the secondary stems have nearly finished flowering has produced good results. Aerial application avoids mechanical damage to the crop. Spraying at the correct rate and time has been as reliable as drilled applications of Mn sulphate. Increasing the rate beyond 4 kg/ha does not significantly increase the effectiveness of the spray.

Other compounds such as mangansol and Mn chelates are effective as sprays if the rate of Mn is equivalent to that in 4 kg of manganese sulphate per ha (i.e. 1 kg/ha of elemental Mn). Manganese chelates tend to be very expensive for broadscale agriculture.

RESIDUAL EFFECTIVENESS

Cereals

The residual benefit of manganese sulphate to subsequent cereal crops was negligible from field trials by Agriculture Western Australia. However, farmers in the Great Southern report a decreasing requirement for Mn after five or six applications of 15 kg of manganese sulphate per ha. This corresponds to half of the residual benefit that was recorded with narrow-leafed lupins two years after Mn was applied to deep pale sands.

Narrow-leafed lupins

Manganese sulphate applications on slightly acidic deep grey sands can have a marked residual effect on a following lupin crop. This means that manganese sulphate rates can be halved with each succeeding crop. On the grey sands in the West Midlands, 60 kg of manganese sulphate per ha applied in 1978 is still fully effective for maximum grain yields and in reducing split seed. It is likely that when a cumulative total of 50 to 60 kg of manganese sulphate per ha has been applied, no more is needed and the split seed problem will be cured.

PADDOCK ASSESSMENT

Cereals

Plant symptoms

Seedlings usually grow normally until early tillering in the sandy gravels and alkaline shallow sandy duplex soils, but earlier symptoms of deficiency have been seen on calcareous deep sands. Pale yellow-green patches with irregular but clearly defined edges develop in the crop. These become paler as the season develops and the plants wilt and droop in warmer weather.

All leaves of wheat, barley and oats become pale, but symptoms appear first on older leaves before extending to newer growth. The pale leaves become limp and soft, eventually dying, producing a dead and wilted basal flag leaf. A very weak head may be produced, or none at all if the plant dies prematurely from severe deficiency.

Similar symptoms can also be induced by copper deficiency, but usually the older leaves are darker green than the younger leaves. Manganese deficiency can also be mistaken for take-all, except that oats are immune to this fungal disease. In addition, Mn deficiency usually occurs in definite patches of characteristic soils, although this may not be true with calcareous deep sands and pale deep sands.

Tissue tests

Precise diagnosis of the Mn status of cereal plants using tissue analysis is not yet possible. In whole shoots, a Mn level below 20 ppm indicates a possible problem, and below 10 ppm indicates almost certain deficiency. In the youngest emerged leaf blades a level of 12 ppm indicates Mn deficiency. Indications gained from tissue analysis can be checked by spraying foliar Mn in strips across the crop.

Soil tests

For many reasons there will probably never be a reliable soil test for Mn, including: the concentration of Mn^{2+} in the soil solution can vary by orders of magnitude within short periods, Mn can be accumulated by the plant far in excess of its requirements and the plant can take up the insoluble Mn-oxide (Reisenauer 1988).

The soil test is regarded as no more than a rough guide. The top 10 cm of problem soils usually contain <2 mg/kg of Mn extractable in hydroxyquinone, or <0.5 mg/kg of Mn that can be extracted in 1M ammonium acetate.

Narrow-leafed lupins

Plant symptoms

Manganese deficient plants tend to drop their lower leaves and remain green long after healthy plants have matured, which can cause difficulties at harvest. In the early growth stages, deficiency symptoms are usually not striking and are easily over-looked. Leaves may show mild mottling, more pronounced at the leaflet tips, which die and turn brown. Unopened new leaves are pale green with small necrotic spots on the tips of the leaflets. After the onset of split-seededness, new growth is dwarfed and bunched. Manganese deficiency is often patchy in paddocks. The symptoms are illustrated in Brennan (1993).

Tissue tests

The Mn level in the main stem of the lupin plant at flowering is a reliable test for Mn deficiency. The test is carried out on the main stem (all leaves and branches are removed as well as the flowers and root system) at mid-flowering. About 20 to 30 stems are required. The critical value for Mn in the main stem is 20 mg/kg* of dry matter.

Whole tops of plants from crops that have subsequently produced split seed have typically contained less than 50 mg/kg of Mn when sampled six to eight weeks after germination. This test appears less reliable than using the main stem.

Split seeds usually contain less than 10 mg/kg dry matter of Mn, although apparently healthy crops have produced seed with as little as 7 mg/kg of Mn. Often the difference in Mn concentration between normal and split seeds from the same plants is negligible.

Soil tests

Not reliable, see comments above for cereals.

Further reading

Graham, R.D., Hannam, R.J. and Uren, N.C. (1988). 'Manganese in soils and plants'. Kluwer Academic Publishers, Dodrecht, The Netherlands.

* *The unit of measurement, milligram per kilogram (mg/kg), is equivalent to parts per million (ppm).*

6.10 BORON

Geoff Moore

Boron (B) *deficiency* has been reported in eastern Australia, but is rare in Western Australia. The isolated cases have usually been associated with lime application (J. Gartrell personal communication). However, B *toxicity* can occur in arid or semi-arid areas where leaching is limited (Gupta *et al.* 1985) and in soils with alkaline, sodic subsoils (Cartwright *et al.* 1986). This section deals with B toxicity.

Riley (1989) found symptoms of toxicity on the leaves of Stirling barley crops in widespread areas of WA receiving less than 550 mm annual rainfall, but not in areas receiving more than 550 mm. On the basis of observations from field and glasshouse studies, he concluded that it was unlikely that B toxicity reduced grain yields of Stirling barley in WA, even though leaf symptoms in spring were quite severe. However, the concurrence of sodicity or salinity with elevated concentrations of B makes it appear that B toxicity has reduced Stirling barley yields in some situations (Riley 1989).

However, Riley's conclusion is not undisputed, and K. Young (personal communication) suggests that the grain yield superiority of some B-tolerant barley cultivars in the Salmon Gums area may be due to B toxicity reducing the yield of Stirling. Grain yield seems to be most affected in dry years when the crop is more reliant on subsoil moisture, while in wet seasons yields are unaffected although leaf symptoms can still appear (Young 1995).

The strong seasonal influence on the response of crops to B toxicity, together with the spatial variability and common association with sodicity and occasionally subsoil salinity, exacerbate the difficulty of quantifying the effect on grain yields. Modern wheat varieties with a yield potential approximately 10% greater than Halberd do not have the genes for B tolerance which are found in that variety (Paull and Ralph 1991). Preference for Halberd by farmers in some parts of the wheatbelt possibly indicates the extent of soils with moderate to high concentrations of B. The breeding of isogenic cultivars, theoretically varying only in genes for B tolerance, may offer an improved technique for the assessment of toxicity in the future.

BORON IN PLANTS

Boron is an essential element for plant growth, however a specific biochemical role is still to be identified conclusively. The evidence generally shows a role in cell division and that B is a necessary part of the cell wall. Boron may also have a role in protein synthesis and the translocation of sugars (e.g. Gupta *et al.* 1985). There is a narrow range of concentrations in the plant between deficiency and toxicity, therefore nutrition needs to be managed carefully in deficient crops. Boron is one of the least mobile nutrients within plants.

The effect of high concentrations of B on plant growth and crop yields is complex and intrinsically linked to the plant-water relations (Bingham and Garber 1970). Plant uptake is predominantly passive over the range from normal to high concentrations of B in the soil, a process greatly influenced by transpiration rates (Raven 1980).

Observations of crop growth suggest that leaf symptoms of toxicity generally develop in the latter stages of plant growth, which appear to be related to the pattern of water extraction by the crop. For instance, a crop may tolerate B at concentrations prohibitive to normal growth in some soil layers, providing most of the root system is growing in soil with a low concentration of B. Conversely, a long dry period during spring may force the crop to extract most of its water from a subsoil layer containing a high concentration of B, when plant transpiration rates are higher and the topsoil is relatively dry. This may partly explain the highly seasonal response of crops to B, although response is also affected by other factors including: species and varietal tolerance, concentration of B in the soil, depth to high concentrations of B (>20 ppm), soil water storage of the layer with low concentrations of B, rainfall and root growth pattern.

There is considerable variation in both species and genotypic tolerance (Eaton 1944; Davis *et al.* 1978; Gupta 1979; Nable *et al.* 1990a). There are thought to be at least two reasons why plants differ in their tolerance to high concentrations of B. The main reason is genotypic variation in the permeability of the cell membrane enabling plants to exclude B (Nable 1988). There are also differences in internal plant tolerance, although no varieties can withstand high concentrations of B in the tissue (Nable 1988; Riley 1989).

BORON IN THE SOIL

High concentrations of B in the soil result from a combination of climatic and soil properties. Concentrations can vary widely within a short distance. In South Australia, toxicity has been found on a range of soils, including red-brown earths, calcareous earths, calcareous sands and grey clays

(Cartwright *et al.* 1986). For high concentrations of B to accumulate in the soil profile, there are three main requirements:

- a source of boron
- high adsorption ability
- limited leaching - low rainfall (also low rainfall per wet day)
  - low soil permeability.

Source of boron

In Western Australia, the source of soil B varies between regions. It includes tourmaline-bearing rocks, marine sediments and aeolian material from salt lakes or from arid zones (B. Cartwright, G. Scholz personal communications). The ocean contains relatively high concentrations of B, so soils derived from marine sediments can have a high content and there may also be some aerosol accession from oceans.

High adsorption ability

In acid soils, B is highly mobile and readily leached (Gupta *et al.* 1985). That B deficiency is rare in WA, despite large areas of acid sandy soils, may be due to the presence of kaolinite, which can adsorb some B under acidic conditions (Hingston 1964). Adsorption increases with pH, especially in the alkaline range. Adsorption is mainly on the clay fraction and illite is the most reactive clay mineral. On a weight basis, adsorption decreases in the order:

illite > montmorillonite > kaolinite (Hingston 1964).

The greater the ability of the soil to adsorb B, the lower the concentration in the soil solution (Gupta *et al.* 1985).

Alkaline soils have a high adsorption capability and adsorption increases with clay content (Gupta *et al.* 1985). Cartwright *et al.* (1986) found statistical relationships between B concentrations and pH, exchangeable sodium, cation exchange capacity and clay content, but the predictive values of these relationships proved negligible between soil profiles. Soils with high concentrations of B are invariably sodic, although not all sodic soils have high concentrations of B (Cartwright *et al.* 1986). Calcium carbonate and/or sodium chloride are common in soils which contain high concentrations of B, although this is a coincidental relationship. The depth to the maximum concentration of B in the soil is usually less than the depth to the maximum content of calcium carbonate (Cartwright *et al.* 1986).

Limited leaching

Boron toxicity in Western Australia is largely confined to areas with an average rainfall of less than 550 mm/annum (Riley 1989). The relationship is related to the degree of leaching of the profile. Boron is adsorbed in soils and the amount of water required to leach it from the soil is approximately three times that required to leach sodium chloride (Bingham *et al.* 1972).

SUMMARY OF SOIL AND CLIMATIC FACTORS ASSOCIATED WITH BORON TOXICITY IN WA

- low rainfall (<550 mm per annum)
- alkaline subsoil (pH_w >8.0)
- sodic subsoil (ESP >6).

Common additional factors are:

- soil formed from marine sediment, aeolian material from arid regions or salt lakes
- high clay content
- dominant clay mineral is illite or montmorillonite
- low rainfall per wet day.

PADDOCK ASSESSMENT

Leaf symptoms are the simplest method of identifying B toxicity, but even when symptoms are present, it is often impossible to accurately gauge the effects (if any) on grain yield.

Soil testing can identify high concentrations of B in the subsoil and therefore soils with a potential to develop toxicity problems. Plant tissue or grain analysis is of limited usefulness in identifying whether grain yields are reduced, although the grain content can be used to identify potential problem sites.

Plant symptoms

Boron toxicity shows first in the older leaf tips and margins, because B is not readily translocated from old to young parts of the plant (Gupta *et al.* 1985). The symptoms in barley appear as dark brown spots frequently surrounded by necrotic (dead) areas on the edges and tips of affected leaves (Eaton 1944; Khan *et al.* 1985). Leaf spotting symptoms in barley are not evident in wheat or oats. Symptoms on cereal crops can appear severe without affecting grain yield (Riley 1989).

Tissue tests

Plant tissue analysis for identifying B toxicity gives erratic results (Paull *et al.* 1988; Riley 1989; Nable *et al.* 1990b). A *critical toxic concentration* (CTC) from plant tissue or grain analysis has yet to be defined because it varies depending on a number of factors. The pattern of B uptake varies with the growth stage and transpiration rate of the plant and on the plant cultivar (Paull *et al.* 1988; Riley 1989; Nable *et al.* 1990a, b; Paull and Ralph 1991). Boron does not accumulate evenly in the plant and significant amounts can be leached from leaves by rain (Oertli 1969; Nable *et al.* 1990b).

The CTC was found to vary greatly depending on the growth stage at the time of sampling, the tissue sampled and the yield parameter chosen; diagnostic or prognostic (Riley *et al.* 1994; Riley and Robson 1994). In whole shoots of Stirling barley the CTC ranged from 50 to 420 μg/g (Riley and Robson 1994).

Boron concentrations in grain can be useful for identifying potential toxicity. If the concentration is 3 ppm or higher, then the crop was grown on soil with a potential to develop toxicity problems.

Soil tests

Soil testing for B is difficult because of high spatial variability and the high B concentrations occur in the subsoil which makes sampling more difficult (Cartwright *et al.* 1984, 1986). Toxic concentrations tend to occur in patches with an irregular distribution across a paddock, therefore soil testing is best correlated with known areas of 'good' and 'poor' growth. Samples should be collected at regular intervals down the profile, especially from the clayey subsoil.

The method for measuring B in soils for many years was the hot water soluble (HWS) extraction method (Berger and Truog 1939) which was developed to diagnose deficiency in acid soils in the United States. The HWS method does not work well with dispersive, alkaline clay soils (Cartwright *et al.* 1983). Extraction in hot $CaCl_2$ is now the preferred method, as it is suitable for a range of soils including sodic and alkaline soils (Spouncer *et al.* 1992). Soils with B values of more than 20 ppm could cause toxicity.

MANAGEMENT OPTIONS

For soils with high concentrations of B in the subsoil, management options are confined to the use of tolerant genotypes. The relative tolerance of crop varieties is described in the *Crop variety sowing guide* (published annually by Agriculture Western Australia). Stirling barley and Eradu wheat are examples of particularly susceptible varieties.

Further reading

Gupta, U.C. (1993). Boron and its role in crop production. CRC Press. Boca Raton, Florida, USA. 237 p.

6.11 OTHER NUTRIENTS

CALCIUM

Mel Mason[*]

Calcium (Ca) is an essential element for plant growth and is involved in the formation of cell walls and in plant structure generally.

Deficiency is rarely, if ever, seen in broadscale crops in Western Australia except for canola, though it is relatively more important in horticultural crops e.g. bitter pit in apples. Although not widespread, calcium deficiency has been observed in some plants in canola crops (Mason 1992), particularly under wet soil conditions. This is despite an adequate supply of calcium in the soil and Ca applied as fertiliser. The problem is transport of calcium within the plant rather than an absolute deficiency; calcium is almost immobile in the plant and sometimes transfer within the plant cannot keep up with rapid growth and stem elongation.

Calcium is essential for structural growth and cell wall formation and is very immobile in plants, therefore symptoms of deficiency usually consist of structural breakdown in young tissue. For example, calcium deficiency in canola results in the collapse of the flowering stem just below the inflorescence (Mason 1991).

Calcium represents a large proportion of the exchangeable cations in most soils and is adsorbed onto the exchange sites. It is obviously present in high levels in calcareous soils, but is also usually high in the finer-textured soils. Levels are lower in coarse-textured soils with a low cation exchange capacity, particularly in the more acid soils in high rainfall areas, where it can be replaced on the exchange complex by hydrogen ions and leached from the rooting zone. Apart from the exchangeable forms, calcium can be present in soils as soft carbonate deposits and carbonate nodules (pedogenic carbonate) and as limestone fragments and gypsum salts. (See Chapter 10 for more information on exchangeable calcium.)

Calcium is applied to the soil incidentally in fertilisers and soil ameliorants e.g. single superphosphate (20% Ca), lime (up to 76% Ca) and gypsum (up to 26% Ca). The residual effect is considerable and generally levels of soil calcium are well above the requirements of crops and pastures.

Lime is often applied to overcome soil acidity problems and will also increase the availability of some nutrients (e.g. molybdenum), but reduces the availability of some potentially toxic elements, notably aluminium and manganese. Overuse of lime can result in deficiencies of nutrients such as iron, manganese and zinc.

Calcium is almost always present in sufficient quantities so there is little need for a soil test. Critical levels are available for plant tissue analysis for many crops (Reuter and Robinson 1986), and can be used for a general diagnostic or monitoring tissue test.

[*] *Formerly Agriculture Western Australia, now Mel Mason and Associates, Perth.*

COBALT

Ross Brennan[*]

The requirement for cobalt (Co) in animals was established in the mid-1930s (Lee 1975). However, an essential role in plants was not discovered until 1960 when it was demonstrated that it was essential for the growth of various legumes reliant on symbiotic nitrogen fixation: soybeans (Ahmed and Evans 1960), alfalfa (*Medicago sativa L.*) by Reisenauer (1960), and subterranean clover (*Trifolium subterranean L.*) by Hallsworth *et al.* (1960). Cobalt is also essential for nitrogen fixation in the non-legume species *Alnus* (Hewitt and Bond 1966) and *Azolla* (Johnson *et al.* 1966).

Cobalt has not been established as an essential nutrient for plants that do not depend on symbiotic nitrogen fixation. However, beneficial effects on growth have been reported in *Hevea* (Bolle-Jones and Mallikarjuneswara 1957), wheat (Wilson and Nicholas 1967) and non-nodulated subterranean clover (Hallsworth *et al.* 1965; Wilson and Hallsworth 1965; Wilson and Nicholas 1967).

Field responses of subterranean clover to Co fertiliser were demonstrated on poor siliceous sands in South Australia and WA soon after the discovery that it was essential for nodulated legumes (Powrie 1960; Ozanne *et al.* 1963). Later work on narrow-leafed lupins (*Lupinus angustifolius*) showed that the seed yield of plants grown on a deep yellow sand without deliberate additions of Co was only 40% of that on plus-Co plots (Gartrell 1974). Narrow-leafed lupins may be particularly sensitive to deficiency, as Gladstones *et al.* (1977) demonstrated a 50% response in vegetative yield in this species at a site where *L. cosentinii, L. luteus*, and subterranean clover showed no significant responses.

Seed of narrow-leafed lupin may vary widely in status from 4 to 730 µg of Co/kg (Robson and Mead 1980; Robson and Snowball 1987), and the magnitude of response to fertiliser is strongly affected by the Co status. Robson and Snowball (1987) concluded that yield responses to fertiliser Co are likely to be small if the seed contains >128 µg of Co/kg. Further work is needed to establish the extent to which fertilising seed crops can be used to overcome the need for more widespread application on production crops, as has been done for molybdenum in corn (Weir and Hudson 1966). However, in the study of Chatel *et al.* (1978) fertilising a Co-deficient sand with $CoSO_4.7H_2O$ at 420 g/ha resulted in seed containing relatively low concentrations of Co (19 to 24 µg/kg).

Most of the interest in Co relates to the health of ruminant animals, which require it for the synthesis of vitamin B12 by their rumen microflora (Underwood 1984). For these animals, inadequate dietary Co leads to wasting diseases characterised by anaemia and loss of appetite. In the case of breeding ewes, Co deficiency can lead to reduced lamb birthweights, elevated neonatal mortality and poor lamb survival associated with depressed milk production by the ewe (Norton and Hales 1976).

It is possible to prevent Co deficiency in livestock by using fertilisers containing Co on the pasture (Adams *et al.* 1969). However, in many situations it is more practical to supply the Co directly, via a bullet administered orally to young animals.

[*] *Agriculture Western Australia, Albany.*

IRON

Ross Brennan[*]

Iron (Fe) deficiency is one of the rarer mineral deficiencies found in crops and pastures in Western Australia. It occurs predominantly in oats grown on the peaty sands in the west and lower south coastal districts but can also be seen in other species.

Iron, an essential micro-nutrient, is an important component of many plant enzyme systems, such as cytochrome oxidase (involved in electron transport) and cytochrome (involved in the terminal respiration step). Iron is a component of the protein ferredoxin and is required for nitrate and sulphate reduction, N_2 assimilation, and energy (NADP) production. It also functions as a catalyst or part of an enzyme system associated with chlorophyll formation. It is thought to be involved in protein synthesis and the growth of root-tips.

There are marked differences in the tolerance of plants to iron deficiency. The symptoms are generally most apparent in oats, less obvious in clover and least obvious in capeweed. Deficiency has also been reported in lupins at Dongara, lucerne at Yanchep and West Harvey, and a number of ornamental garden plants in the Perth metropolitan area.

Iron exists in the soil as both the *ferric* (Fe^{3+}) and *ferrous* (Fe^{2+}) cations. The availability of the *ferrous* form (the form taken up by plants) is affected by the degree of aeration. Iron-sufficient plants can acidify the rhizosphere as well as release Fe-complexing substances which enhance availability and uptake.

Most WA soils have abundant iron. This is evident from their yellow to red colours (caused by the goethite and haematite) and/or the presence of ironstone gravel and ferricrete (Section 2.3).

Deficiency is common on the grey-black coastal sands overlying limestone and on the peaty sands and grey siliceous sands in the west and lower south-west coastal districts. The coastal sands overlying limestone are found as a narrow belt from Dunsborough in the south to north of Geraldton; the largest area is between Dongara and Geraldton. The dominant native vegetation on these soils is wattle. The Plantagenet peaty sands originally carried a low heath containing tea-tree, paperbark, grass trees, kangaroo grass, bottlebrush and, in some instances, stunted jarrah. These soils are high in organic matter, often waterlogged in winter and are naturally acidic.

Elsewhere, Fe deficiency occurs only spasmodically, as a result of periodic waterlogging, over-liming or on old windrows as a result of the alkaline ash bed.

High levels of phosphorus decrease the solubility of Fe in the plant, a P:Fe ratio of 29:1 being average for most plants. Potassium increases its mobility and solubility, while nitrogen accentuates deficiency due to increased growth. The bicarbonate anion is believed to interfere with Fe translocation.

Deficiency can be overcome by using foliar sprays or by topdressing with iron sulphate. A limited number of trials on oats and clover have shown that a 2 to 4% iron sulphate spray (2 to 4 kg iron sulphate per 100 L water) should be applied at 100 L/ha immediately the plants become pale and should be repeated whenever the symptoms reappear. A single spray, although increasing yields markedly, is rarely enough to overcome a severe deficiency.

Topdressing also effectively controls the problem. In trials, 50 to 100 kg/ha of iron sulphate removed the symptoms and increased yields. There is some residual value.

For a marginal deficiency or where the deficiency develops late in the season it is doubtful whether treatment is economical.

Paddock assessment

Plant symptoms

In shoots, the first visible symptom of Fe deficiency is chlorosis of young leaves. In spite of a decrease in chlorophyll concentration, the leaves expand normally. The chlorosis in young leaves is reversible unless necrotic spots occur with severe deficiency. When deficiency becomes severe, cell division is also inhibited, so leaf growth is impaired.

Severe iron deficiency is distinctive, particularly in oats. In the early stages, symptoms for oats are pale leaves. The area between the veins gradually becomes yellow while the veins are still green, giving a striped appearance. In more severe cases, the leaf becomes uniformly yellow, turns white and finally dies. In paddocks, deficient areas are white to yellow, depending on the severity, in an otherwise green crop.

Iron-deficient clovers have the same colour as oats but instead of a striped appearance there is a fish-bone pattern. There is also a marked marginal chlorosis. Affected clover paddocks lack the dark green appearance so typical of healthy pastures. On the same soil, symptoms in oats are more severe than those in subterranean clover.

[*] *Agriculture Western Australia, Albany.*

Weeds such as capeweed are generally unaffected by Fe deficiency. Among the cereals, oats are more susceptible than barley. Healthy barley plants may grow among iron-deficient oats. In lupins, *Lupinus angustifolius* is the most affected and *L. albus* the most resistant.

Tissue tests

Leaf contents range from 10 to 1000 ppm in dry matter with sufficiency ranging from 50 to 75 ppm, although total Fe may not relate to sufficiency. Most plant Fe is in the *ferric* (Fe^{3+}) form as a ferric phospho-protein, although the *ferrous* (Fe^{2+}) iron is believed to be the metabolically active form.

The critical deficiency concentrations (CDC) in leaves range from 30 to 50 mg/kg dry weight basis (Wallihan 1978). The CDCs are presumably considerably higher in fast-growing meristematic and expanding tissues, presumably in the range of 200 mg/kg for total Fe and 60 to 80 mg/kg for HCl extractable-Fe. Extractable-Fe, often defined as 'active Fe', is usually a better parameter than total Fe for nutritional status, especially for field-grown plants.

Soil tests

There are a few methods for extracting Fe from soils, but they have not been widely calibrated with plant growth.

MAGNESIUM

*Noeleen Edwards**

In WA, magnesium (Mg) deficiency has been reported in fruit crops (Halse 1959) and vegetable crops (McKay and Galati 1993), but not in commercial cereals or narrow-leafed lupins. Deficiencies can be controlled with dolomite (calcium-magnesium carbonate) or epsom salts (magnesium sulphate).

Deficiencies are rarely seen in agricultural plants and low levels are more of a problem for stock health. Magnesium uptake may be reduced by high levels of NH_4^+ or potassium (K) in the soil. Heavy applications of K fertilisers may restrict Mg uptake in pastures and induce hypomagnesaemia in grazing animals.

Magnesium in plants

Magnesium is an essential nutrient in plants. A high proportion of the total plant Mg is involved in regulating cellular pH and the cation-anion balance. It also has an essential role in protein synthesis and many enzyme reactions. A major function of Mg is as the central atom of chlorophyll molecules. This accounts for 15 to 20% of the total plant Mg. Plants require less Mg than calcium or potassium, and tissue content averages 0.5% of dry matter. Impaired transport of photosynthates from the leaves to roots, fruits or tubers is a major consequence of Mg deficiency. Root growth is affected to a much greater extent than shoot growth.

Deficiency symptoms in wheat start with a faint mottling of the leaves with beads of darker green on a light green to yellow background. The mottling and yellowing becomes more marked with time and the beading persists throughout the life of the leaf. The young leaves remain unopened resulting in a twisted appearance that makes the plant appear water-stressed. Symptoms in the glasshouse appear on the mid or youngest leaves first, indicating that Mg is variably mobile.

In narrow-leafed lupins, symptoms are expressed in both new and old leaves. In old leaves, reddish brown oval spots develop randomly over the entire leaflet. The leaves slowly turn a dull greyish green. The new leaves form a cluster, with leaflets becoming thin and spiky, and the tips rapidly dying. The whole plant becomes a dull green colour.

Magnesium in soil

The Mg content of soil generally ranges from 0.05 to 0.5%. Magnesium is found in ferromagnesian minerals such as biotite, serpentine and hornblende; in secondary clay minerals including chlorite, vermiculite, illite and montmorillonite; and as dolomite. Its distribution can be considered in the same way as for potassium and divided into non-exchangeable, exchangeable and water soluble forms. Most Mg is found in the non-exchangeable form and is only released slowly. Exchangeable and water soluble Mg are of greatest importance to plants.

Exchangeable Mg represents about 5% of the total soil Mg and normally occupies between 4 and 20% of the cation exchange capacity (CEC), while Ca usually occupies about 80% of the CEC. However, in WA soils, there is a general trend for the exchangeable Ca:Mg ratio to decrease down the profile (McArthur 1991). WA subsoils are relatively rich in exchangeable Mg. On some soils, this has been shown to exacerbate the adverse effect of exchangeable Na on the structural stability of clay subsoils, increasing dispersion. In the Jerramungup district, there are minor but widespread areas of permeability contrast soils where the exchange complex in the subsoil is dominated by Mg and Na.

Magnesium is leached relatively easily from soil. Nitrification processes increase leaching. The rate of removal depends on the rainfall and properties such as texture and structure. Highly leached soils are generally low in Mg, whereas soils formed in depressions where leached nutrients may accumulate tend to be high in Mg. Atmospheric accessions of Mg in rainfall are 2 to 10 kg/ha/year (e.g. Hingston and Gailitis 1976).

Further reading

Mengel, K. and Kirkby, E.A. (1987). 'Principles of plant nutrition'. 4th edition. International Potash Institute, Berne, Switzerland.

* *Agriculture Western Australia, South Perth.*

CHAPTER 7

SUSTAINABLE SOIL MANAGEMENT

7.1 WIND EROSION

Geoff Moore, Paul Findlater* and Dan Carter**

Wind erosion has long been recognised as a major land degradation issue in Western Australia because of the large area of coarse-textured surface soils and windy environment. There is considerable evidence that aeolian action contributed to the formation of many landscapes in south-western Australia (Jutson 1934; Stephens and Crocker 1946; Harper et al. 1988), but this does not mean that contemporary wind erosion is something to be tolerated.

Wind erosion has many adverse effects including sand blasting damage to crops, the loss of macro and micro-nutrients (Leys and Heinjus 1992), the long-term loss of productivity and atmospheric pollution. There are also off-site costs to both individuals and the community. The dust lost from paddocks is rich in nutrients and is carried high into the atmosphere before being deposited, possibly thousands of kilometres downwind. Significant dust loss can be caused by short episodes of non-extreme winds which do not result in appreciable accumulation of drift material (Marsh 1984). Wind erosion can be minimised by careful management.

PRINCIPLES OF WIND EROSION

The principles of wind erosion are well established (Bagnold 1941; Chepil and Woodruff 1963; Middleton and Southard 1984; Greeley and Iverson 1985; Lancaster and Nickling 1994). For wind erosion to occur, three conditions have to be met:

(i) Insufficient ground cover (trees, plants, stubble, plant residue) to protect the soil surface;

(ii) Loose, dry soil on the surface;

(iii) Erosive winds e.g. more than 8 m/s (28 km/h) at a height of 2 m, which is approximately 1 km/h at the soil surface. For winds above the threshold velocity blowing over a loose dry soil, soil loss is approximately proportional to the cube of the wind speed.

Plant cover

The physics of air moving across vegetation has been studied extensively in controlled laboratory experiments, mathematical models and field experiments (Bagnold 1941; Chepil and Woodruff 1963; Marshall 1971; Lyles and Allison 1976; Fryrear 1985; Findlater et al. 1990).

Roughness elements such as vegetation (e.g. standing stubble, growing plants), clods and gravel affect the wind profile by altering the roughness of the surface. A barrier of slow moving air develops near the surface, within the roughness elements, caused by the increased friction. This barrier protects the soil from erosion by preventing winds of an erosive velocity reaching the surface (Chepil and Woodruff 1963). This sheltering effect may cover an area several times greater than the area the roughness elements occupy (Marshall 1971; Raupach 1992; Raupach et al. 1992). Standing stubble is more effective than an equivalent weight of prostrate stubble (Carter 1984). The critical cover percentage for standing stubble to minimise wind erosion is 30% (i.e. when viewed from above, more than 30% of the ground surface would be covered).

Prostrate ground cover not only increases surface roughness but also physically covers the soil surface, which reduces the amount of soil exposed to the wind. Research in Western Australia has demonstrated that ~50% prostrate ground cover is adequate to minimise wind erosion (Figure 7.1.1). With ~50% of the surface covered, the predicted soil loss is 10% of that predicted from a bare surface (Fryrear 1985; Findlater et al. 1990). The stubble should be anchored in the ground. The relationship between prostrate ground cover and relative soil loss applies to all soils and plant covers, unless the vegetation is easily detached (e.g. field pea stubbles), where much higher levels are required to control wind erosion.

Extensive wind erosion in the South Stirlings, February 1988 (D. Carter).

* Agriculture Western Australia, Geraldton.
** Agriculture Western Australia, Albany.

The benefits of retaining stubble levels higher than 50% are small (Figure 7.1.1), but may be necessary to ensure adequate paddock protection following seeding. The traditional seeders (e.g. off-set disc, culti-trash) leave as little as 20% of the stubble in the first pass, while no-till seeders can leave between 45 and 100% (Bligh and Findlater 1996).

Windbreaks absorb or deflect the force of the wind causing an extensive downwind effect and some effect upwind. They are most effective when the wind is at right angles to the windbreak and have only a small effect if they are parallel to the wind direction (Chepil *et al.* 1964a). They should not have big gaps and should be relatively long. Windbreaks should have significant porosity (~40%) to reduce turbulence on the leeward side (Hagen and Skidmore 1971).

For a susceptible soil, either bare or carrying short pasture, wind speeds need to be lower than 30 km/h at a height of 10 m over a 'smooth' natural surface or approximately 28 km/h at 2 m. The structure of a tree windbreak affects the extent to which wind speed is reduced, but the following rule of thumb is useful:

> *At 10H downwind (H = height of windbreak), the wind speed is reduced by 50% and between 15H and 20H the wind speed is reduced by only 20%.*

The distance that windbreaks can protect the soil therefore depends on wind speed. Assuming that winds are up to 60 km/h, protection extends to about 10H. Protection to 20H is only likely with wind speeds of less than 38 km/h. On slopes away from the wind, protection is increased; on slopes facing the wind, protection is decreased.

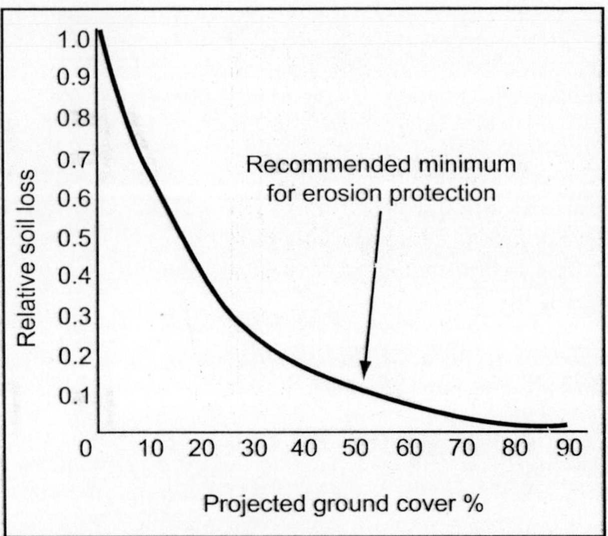

***Figure 7.1.1** The effectiveness of prostrate ground cover in reducing the amount of soil lost by wind erosion, relative to bare ground (from Findlater* et al. *1990).*

Soil movement

A typical soil consists of particles with a wide distribution of sizes and a correspondingly wide variation in erodibility. Movement is initiated when the aerodynamic uplift of the wind overcomes the force of gravity on the loose surface grains. Particles of fine sand with a diameter of 0.1 mm (100 µm) are the most erodible. With particles smaller than 0.1 mm, cohesion between particles and aerodynamic effects begin to replace gravity as the main forces the wind has to overcome. A small amount of soil moisture is sufficient to stick the particles together, and soils have a low erodibility when the water content is above the lower storage limit (Chepil and Woodruff 1963). Particles larger than 0.1 mm become progressively harder to move because of their increasing mass.

Soil particles are sorted by wind action. There are three distinct mechanisms of soil movement, which are usually associated with certain particle sizes: saltation, suspension and surface creep.

If the wind velocity is gradually increased, a point is reached that starts the most erodible particles (fine to medium sand, 0.1 to 0.5 mm) in motion. These particles are lifted and carried a short distance in a jumping or bouncing motion called *saltation*. The saltating grains strike other sand grains triggering their saltation, thus initiating an acceleration of soil movement downwind in a process called avalanching. Saltating grains are transported for hundreds of metres, usually being deposited along fence lines where the roughness increases. The saltating sand grains dislodge other particles both smaller and larger to initiate the two other mechanisms of soil movement: *suspension* and *surface creep*. Particles smaller than 50 µm (very fine sand, silt and clay) are not easily lifted by wind, but become airborne through saltation bombardment (Shao *et al.* 1993) and in some soils as coatings on sand grains. Once airborne they are light enough to be easily carried by the wind and become airborne dust (or suspension), which is transported considerable distances (McTainsh and Leys 1992). The larger particles (>0.5 mm) are too heavy to be lifted by the wind, but are rolled along the surface by the impact of saltating sand grains, in a mechanism called surface creep (or creep). These particles generally travel short distances only, and are confined to the eroded area (Bagnold 1941; Chepil and Woodruff 1963; Middleton and Southard 1984; Greeley and Iverson 1985; Lancaster and Nickling 1994). Particles larger than 0.84 mm (very coarse sand and gravel) are essentially non-erodible (Chepil 1950, 1953) unless there is an extreme wind event (Figure 7.1.2).

Large particles and aggregates

Wind erosion on a bare, loose soil will continue as long as the wind blows or until the large particles (coarse sand, gravel etc.) and aggregates (clods) are concentrated on the surface. These large particles and aggregates stabilise the surface by: (i) reducing the wind speed at the soil surface, (ii) shielding the underlying soil and (iii) capturing saltating materials (Gillette and Stockton 1989). The point at which the surface is stabilised depends on the size and coverage of the particles and aggregates at the surface.

Fryrear (1984) found that a 60% coverage of artificial aggregates (45 x 25 mm) on a bed of dry erodible soil (particle size <0.42 mm) resulted in a 90% reduction in soil loss compared with a bare surface. Chepil (1953) and Leys *et al.* (1997), using natural aggregates with a wide distribution of particle sizes, report that 40% of non-erodible particles (>0.84 mm) gives adequate erosion control. The shape and smoothness of the aggregates (or gravel) are also important (Logie 1982). (In the sections 'Potential soil loss' and 'Assessing the current risk of wind erosion', below, aggregate sizes (>0.85 mm) covering 50% of the surface are used to accommodate the variation observed in experiments.)

There are two straightforward methods for assessing the proportion of soil particles in the non-erodible category (>0.84 mm):

- Sieve the dry soil in the field followed by visual assessment (volume of soil in the pan relative to volume retained in a 0.85 mm sieve). For convenience, 0.85 mm is generally used because it coincides with a commonly available sieve size. Use a dustpan and brush to sweep up all the loose material on the surface.
- A second, more rigorous hand sieving method is described in Semple and Leys (1987) or Leys *et al.* (1997).

Limited protection may be provided by smaller particles, but at higher concentrations. Marsh (1983) observed that under conditions of extended erosion, a protective layer of the creep fraction (>0.5 mm) develops on the levelled surface of an eroded paddock. A 10 mm thick armouring layer appears to reduce the soil lost in suspension, even though the coarse sand particles are loose and form ripples on the soil surface. In most coarse-textured soils in WA, less than 35% of particles are larger than 0.5 mm, while in the fine sands on the Esperance sandplain, less than 5% are larger than 0.5 mm.

Potential soil loss

The potential soil loss from a site with inadequate surface protection depends on the depth of loose material and the proportion of particles larger than 0.85 mm. As successive layers of soil particles are stripped from the surface, the non-erodible fraction remains and contributes to the underlying layer.

For example, in a soil with 10% of the particles larger than 0.85 mm (average size 1 mm), the removal of the top mm by wind erosion will increase the non-erodible fraction on the surface to about 20%. The removal of successive layers will increase the non-erodible fraction by about 10%, a process that will continue until ~50% is covered by the non-erodible fraction. Soil loss would be limited to about 4 mm. A guide to potential soil loss for a range of soils is given in Table 7.1.1.

Field observations of wind speed

The Beaufort Scale (Table 7.1.2) provides a means of assessing wind strength from field observations. Wind measurements at climate stations may be in terms of peak (gusting) and average velocity (km/h), or wind-run over a time period (km). Extreme wind events (e.g. gale force to storm winds) may be unlikely, but winds strong enough to cause significant dust loss are almost certain. A Beaufort Scale reading of 5, which corresponds with a fresh breeze (small trees in leaf begin to sway), will result in soil movement on a susceptible paddock. A measurement of 4 actually mentions that dust and loose paper are raised, but these winds are unlikely to cause significant soil movement on agricultural soils. Wind erosion will start on beach sands at lower velocities (~20 km/h at a height of 10 m) than the accepted threshold of 30 km/h (at 10 m) for agricultural soils.

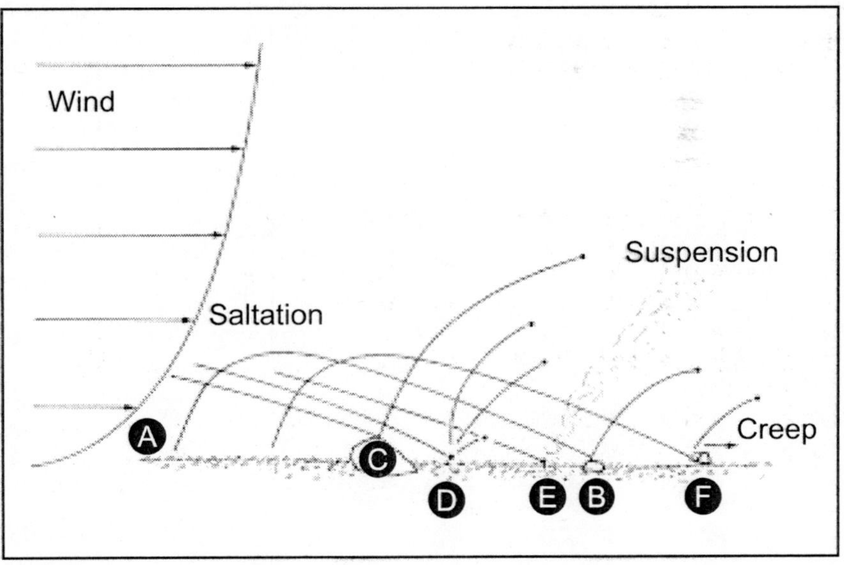

Figure 7.1.2 *The three principal modes of particle movement in wind erosion (from Figure 1.12, Greeley and Iverson 1985).*

*Surface shear stresses exerted by the wind cause sand grains at (A) to be lifted, carried downwind and bounced back into flight (*saltation*). Sand grain at (C) hits a large rock and rebounds to a high saltation trajectory; grain at (D) strikes the surface and triggers other grains into saltation; grain at (E) strikes the surface and causes very fine particles to be lifted from the surface and carried by turbulence in* suspension*; the grain at (F) strikes a larger sand grain and pushes it a short distance downwind (*surface creep*).*

Table 7.1.1 A guide to the potential soil loss (mm) from a bare, loose, unconsolidated dry surface under extended erosive winds*.

Proportion of particles >0.85 mm	20%	10%	5%	1%
Maximum soil loss (mm)	2	4	9	49

** Assumes the particles (>0.85 mm) have an average diameter of 1 mm.*

Table 7.1.2 Wind speeds (Beaufort Scale) related to readily observed field conditions (from Meteorological Office 1969).

Beaufort number	Description	Wind speed equivalent*		Specifications for estimating speed over land
		(m/s)	(km/h)	
0	Calm	0-0.2	<1	Calm, smoke rises vertically.
1	Light air	0.3-1.5	1-5	Direction of wind shown by smoke drift, but not by wind vanes.
2	Light breeze	1.6-3.3	6-11	Wind felt on face; leaves rustle; ordinary vane moved by wind.
3	Gentle breeze	3.4-5.4	12-19	Leaves and small twigs in constant motion; wind extends light flag.
4	Moderate breeze	5.5-7.9	20-28	Raises dust and loose paper; small branches move.
Erosive winds				
5	Fresh breeze	8.0-10.7	29-38	Small trees in leaf begin to sway; crested wavelets form on inland waters.
6	Strong breeze	10.8-13.8	39-49	Large branches in motion; whistling heard in telegraph wires; umbrellas used with difficulty.
7	Near gale	13.9-17.1	50-61	Whole trees in motion; inconvenience felt when walking against wind.
8	Gale	17.3-20.7	62-74	Twigs break off trees; generally impedes progress.
9	Strong gale	20.8-24.4	75-88	Slight structural damage occurs (chimney-pots and slates removed).
10	Storm	24.5-28.4	89-102	Seldom experienced inland; trees uprooted; considerable structural damage occurs.
11	Violent storm	28.5-32.6	103-117	Very rarely experienced; accompanied by widespread damage.
12	Hurricane	>32.7	>118	Severe and extensive damage.

** Wind speed equivalent at a standard height of 10 m above the ground.*

SUSCEPTIBILITY TO WIND EROSION

The effect of soil type

All soils are subject to wind erosion given certain conditions. The key question is not the soil type, but the level of disturbance (by mechanical or animal action) required to bring a soil to an erodible condition. The erodibility of a soil changes with both time of year and with different management practices.

In this context, susceptibility is the level of disturbance required to bring a soil to an erodible condition.

Soils can be classified into five categories of susceptibility to wind erosion (Table 7.1.3). Soils in category (v) start with a loose surface and have an inherently low stability. On these soils, control of erosion must rely on windbreaks and/or maintaining adequate vegetative cover. Categories (iv) through to (i) have a decreasing susceptibility to wind erosion. These soils are less fragile and require some disturbance by machinery or stock to loosen the soil.

The susceptibility of a soil can be assessed from a simple matrix of the surface texture and surface condition (Table 7.1.3). The data in this table are for a dry soil; repeated field observations can accommodate the dynamic nature of erodibility in the field. It is unlikely that a moist soil will erode even if it is loose. However, under windy conditions the surface can dry to an erodible condition within a few hours.

The susceptibility rating does not predict the amount of soil lost if the surface is loose and bare. Soil loss will vary with the aggregate size distribution which is partly determined by the primary particle size distribution (texture). The textures in Table 7.1.3 are in descending order of expected soil loss, from very fine sand (highest) to silty clays (lowest). The surface texture groups (Table 7.1.3) are based on the 'wind erodibility groups' (WEG) developed by the United States Department of Agriculture using wind tunnel measurements on field plots (Chepil and Woodruff 1963; Gillette 1988). For the purpose of the classification, some WEGs have been grouped, because they have the same wind erodibility index.

The concepts behind the susceptibility ratings are illustrated in Figure 7.1.3. The level of disturbance required to bring a soil to a loose condition depends on the surface condition in ascending order from soft < firm < hardsetting for a given soil texture. The erodibility of self-mulching soils is inversely related

Table 7.1.3 Assessing the susceptibility of a dry soil to wind erosion using surface texture and surface condition.

Surface texture	Susceptibility rating				
	*Surface condition**				
	Loose	*Soft*	*Firm*	*Hardsetting*	*Self-mulching*
Very fine sand Fine sand Medium sand	**(v)**	**(iv)**	**(iii)**	-	-
Loamy sand Clayey sand Coarse sand	**(v)**	**(iv)**	**(iii)**	**(iii)**	-
Sandy loam Light sandy clay loam Loam	**(v)**	**(iii)**	**(ii)**	**(i)**	-
Sandy clay loam Sandy clay Clay loam Clay Silty clay	**(v)**	**(iii)**	**(ii)**	**(i)**	**(i)–(iv)**

* *Assessing the surface condition of a dry soil based on McDonald* et al. *(1990):*

Loose. *Incoherent[1] mass of individual particles or very small aggregates. Surface easily disturbed by pressure of forefinger.*

Soft. *Coherent[2] mass of individual particles or aggregates. Surface easily disturbed by pressure of forefinger.*

Firm. *Coherent[2] mass of individual particles or aggregates. Surface disturbed or indented by moderate pressure of forefinger.*

Hardsetting. *Compact, hard, apparently apedal condition forms on drying. Surface not disturbed or indented by pressure of forefinger.*

Self-mulching. *Highly pedal, loose surface mulch forms on drying.*

1. Incoherent means that less than two-thirds of the soil material, whether composed of peds of not, will remain united at the given moisture state without significant force (very weak or less) having been applied.

2. Coherent means that two-thirds of the soil material, whether composed of peds or not, will remain united at the given moisture state unless force is applied.

to their aggregate size. The probability of wind erosion on bare soils with varying surface conditions is intrinsically linked to the threshold wind speed required to initiate movement and the likelihood of that wind occurring. For example, there is a high probability (1:1) of wind speeds ~30 km/h, which will result in erosion on a loose sand. Conversely, there is a low probability (1:200) that a high velocity wind will erode a hardsetting loam.

Wind erosion of a cultivated paddock with inadequate stubble protection (G. Moore).

The effect of landscape

The landscape does not directly affect the inherent susceptibility of a soil to wind erosion, but influences the third requirement, wind speed. Streamlines of air moving over a rise converge or are constricted, so the wind velocity increases. The relative increase in wind speed depends on the length of slope and the slope angle. This concept is illustrated in Figure 7.1.4.

When the slope is <1.5% there is minimal effect on the wind speed and the wind profile generally conforms with the pattern experienced over level ground. With slopes >1.5% the wind profile is compressed, resulting in increased velocities unless the threshold slope length is exceeded. When a slope exceeds a certain length, which varies with the angle, the wind profile tends to follow the contours of the land. The wind speed on the slope is then comparable to level terrain (Doughty *et al.* 1943; Chepil *et al.* 1964b).

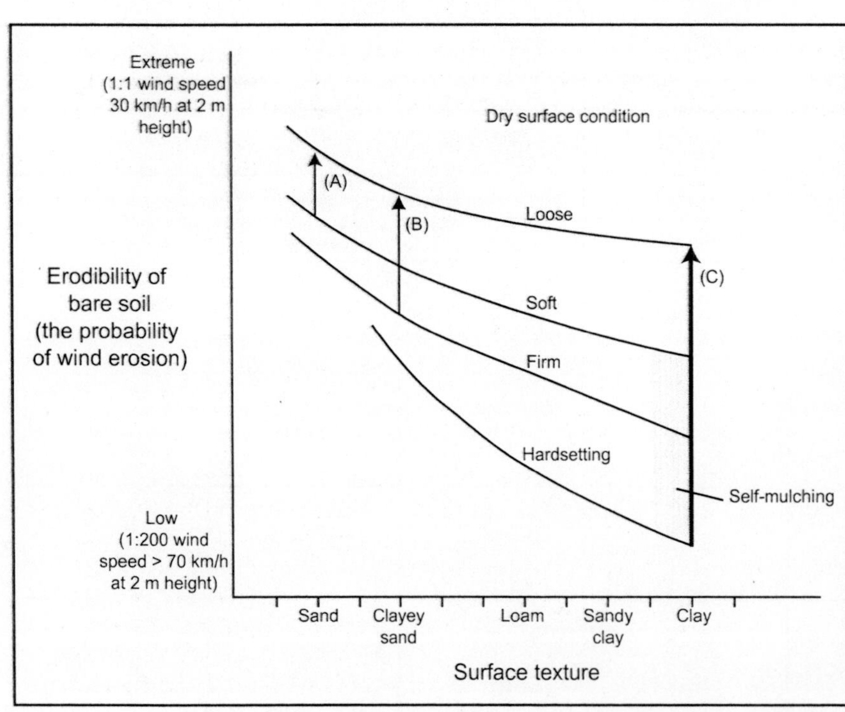

Figure 7.1.3 *The conceptual relationship between surface texture, surface condition and wind erodibility. The vertical lines suggest how management may affect surface condition and erodibility.*

(A) The effect of minimum tillage or careful summer grazing on sands initially in a soft condition.

(B) The effect of two cultivations or moderate summer grazing on a clayey sand, which was initially in a firm condition.

(C) The effect of rotary hoeing (maximum disturbance) or very heavy summer grazing on a hardsetting clay soil.

The rate of soil loss varies with the cube of the wind speed (Bagnold 1941). Therefore, small changes in the wind speed can result in large increases in erosion if the speed exceeds the threshold velocity for erosion (~8 m/s at a height of 2 m). The effect of the landscape can be considered in terms of the soil loss relative to level ground (Table 7.1.4). Soil loss will be greatest from crests and then from relatively short slopes. The amount of soil lost from the windward and leeward slopes (<10%) is comparable. Wind speed may be higher on the windward side, but on the leeward slope the sheltering

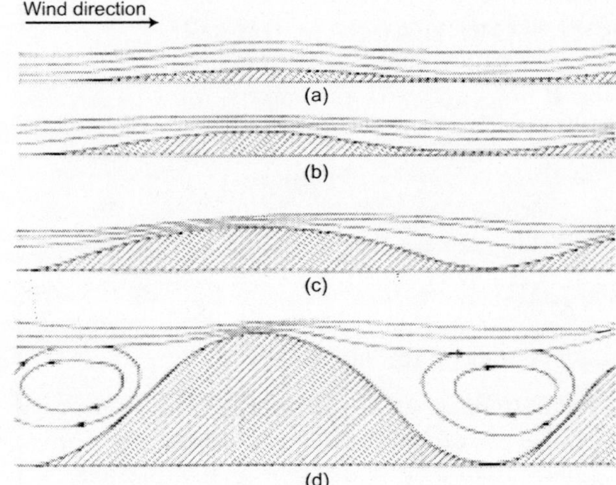

***Figure 7.1.4** Lines of equal wind velocity over:* (a) *Low rise (slope 1.5%); (b) Low rise with 3% slope, length <130 m; (c) Rise with 6% slope, length <130 m; (d) Ridges with slope >25%. Diagrams are not to scale (from Chepil* et al. *1964b).*

Note the increased turbulence and upward lift on the mid-slope (leeward side), which can initiate erosion. Where the lines of equal wind velocity converge and are close to the surface, higher wind velocities are experienced.

Table 7.1.4 The effect of the position in the landscape on potential soil loss. It includes the effect of slope, slope length and landform element on wind speed and consequently potential soil loss. The percentage values are relative to level ground, which has been given an arbitrary value of 100% (adapted from Chepil *et al.* 1964b).

Landform element	Slope (%)	Length of slope (m)	Soil loss relative to level ground (%)
Closed depression	–	–	<100
Flat	<0.5	–	100
Slope	0.5-1.5	–	100
	2	<80	110
		>80	100
	3	<100	130
		>100	100
	5	<150	190
		>150	100
	10	<250	370
		>250	100
Crest	<1.5	–	100
	2	<80	120
		>80	100
	3	<100	150
		>100	100
	5	<150	250
		>150	100
	10	<250	660
		>250	100

effect is off-set by the impact of soil material from the crest moving downslope readily dislodging soil particles because of a higher angle of repose of the grains (Chepil *et al.* 1964b).

An alternative way of considering the influence of the landscape on wind erosion is that it changes the probability of erosive winds. For instance, the probability of erosive winds on a crest is higher than on a flat plain.

CURRENT RISK OF WIND EROSION (PADDOCK STATUS)

To prevent wind erosion it is important to monitor paddock conditions, especially when land use significantly reduces the amount of ground cover. A monitoring system consists of a flow diagram with six steps (Figure 7.1.6). Steps 1 and 2 refer to physical features that cannot be altered in the short-term, while steps 3 to 6 are controlled by management.

If the planned use of the land will reduce the amount of plant cover (e.g. summer grazing, cultivation), the monitoring system can be used to identify which step is critical for wind erosion control.

STEPS USED IN THE MANAGEMENT DECISION FLOW DIAGRAM (FIGURE 7.1.6)

Step 1: Windbreaks

The percentage of the paddock effectively protected by windbreaks can be estimated from Table 7.1.5. The prevailing winds from the appropriate direction for the time of the year in question should be used.

Step 2: Gravel/stones

Gravels and stones on the surface reduce susceptibility to wind erosion in two ways. First, they physically cover the surface and second, they increase the roughness. The increased roughness results in a layer of slow moving air near the surface. This air layer can act as a barrier and protect the soil surface by excluding winds of an erosive velocity.

Step 3: Plant cover

These cover levels assume that at least one third of the stubble is anchored and do not apply to stubble which is fragile or easily detached, such as field pea stubble. Estimate using photo standards and Figure 7.1.5 (Carter 1994).

Step 4: Soil surface condition

For erosion to occur, the soil surface generally has to be loose or have been cultivated recently. The whole surface does not have to be loose, as any loose soil on the surface is likely to erode. Self-mulching soils (rare in WA) can also erode, depending on the size of the individual clods or peds of soil.

Step 5: Clods on the surface

The clods (or aggregates) on the soil surface reduce the wind speed because they increase surface roughness and they also shield smaller particles. Particles larger than 1 mm cannot be moved by the wind unless there is an extreme event (e.g. cyclone). If soil moves off an adjacent paddock it can abrade and break down clods to an erodible size.

Step 6: Combined effects

Wind erosion can be controlled by the combined effects of gravel, stones, anchored plant material and clods on the surface (assess from plan view).

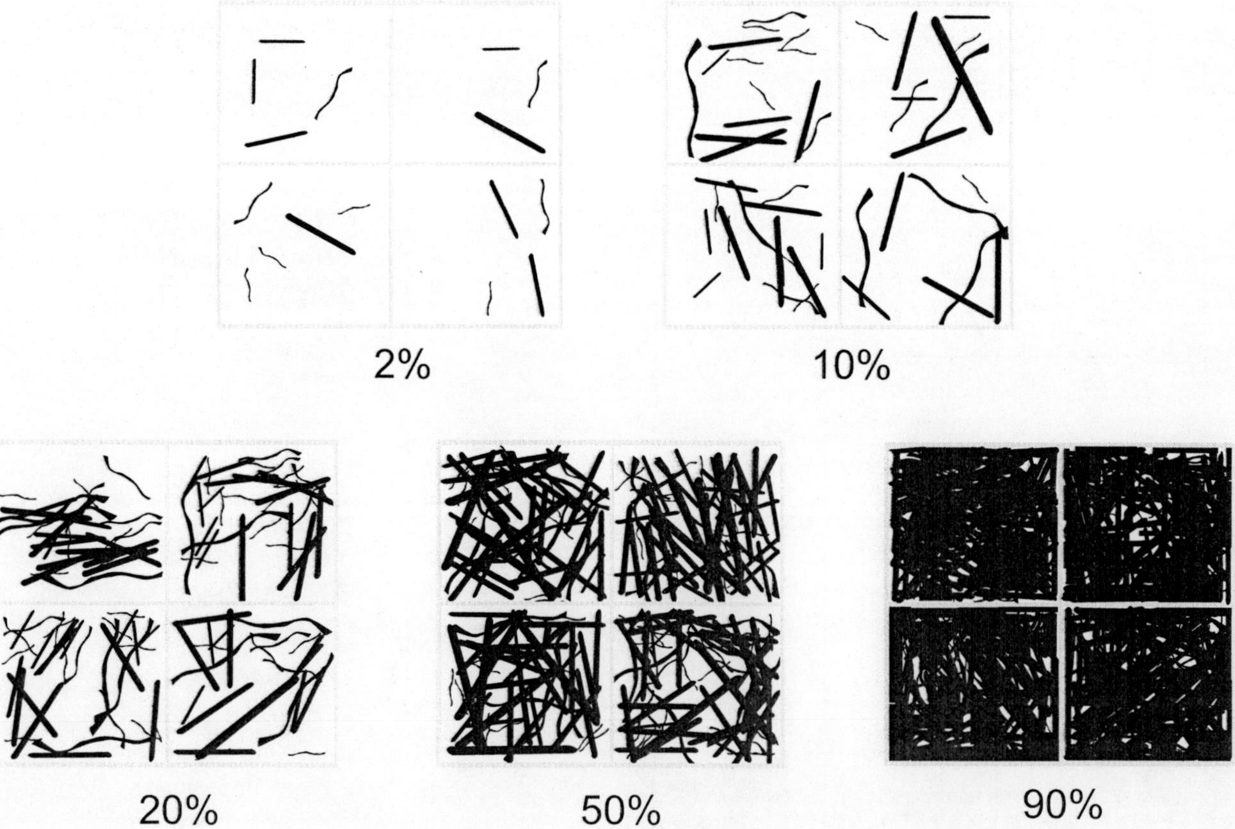

Figure 7.1.5 Estimating percentage of ground covered by prostrate plant residues. Each quarter of any one square has the same area of black.

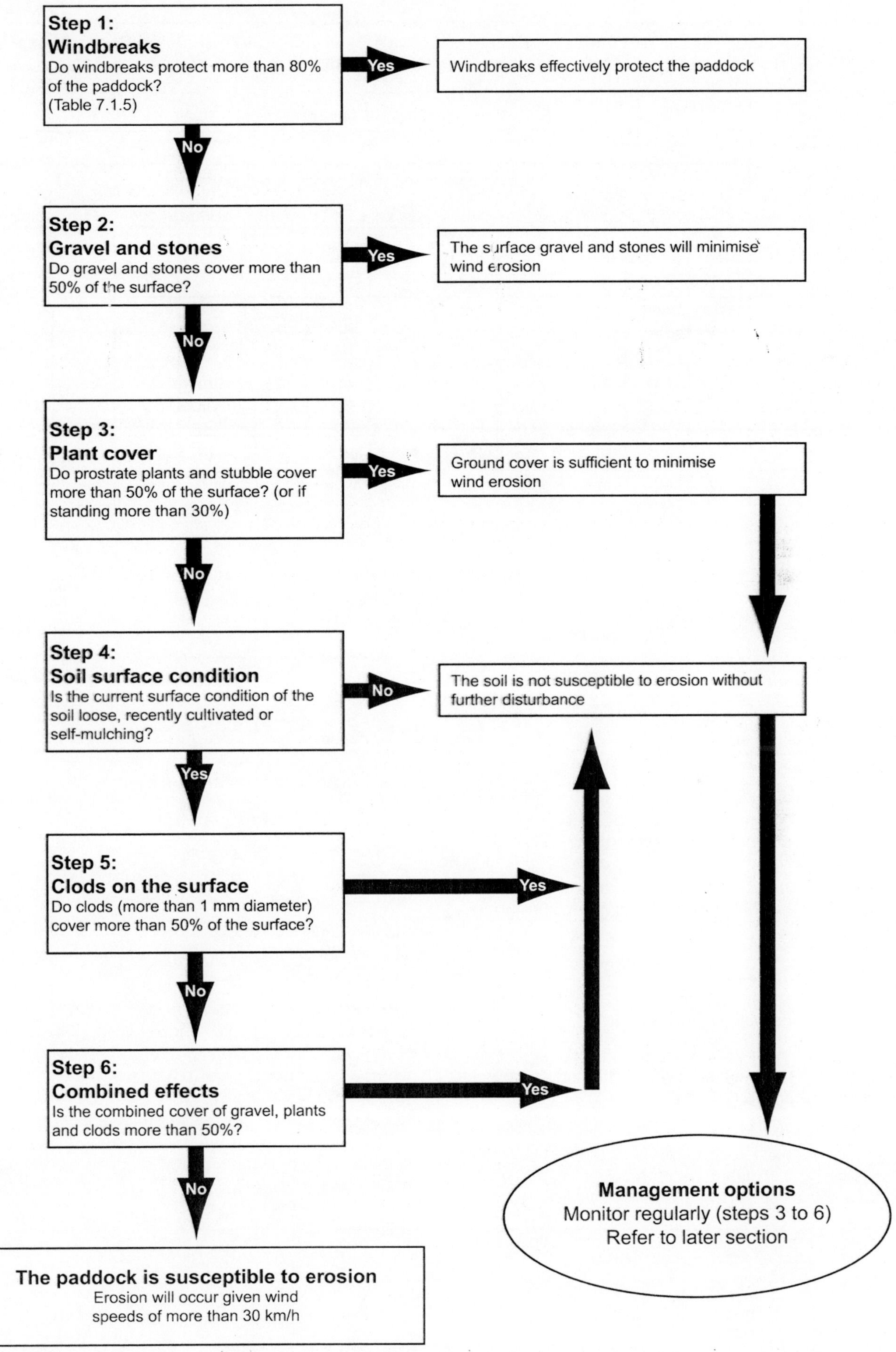

Figure 7.1.6 Management decision diagram for wind erosion control.

Table 7.1.5 Estimating the percentage of a paddock effectively protected by windbreaks at right angles to the wind direction. The direction of prevailing winds for the time of year should be used.

Height of windbreak (m)	Slope of paddock											
	<1%				~3%				~6%			
	Distance downwind or between windbreaks (m)											
	200	**300**	**500**	**800**	**200**	**300**	**500**	**800**	**200**	**300**	**500**	**800**
3	15%	10%	6%	4%	12%	8%	5%	3%	9%	6%	4%	2%
4	20%	13%	8%	5%	16%	10%	6%	4%	12%	8%	5%	3%
5	25%	16%	10%	6%	20%	13%	8%	5%	15%	10%	6%	4%
6	30%	20%	12%	8%	24%	16%	10%	6%	18%	12%	7%	5%
10	50%	33%	20%	13%	40%	27%	16%	**10%**	30%	20%	12%	8%
15	75%	50%	30%	19%	60%	40%	24%	15%	45%	30%	18%	11%
20	100%	67%	40%	25%	80%	53%	32%	20%	60%	40%	24%	15%
30	100%	100%	60%	38%	100%	80%	48%	30%	90%	60%	36%	23%

** The shaded example is for an 800 m wide paddock (equivalent to 800 m between windbreaks) on a 2-3% slope, with a 10 m high shelterbelt of trees on the eastern fenceline. For prevailing winds from the east, approximately 10% of the paddock is protected. The effect of slope on the area protected by windbreaks was derived from United States Department of Agriculture data. The effective protection is reduced when the prevailing winds are at an oblique angle, not perpendicular to the windbreaks (refer to Bird* et al. *1992).*

EXAMPLE

Paddock description: An 80 ha, gently sloping (2-3%) paddock (1000 m north-south x 800 m east-west) has a 10 m high shelterbelt of eucalypts along the northern and eastern fencelines. The soil is a red-brown gravelly sandy loam. A flock of wethers is grazing the annual pasture. The prevailing winds are from the east during January and February.

Assessment of wind erosion status (mid-January)

Step 1: Windbreaks
The shelterbelt on the eastern fenceline protects only 10% of the paddock (Table 7.1.5).

Step 2: Gravel/stones
There is gravel in the surface, although the amount exposed is less than 5%.

Step 3: Plant cover
The paddock has been grazed since early December and the estimated cover level from the photo-standards and chart (Figure 7.1.5) is about 20%.

Step 4: Soil surface condition
The surface is firm, although the sheep have loosened the top 1 to 2 mm of soil.

Step 5: Clods on the surface
The soil has been under pasture for two years and there are no clods on the surface.

Step 6: Combined effects
The gravel and plant residues provide about 25% surface cover.

Conclusion: The shelterbelts and the plant cover do not protect the paddock adequately. The loose soil on the surface is very likely to erode during winds of more than 30 km/h (high probability). The firm surface would then prevent any further erosion. If the wethers are left in the paddock they will continue to loosen soil and increase the potential soil loss. The recommendation would be to remove the stock.

Lupin stubble. (a) A paddock with sufficient lupin stubble to reduce wind erosion after cultivation. The stubble (6,852 kg/ha) consists of 2,812 kg/ha which is effective stubble and 4,040 kg/ha of small, loose ineffective stubble. (b) Inadequate lupin stubble to minimise wind erosion. The stubble (816 kg/ha) is all loose and the paddock is highly susceptible to wind erosion.

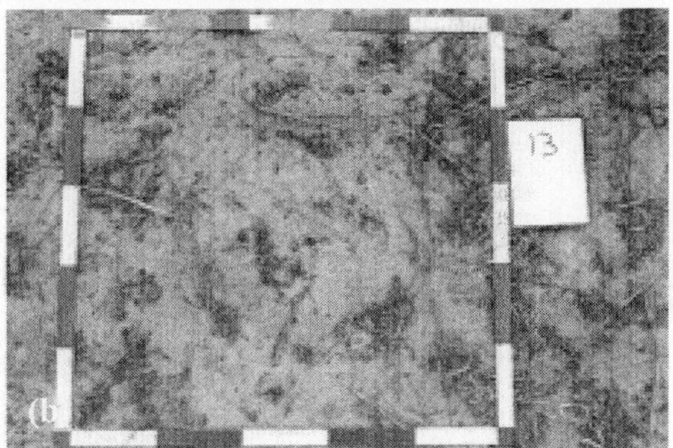

Cereal stubble. (a) More than enough cereal stubble to minimise wind erosion. The 0.25 m² quadrat represents 1,400 kg/ha. (b) Inadequate stubble levels to protect the paddock from wind erosion. The 0.25 m² quadrat represents 400 kg/ha.

MANAGEMENT OPTIONS FOR CONTROLLING WIND EROSION

The susceptibility of a soil to wind erosion has been defined in terms of the amount of disturbance necessary to bring that soil to an erodible condition. *Use the monitoring system (Figure 7.1.5) to determine the 'critical' step for wind erosion control in each paddock.*

Windbreaks

- Windbreaks can be an effective means of minimising wind erosion and provide benefits by reducing groundwater recharge and increasing crop yields.
- The size of most paddocks in the WA wheatbelt precludes the complete control of erosion by growing trees along the fencelines. Windbreaks can be used in conjunction with other management methods or as part of an alley farming system.

Problem areas

- If the susceptibility of the paddock varies widely, a long-term plan to stabilise the paddock may include changing fencelines to follow land class boundaries (e.g. if part of the paddock is susceptibility category **(v)** but the remainder is in categories **(i)**, **(ii)** or **(iii)**, then the land with a high risk of wind erosion **(v)**, should be fenced off from the rest of the paddock so it can be managed separately.)
- Stabilise blowouts by fencing to allow native vegetation to regenerate, or by sowing to cereal rye.
- Replace trace elements when a significant proportion of the cultivation layer is removed by erosion.

Land use

Management options to minimise soil erosion from the different grazing and cropping regimes are summarised in Table 7.1.6.

Table 7.1.6 Summary of the management options for controlling wind erosion.

<table>
<tr><th rowspan="2">Land use</th><th colspan="5">Wind erosion susceptibility class
(from Table 7.1.3)</th><th rowspan="2">Further information</th></tr>
<tr><th>(v)</th><th>(iv)</th><th>(iii)</th><th>(ii)</th><th>(i)</th></tr>
<tr><td rowspan="4">Grazing (general)</td><td colspan="2">Essential to maintain >50% ground cover. Monitor regularly.</td><td colspan="3">Desirable to maintain >50% ground cover. Monitor regularly.</td><td rowspan="4">Carter et al. (1991, 1992).</td></tr>
<tr><td colspan="5">Plan ahead by assessing the amount of feed available from stubbles and pasture in October/November to determine the carrying capacity over summer.</td></tr>
<tr><td colspan="5">Soil detachment by sheep grazing in summer (t/sheep/week)</td></tr>
<tr><td>Loose surface</td><td>2.0</td><td>0.6</td><td>0.3</td><td>0.1</td></tr>
<tr><td>Grazing lupin stubbles</td><td colspan="5">Use movable watering points to increase uniformity of grazing.
Remove stock if cover levels <50%, or <70 kg/ha of lupin seed on the ground because otherwise stock will lose condition (C. McDonald personal communication).</td><td>Carter and Findlater (1989)</td></tr>
<tr><td rowspan="2">Grazing field pea stubbles</td><td colspan="2" rowspan="2">Not recommended
Extreme wind erosion hazard.</td><td>High hazard, unless adequate protection from summer weeds. If grazed use management options under (i), (ii).</td><td colspan="2">Minimise harvest grain losses by thorough paddock preparation prior to seeding. Graze heavily, but for a maximum of 4-6 weeks, or until seed on the ground is <50 kg/ha (which ever comes first).</td><td rowspan="2">Russell (1993)</td></tr>
<tr><td>Grazing should be left until autumn.</td><td colspan="2">Grazing should be immediately after harvest, to allow for summer rain to form a crust on the surface or for cultivation to increase surface roughness.</td></tr>
<tr><td rowspan="3">Cultivating/sowing</td><td colspan="5">The amount of disturbance to make a soil erodible</td><td rowspan="2"></td></tr>
<tr><td>None (loose surface)</td><td>no-till sowing</td><td>direct-drill</td><td>2 passes prior to sowing</td><td>Maximum disturbance (i.e. rotary hoe)</td></tr>
<tr><td colspan="4">Stubble retention with ~1 t/ha of straw (incorporate into the soil to anchor if necessary), or ~50% ground cover. High levels of stubble may be necessary depending on the amount of stubble left by the seeding machinery.
Minimise cultivation, no-till or direct-drill sowing is desirable.
Reduce the cultivation speed on drier soils; stop if clouds of dust form.</td><td>Low erosion hazard</td><td>Leonard (1993)
Bligh and Findlater (1996)</td></tr>
</table>

7.2 RUNOFF AND WATER EROSION

Neil Coles and Geoff Moore*

The land surface of south-western Australia has been subjected to long and complex weathering and denudation processes which are expressed today as variations in topography, soils and hydrology (Churchward 1986). The physical degradation of the soil caused by agricultural practices has affected the movement of water in the soil and through catchments. These changes have altered runoff generation and erosion patterns in the agricultural and pastoral areas. For these reasons, soil management and the effect on runoff and erosion need to be examined.

Gullies remove land from production, add large amounts of sediment to river systems and divide paddocks into small areas which are difficult to manage (G. Moore).

Water erosion affects the long-term sustainability of agriculture in Western Australia. The fertility of wheatbelt soils is susceptible to rapid decline as a result of water erosion. For example, the loss of only 4 mm of topsoil from a pasture paddock may result in a 10 to 20 percent decline in the following cereal crop (Marsh 1982). Water erosion is highly variable, both temporally and spatially, with most soil loss occurring from a small proportion of the agricultural area, and is usually associated with extreme rainfall events (e.g. NSW, Edwards 1980, 1987; WA, McFarlane *et al.* 1992a).

Significant soil losses are associated with infrequent but extreme events. Therefore, to be effective the assessment of erosion risk and erosion management must focus on these events.

Variability in runoff in south-western Australia is related to the unique nature of the soil and landforms and the way in which the landscape has been developed for agriculture. Quantifying the soil-landform response to rainfall is therefore highly desirable to identify the sources of runoff, potential erosion sites and areas prone to waterlogging or flooding (Coles 1993).

Although runoff causes water erosion and is therefore viewed as a problem, the collection and storage of runoff is necessary to maintain on-farm (surface) water supplies. Reliable water supplies may require an improvement to water shedding characteristics of a catchment. Surface water can be collected from water courses (either perennial or ephemeral) or from overland flow. The objective of any farm water supply plan or design is for the catchment to generate sufficient water to provide a continuous supply of water and to ensure that excess runoff is redistributed or conveyed in a manner that minimises soil erosion. Catchment runoff estimation is therefore made with three outcomes in mind: to determine the expected water yield for dam design, to determine flood potential and to assess the erosion potential.

FACTORS AFFECTING RUNOFF GENERATION

Climate determines the broad upper limit to the amount of runoff that can be generated from a catchment. Meteorological factors that can be correlated with runoff and measured with reasonable accuracy are:

- precipitation (amount, intensity, frequency and duration);
- spatial variability of precipitation;
- other climatic factors which affect evaporation and transpiration (e.g. wind velocity, solar radiation, temperature and air humidity).

* *Agriculture Western Australia, South Perth.*

The relationships between runoff, rainfall and evapotranspiration are modified by short-term changes in catchment management. Factors such as catchment shape, topography and soil distribution remain fairly constant over long periods while others associated with land management may alter rapidly (Coles 1988). Catchment factors that influence runoff generation include:

- catchment shape and topography;
- underlying geology and hydrogeology;
- vegetation cover;
- soil properties and soil distribution.

The effect of some of these factors is described by Laing (1988):

- forest areas yield less runoff than crop and pasture areas;
- dense, vigorous forest yields less runoff than sparse forest;
- cultivated soil generates less runoff than uncultivated soil;
- vegetated and mulched areas yield less runoff than unvegetated and bare areas;
- well-structured and self-mulching soils generate less runoff than massive, hardsetting soils;
- loose sandy soils generate less runoff than compacted, hard clayey soils;
- deep soils generate less runoff than shallow soils with significant rock outcrop.

RUNOFF MECHANISMS

The main mechanisms that produce runoff in WA are *infiltration excess runoff* and *saturation excess runoff* (Coles 1993). They are not mutually exclusive and may generate runoff at different times during a storm or from different locations within a catchment. At a particular site, runoff can either be generated primarily by rainfall excess or by saturation excess depending on the rainfall intensity, surface conditions and soil moisture status (Hauck and Coles 1991).

Infiltration excess runoff (also called Hortonian flow or rainfall excess) occurs when the rainfall intensity exceeds both the infiltration capacity of the soil and it's surface storage (Horton 1945). This runoff mechanism is normally associated with high intensity storms and/or soils that have slow infiltration rates. In a flood situation, this process is likely to be a large-scale or 'catchment-wide' phenomenon with all sections of the catchment contributing to runoff more or less simultaneously.

Infiltration excess is an important mechanism on the following soils:

- fine-textured soils with low infiltration capacities;
- surface sealing and hardsetting soils (depends on the dispersive nature of the clay component);
- water repellent soils (area affected will vary with season and vegetation cover);
- surface-compacted soils with reduced surface infiltration and surface storage capacities.

Saturation excess runoff (also called saturated overland flow) is generated when rain falling on saturated soils results in ponding, and runoff is initiated once surface storages are exceeded. This is an important mechanism on wet or waterlogged soils. The high soil water content before the rain can be associated with a perched watertable, high groundwater table or lateral (subsurface) flow.

Saturation excess is commonly associated with the following soils and conditions:

- a permeability contrast within the profile (e.g. permeability contrast soils, Section 1.2) that results in the topsoil becoming waterlogged due to: relatively higher infiltration and low evaporation characteristics of the topsoil, low permeability of the subsoil and low soil water storage in the coarse-textured A horizons.
- saturated overland flow can develop where perennial groundwater rises to the surface because of vertical percolation or where there is lateral downslope movement of subsurface water.
- soils in groundwater discharge areas that may remain saturated throughout the year, or areas at the base of slopes or along rivers/creeks where the watertable is near the surface.
- fine-textured soils on valley floors where water may pond due to slow internal and external drainage.

The amount of runoff generated by a particular rainfall event can be estimated using five key factors:

- rainfall intensity,
- surface storage,
- surface infiltration rate,
- storage capacity of the topsoil,
- topography.

Significant soil losses are usually associated with heavy rain on recently cultivated soils. However, the rainfall intensity required to initiate runoff on a saturated soil may be as low as 1 mm/h. Susceptibility to erosion may also depend on the frequency and spatial extent of waterlogging (Coles 1993).

Contributing area

Areas contributing to runoff can comprise 5 to 85% of small agricultural catchments. This variation in contributing area can be compared using two mechanisms at the extreme ranges of the scale, 100% infiltration excess runoff or 100% saturation excess runoff. During 100% infiltration excess runoff, water commonly reaches a drainage line or stream channel shortly after a rainfall event, often within an hour. For storms of longer duration, the percentage of contributing area (C_A) is equal to the percentage of the catchment where rainfall intensity exceeds the infiltration rate.

$$C_A\,(\%) = (\Sigma \text{Area (Rainfall intensity} > \text{(infiltration and surface losses))} \times 100) / \text{Total area} \quad (7.2.1)$$

Where vegetation and soil are thin or absent, the contributing area may be large. For example, the area contributing runoff (through infiltration excess runoff) was 85%, for an abandoned mining site with less than 36% vegetation.

Areas that are associated with 100% saturation excess runoff include:

- stream channels;
- areas of saturated or near-saturated soil adjacent to the channels, which respond rapidly to the changes in rainfall intensity; and
- narrow strips of hillside around the saturated areas, the width of which is determined by the extent of unsaturated soils with low hydraulic conductivity (Kirkby 1969, 1978).

The changing size of the contributing area plays a dynamic role in generating runoff from hillslopes and catchment basins. The contributing area varies with season, antecedent soil moisture and climatic conditions during and between storms (Chorley 1978).

ESTIMATING RUNOFF

In agricultural catchments, *peak flow* (or peak discharge) is required for earthworks designed to convey water. Both peak flow and *total volume* (yield or total discharge) are required for the design of structures used to harvest or store runoff. Runoff estimation methods that consider soil and land use are desirable. Estimates of catchment water balance also require estimation of catchment yield with reasonable accuracy.

As runoff is influenced by a complex interaction of topographic, soil and climatic factors, there is no universal method to predict the amount or frequency of runoff from catchments with variable characteristics (Sivapalan *et al.* 1990; Coles and Sivapalan 1991). Localised parameters can be incorporated into deterministic models, or empirical equations can be used to improve estimations.

Relationships between peak runoff rates and catchment characteristics for south-western Australia were developed by Flavell *et al.* (1987) and are described in *Australian rainfall and runoff* (IEA 1987). These runoff estimation methods are described for different regions and soil textures in the *Soil conservation earthworks design manual* (Bligh 1989).

The frequency of a rainfall or flood event is expressed in terms of the *average recurrence interval* (ARI) or the *probability of exceedence* (or average exceedence probability). The ARI is the number of years on average, between hydrologic events which exceed a given magnitude. The *probability of exceedence* is the reciprocal of the ARI.

Comparable average recurrence intervals and probability of exceedence:

ARI (years)	Probability of exceedence (%)
2	50
5	20
10	10
20	5
50	2

Statistical Rational Method

The Statistical Rational Method is based on the probabilistic (or statistical) interpretation of the catchment factors and rainfall and runoff data. The equation is expressed as:

$$Q = F\,C\,I\,A \quad (7.2.2)$$

and rewritten as:

$$Q_y = 0.278\,C_y\,It_{c.y}\,A \quad (7.2.3)$$

where,

Q_y is the peak flow rate (m^3/s) for an average recurrence interval (ARI) of y years;

0.278 is a factor of proportionality (F) related to the units used;

C_y is the runoff coefficient (dimensionless) for an ARI of y years;

$It_{c.y}$ is the average rainfall intensity I (mm/h) for design duration of t_c hours and y years;

A is the area of the catchment (km^2);

t_c is the time of concentration (hours).

Runoff coefficient. The calculation of a runoff coefficient (C) is based on the dimensionless ratio of the total volume of runoff to the total volume of rainfall, where both rainfall and runoff are expressed in the same units (Wanielista 1990). Therefore, if 30% of the rainfall appears as runoff for a storm with a 5 year ARI, then $C_5 = 0.30$. In using the Statistical Rational Method, runoff estimates depend on rainfall estimates of a given duration with the same average recurrence interval. The runoff coefficient C_y is based on regional streamflow and rainfall data. The derived value of C_y is given by:

$$C_y = Q_y/(0.278\ It_{c.y}\ A) \qquad (7.2.4)$$

with values for $It_{c.y}$ obtained from the regional data for the catchment of interest and derived from the appropriate rainfall intensity-frequency-duration (IFD) curves. The IFD curves were designed to provide accurate, temporally and spatially consistent values (Canterford *et al.* 1987). The IFD curve estimation procedures use the annual maximum rainfall data for a duration between 5 minutes and 72 hours and are fitted to Log Pearson Type III distributions. Curves have been derived for ARIs between 1 and 100 years and are presented in *Australian rainfall and runoff* (IEA 1987).

Time of concentration (t_c) is defined as the longest time taken for water to travel by overland flow from any point in the catchment to the outlet (Hudson 1981; IEA 1987).

Estimating the time of concentration is important and can be derived using a formula based on either the region studied or an area considered to be representative of the study site (Flavell *et al.* 1987). If recorded data are unavailable, other methods can be used to compute t_c which are based on the relationship between the catchment area, slope, stream length, surface cover and rainfall intensities. For Australian conditions, the Bransby-Williams method is recommended by Pilgrim (1987) as "an arbitrary but reasonable approach" and is:

$$t_c = 58L / A^{0.1} S_e^{0.2} \qquad (7.2.5)$$

where,

L is the catchment length (km)

A is the catchment area (km^2) and

S_e is the equal-area slope (m/km).

To apply the Statistical Rational Method to a catchment the following procedure is used:

- determine the values for the catchment characteristics (L, A, etc.);
- estimate the time of concentration, t_c;
- obtain the rainfall intensity, I for the expected duration (from the intensity-frequency-duration curve in *Australian rainfall and runoff*);
- determine the runoff coefficient, C_{10};
- adjust C_{10} to the required average recurrence interval for *y* years by multiplying by C_y/C_{10} (the desired frequency factor);
- calculate the design discharge, Q_y.

Pilgrim (1987) recommended that the deterministic interpretation of the Statistical Rational Method should only be applied to small catchments (i.e. <25 km^2) and suggested the method failed to adequately account for physical factors that may influence the runoff potential. These factors include temporary surface storage areas, antecedent soil moisture and the temporal and spatial variation inherent in rainfall events.

The Statistical Rational Method employed in the probabilistic sense was considered by Pilgrim (1987), Flavell *et al.* (1987) and Sallaway *et al.* (1989) to be a relevant alternative to the Index Flood Method, where the *physical considerations are incorporated* into both the frequency statistics and the derived value of C_y. Its use in the physical sense, although modified by incorporating soil factors into the runoff coefficient (C_y), was not considered applicable in conditions of non-wetting sands, duplex soils and saline seeps.

Table 7.2.1 Comparative values for the Statistical Rational Method and Index Flood Method using Flavell's regional equations for individual catchments* in south-western Australia.

Statistical Rational Method						
Soil	**ARI**	2	5	10	20	50
Loamy	C_y/C_{10}**	0.41	0.65	1.00	1.54	2.20
Loamy and lateritic	C_y/C_{10}	0.43	0.67	1.00	1.45	1.98
Index Flood Method						
Soil	**ARI**	2	5	10	20	50
Loamy	Q_y/Q_5***	0.48	1.00	1.84	3.23	6.10
Loamy and lateritic	Q_y/Q_5	0.50	1.00	1.76	3.05	5.65

* *cleared area of catchment (100% in all the cases, but may be modified to suit other catchments; values given in Flavell* et al. *1987 and IEA 1987).*

** C_y/C_{10} *is the ratio of the runoff coefficient for* y*-years to the runoff coefficient for 10 years.*

*** Q_y/Q_5 *is the ratio of* y*-year ARI peak flow to the 5-year ARI peak flow.*

Index flood method

The Index Flood Method (also called the Regional Flood Frequency Method) estimates floods of a particular average recurrence interval (ARI) at a given location within a region. The method is based on the available flood data which are then related to catchment physical parameters and rainfall characteristics. This method is outlined for different regions of south-western Australia in Flavell *et al.* (1987) and *Australian rainfall and runoff* (IEA 1987).

The steps for applying the Index Flood Method are:

1. Determine the region applicable to the catchment of interest, and acquire the catchment and climatic data required for the regional estimation equation.
2. Calculate the peak discharge for Q_2 using the formula recommended for the region.
3. Adjust the Q_2 to the design discharge Q_y for the required ARI of y years by multiplying by the appropriate frequency factor Q_y/Q_2.

Flood frequency values for C_y/C_{10} derived from Flavell's regional equations are given in Table 7.2.1.

Flood hydrograph models - contributing area concept

There are many types of models used to estimate flood hydrographs from rainfall excess. The most common of these are storage routing models such as RORB and RAFTS. Models represent runoff processes in more detail and generally consider topography, hydraulics of runoff, the routing of runoff and temporary storage in the drainage basin. The modelling approach can lead to an improved prediction of surface runoff if data inputs accurately represent the modelled catchment.

Concepts such as *contributing area* (Wanielista 1990) are used in most models. This enables the runoff response to be characterised according to sub-areas that represent different land management, soil types, groundcover, soil moisture or any other division that can be represented with a set of catchment parameters.

Design rainfall events, or actual rainfall data, can be used to predict rainfall excess runoff. Flood hydrographs are computed by representing rainfall excess in a series of temporary storages. The volume of water in the temporary storage is considered in transit to the catchment outlet. Hydrographs are constructed in a time series that combines runoff responses at a selected point in the catchment. The highest flow in the modelled hydrograph is used as an estimate of peak flow for the average recurrence interval of the design storm. The total flow represented by the hydrograph provides an estimate of the flood volume or total discharge.

Water yield classification

A study by Davies and McFarlane (1987) developed estimation equations based on mean annual flows for flood volumes that used catchment and climatic parameters. Flood frequency factors were developed to allow estimation of the discharge Q, for ARIs of up to 50 years for flow durations of one, two and five days.

In many areas, the volumes of runoff or the frequency of runoff events cannot be determined because data are not available. A practical guide for estimating water yield (or runoff volume, total discharge) for a particular catchment is a useful tool for conservation and water use planning.

A general relationship exists in south-western Australia between the water yield and the landforms, soils and original vegetation. This relationship has been described for the lateritic profile and various erosional and depositional derivatives (Bettenay and Hingston 1961; Mulcahy and Hingston 1961; Churchward and McArthur 1980; McArthur 1980). The application of soil-landform associations in runoff volume estimation is summarised for catchments with differing runoff potential (i.e. high, medium or low; Table 7.2.2).

This method is applicable to the Zones of Ancient Drainage and Rejuvenated Drainage. Some catchments and landforms are expected to give runoff values outside the estimates quoted, including water repellent sands, cracking clays and medium to deep duplex soils. In the Zone of Ancient Drainage the classification system should also be modified to account for: the amount of lime and/or gypsum in the profile (affects soil permeability); the occurrence of gilgai (affects surface roughness); catchment slope and the presence and extent of shallow watertables. These factors interact with the rainfall, soil texture, soil profile type, and depth to an impeding layer (if present) to influence the total water yield of a catchment. The likely runoff mechanisms can also be identified.

The relative yields of agricultural catchments described in Table 7.2.2 and roaded catchments were used to produce a series of tables (Laing 1988) that provide guidelines for estimating annual runoff (Tables 7.2.3 to 7.2.5 below). To estimate the catchment yield the following formula was recommended:

$$\text{Catchment runoff } (m^3) = 10 \times A \times R \qquad (7.2.6)$$

where,

A=catchment area (ha),

R = annual runoff (mm) - with nominated probability of exceedence.

This equation is a simplification of the complex processes involved in the generation of runoff and because the soil-landform associations have been grouped into three general classifications, a wide range in runoff can be expected.

Table 7.2.2 General soil-landform associations and potential water yield classifications for south-western Australia (after Bettenay and Hingston 1961; Mulcahy and Hingston 1961; Salim 1982 and Laing 1988).

Type	General landform description	Runoff generating potential	Surface slopes	Surface texture and profile type	Soil-landform associations	Depth to restricting layer*
A	Incised valleys, valley floor, breakaways, pediments.	High**	Steep**	Loam, clay loam, clay or shallow duplex soils	Murray, Williams, Birberkine, Popanyinning, Michibin, Coolakin, Booraan, Danberrin, Stirling, Baandee, Merredin, Hines Hill, Balkuling, Avon, York, Malebelling, Mortlock, Calje, Belmunging	Shallow (<15 cm)
B	Intermediate slopes	Medium**	Medium**	Sandy loams or medium-depth duplex soils	Pindalup, Yarragil, Mobedine, Kauring, Collgar, Nangeenan	Intermediate (15-40 cm)
C	Lateritic plateau	Low**	Low**	Loamy sand, clayey sand, gravelly sands and/or massive laterite or deep duplex profiles	Dwellingup, Norrine, Quailing, Ulva	Deep (>40 cm)

** Restricting layer - usually subsoil sandy clay.*

*** These descriptive terms are relative for any particular region or zone.*

Table 7.2.3a Annual runoff estimates for landforms of Type A, in the >600 mm average annual rainfall zone (after Laing 1988).

Mean annual rainfall (mm)	Annual runoff (mm) from forested catchments			Additional runoff (mm) generated following clearing			Annual runoff (mm) from cleared catchments in pasture/crop		
	Annual exceedence probability								
	90%	50%	10%	90%	50%	10%	90%	50%	10%
600	5	20	80	20	45	90	25	65	170
800	20	55	140	50	70	90	70	125	230
1000	55	130	260	80	90	100	135	220	360
1200	120	230	430	100	110	120	220	340	550
1400	240	330	530	110	150	220	350	500	750

Table 7.2.3b Annual runoff estimates for landforms of Type B, in the >600 mm average annual rainfall zone (after Laing 1988).

Mean annual rainfall (mm)	Annual runoff (mm) from forested catchments			Additional runoff (mm) generated following clearing			Annual runoff (mm) from cleared catchments in pasture/crop		
	Annual exceedence probability								
	90%	50%	10%	90%	50%	10%	90%	50%	10%
600	1	5	40	14	40	90	15	45	130
800	12	40	120	50	70	100	62	110	220
1000	30	80	220	80	100	110	110	180	330
1200	90	180	400	100	120	130	190	300	530
1400	180	290	500	110	150	220	290	440	720

Table 7.2.3c Annual runoff estimates for landforms of Type C, in the >600 mm average annual rainfall zone (after Laing 1988).

Mean annual rainfall (mm)	Annual runoff (mm) from forested catchments			Additional runoff (mm) generated following clearing			Annual runoff (mm) from cleared catchments in pasture/crop		
	Annual exceedence probability								
	90%	50%	10%	90%	50%	10%	90%	50%	10%
600	0	1	8	2	10	50	2	11	58
800	3	15	75	30	50	70	33	65	145
1000	15	50	180	70	90	100	85	140	280
1200	50	130	350	90	110	130	140	240	480
1400	110	230	470	100	150	220	210	380	690

Table 7.2.4 Annual runoff estimates for cleared agricultural land in the 300 to 600 mm annual rainfall zone (after Laing 1988).

Mean annual rainfall (mm)	Annual runoff (mm) from landform Type A			Annual runoff (mm) from landform Type B			Annual runoff (mm) from landform Type C		
	Annual exceedence probability								
	90%	50%	10%	90%	50%	10%	90%	50%	10%
300	0.1	2	20	0	1	15	0	0	20
400	0.5	5	40	0.2	3	25	0	0.5	30
500	5	25	100	2	15	75	0	3	45
600	25	65	170	15	42	130	2	10	60

Table 7.2.5 Annual runoff estimates for roaded catchments (after Laing 1988).

Mean annual rainfall (mm)	Annual runoff (mm) from sandy clay - massive			Annual runoff (mm) from sandy clay, moderately structured with lime nodules			Annual runoff (mm) from loamy sand - massive, and compacted to >1900 kg/m^3		
	Annual exceedence probability								
	90%	50%	10%	90%	50%	10%	90%	50%	10%
300	50	80	125	30	60	100	40	70	110
400	95	150	230	60	110	190	80	135	205
500	145	230	330	95	170	270	130	210	300

WATER EROSION

Water erosion is a complex process, it involves runoff and soil detachment. In an agricultural context, it is influenced by a number of factors including rainfall erosivity, soil erodibility, slope angle, slope length and management practices (Wischmeier and Smith 1978). The single most important factor affecting soil loss at a site is the amount and type of groundcover (Emerson 1991).

Soil is essentially a non-renewable resource. Under the present climate of Western Australia, soil formation from weathered rock is negligible, only 1 mm per 100 years (McFarlane and Ryder 1987). Given this low rate of soil formation, water erosion is considered to be an irreversible form of land degradation. The acceptable level of erosion approaches zero for sustainable production on many soils.

TYPES OF WATER EROSION

Sheet erosion (or inter-rill erosion) is the removal of a fairly uniform layer of soil over a relatively wide area, with no perceptible channels where flow has been concentrated (Houghton and Charman 1986). Raindrop splash is frequently the main mechanism of soil detachment, although entrainment can also be involved on steep or long slopes (Proffitt and Rose 1991).

Rill erosion is the removal of soil by runoff, whereby numerous small channels (up to 30 cm deep) are formed (Houghton and Charman 1986). These channels are usually small enough to be obliterated by normal tillage and generally only affect a small proportion of the catchment surface, but they are more visible than sheet erosion. Rill erosion mainly involves the detachment of soil particles by concentrated runoff and entrainment. The runoff tends to concentrate in slight depressions or cultivation furrows to form rills, but erosion will only occur when the shear force of the flowing water exceeds the soil's strength. As a consequence, the water can flow in the rill for some distance before erosion starts (Rosewell *et al.* 1991).

Gully erosion is the removal of soil by a concentrated flow with sufficient volume and velocity to cut large channels, generally more than 30 cm deep. Gullies disrupt farming operations and are too large to be removed during normal cultivation (Houghton and Charman 1986).

The hydrology of a catchment is altered when the land is cleared for agriculture. Gullies tend to form in natural watercourses as they adjust to the increased runoff normally generated after clearing. Soils with dispersible subsoils are particularly susceptible to gully erosion if the subsoil is exposed.

Controlling gully development is important because gullies disrupt machinery operations and are a major source of stream sediment. In general, gullies do not cover large areas within a catchment, but can result in dramatic soil losses in areas with a marginal climate for agriculture, or on soils with adverse physical or chemical properties (Hudson 1981).

Tunnel erosion (or piping) is the removal of the subsoil, while the surface soil remains relatively intact, so that cavities are produced under the soil. Eventually, the surface collapses and can form a gully. Tunnel erosion is uncommon in the agricultural area of WA.

Effects of water erosion

The effects of erosion can be catastrophic with an immediate impact on productivity, but can also be subtle. A slight decrease in crop yields (for the inputs) or increase in fertiliser costs may be the only noticeable short-term effect. If erosion does not affect yields in the short-term, it can still severely limit long-term productivity, principally through reduction in soil depth and therefore water storage capacity and fertility. The relationship between soil loss and potential productivity for three contrasting soil profiles is illustrated in Figure 7.2.1.

Water erosion not only removes soil physically from a site, but also changes the composition of the remaining soil. It is normally a selective process, so the sediment removed from catchments is enriched in clay and silt particles, while the remaining soil is depleted (Swanson *et al.* 1965; Miller and Baharuddin 1987).

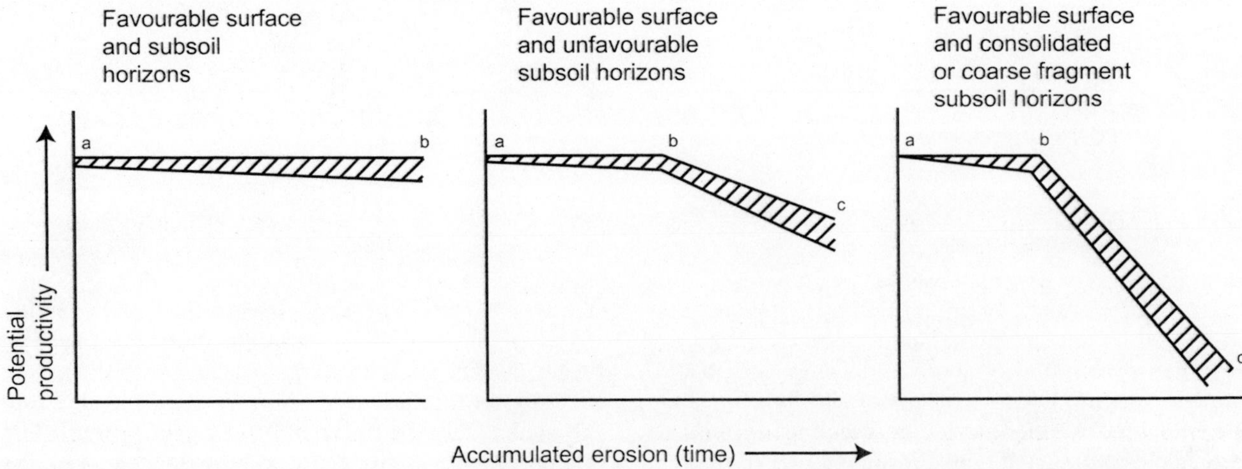

Figure 7.2.1 *The effect of erosion on long-term productivity for three contrasting soil profiles (after Figure 7, Pierce* et al. *1983).*

The degree of sediment enrichment or enrichment ratio (Stocking 1984) varies greatly between soils depending on how evenly the clay and silt fraction is distributed within the profile. In strongly aggregated clayey soils, where the silt and clay is relatively evenly distributed, the enrichment ratio is lower than for a sandy soil where the clay and silt are predominantly in fine particles which are readily detached. As many nutrients are associated with the clay-sized fraction, which includes the organic matter, the enrichment ratio provides a direct link between total soil loss and nutrient losses.

Estimating potential soil losses

The three fundamental processes of water erosion are the detachment of soil particles, their transport and their deposition. Detached soil can result from raindrop impact (terminal velocity of raindrops can reach 9 m/s), tillage (rill erosion often occurs to the depth of cultivation) and the action of animal hooves. Eroded soil is termed *sediment*, and *sedimentation* occurs when soil particles are deposited. The principles of water erosion have been studied extensively from the formative work of Ellinson (1944-47) to the development of the 'Universal Soil Loss Equation' (Wischmeier and Smith 1965, 1978; Renard *et al.* 1991), to the more recent process-based models (i.e. Water Erosion Prediction Program, WEPP, Laflen *et al.* 1991).

The following paragraphs outline a series of steps that can be followed to determine the potential of a land unit to generate runoff under different rainfall intensities and management options. These steps are summarised in flow diagrams (Figures 7.2.6 and 7.2.7) at the end of this section.

Step (i) Rainfall and infiltration

- Rainfall erosivity

Rainfall erosivity is a measure of the potential of rainfall to cause erosion. It is a function of the physical characteristics of rainfall, which include the drop size, their velocity and the intensity of the rainfall. Following the analysis of long-term rainfall events and soil loss, an index of erosivity was developed to represent this relationship (Wischmeier 1959). The index (EI_{30}) represents the multiple of the storm *energy* (E) and the maximum rainfall *intensity* for 30 minutes (I_{30}). The relevance of the EI_{30} index to soil loss under Australian conditions has been demonstrated by Rosewell (1983). The spatial distribution of rainfall erosivity (EI_{30}) for Western Australia was determined by McFarlane *et al.* (1986).

The rainfall erosivity for a storm with a 10 year average exceedence probability (AEP) is illustrated in Figure 7.2.2 (summer, November-April) and in Figure 7.2.3 (winter, May-October). The higher the value the greater the erosivity. It is useful to know whether erosive storms are more likely in winter, or in summer, when designing earthworks. Erosive storms in winter can result in large soil losses if the area is cropped (i.e. erosion of bare, cultivated soil), while erosive storms in summer can also result in large soil losses if the land is heavily grazed (i.e. minimal vegetation cover, soil loosened by stock). In general, winter erosivities are less variable between years than summer erosivities (Bligh 1989).

Figure 7.2.2 The distribution of summer rainfall erosivity (EI_{30} metric units) with a 10 year average exceedence probability for south-western Australia (from McFarlane et al. *1986).*

Figure 7.2.3 The distribution of winter rainfall erosivity (EI_{30} metric units) with a 10 year average exceedence probability for south-western Australia (from McFarlane et al. *1986).*

Rill erosion on soils formed on greenstone in the Zone of Ancient Drainage.

- Surface storage

Surface storage clearly has a direct relationship with runoff; the greater the surface storage the longer the time lag before runoff occurs. The relationship between infiltration, rainfall, surface storage and runoff is represented in Equation 7.2.7.

$$R = P - I - dE/dt \qquad (7.2.7)$$

(from Hairsine *et al.* 1992)

where R is runoff; P is precipitation; I is infiltration rate, and dE/dt is the rate of change of water held in surface storage (which is related to the surface roughness).

The rougher the surface, the higher dE/dt. Aggregate stability is an additional factor. Soils with low aggregate stability usually have a low surface storage, because the aggregates disperse rapidly during rainfall and resettle to form a relatively smooth surface (Hairsine *et al.* 1992).

- Infiltration

Infiltration rate is not an intrinsic value for a particular soil and will vary depending on the surface condition, surface roughness and management. For instance, when a soil is cultivated there is a large increase in the infiltration rate which then declines during the growing season (Ross 1989). The rate of decrease in infiltration will depend on soil stability and prevailing environmental conditions.

The soil's infiltration rate can be measured, but its usefulness is limited because of high spatial and temporal variability associated with seasonal conditions and management. To measure infiltration rates under rainfed agriculture, a rainfall simulator is used, however this can be very time consuming (Grierson and Oades 1977). The disc permeameter, a common alternative, is best confined to comparing different treatments within an experimental plot or catchment. Both methods only provide a measure of infiltration rate and storage capacity at a particular point in time.

Step (ii) Topography

Topography influences whether excess water flows off a site or ponds on the surface. There are four factors associated with topography: slope angle, slope length, slope form (for non-uniform slopes) and runon from higher ground. The slope angle controls the velocity of the runoff; the length affects the volume of runoff and is highly correlated to catchment area; while adjacent rising ground or rock outcrops control the progressive volume of water that accumulates downslope (i.e. runon).

- Slope angle

In general, the steeper the slope the greater the erosion, because there is more splash downhill, more runoff and it will flow faster. However, runoff experiments do not always show an increase in soil loss with increasing slope, especially when the slope exceeds 10%. Slope angle is an important factor influencing soil detachment and erosion by raindrop splash. As the angle increases, there is more splash downhill which promotes the displacement and entrainment of soil particles in runoff (Ellinson 1944). It has been suggested that downslope splash (%) = 50 + slope% (Hudson 1981).

In general, soil erosion (E) increases rapidly with slope and is thought to vary exponentially with the slope angle (S) (Equation 7.2.8). The value of the coefficient *a* varies from 1.35 to about 2 (Zingg 1940; Hudson 1981).

$$E \propto S^a \text{ (from Zingg 1940)} \qquad (7.2.8)$$

- Slope length

Soil erosion is considered to increase as slopes lengthen. As the slope length is correlated to the contributing catchment area, an increase in slope length tends to increase erosion due to an increase in runoff volume (Davies and McFarlane 1987; Hauck and Coles 1991; Coles 1993). On a long slope there is a bigger build-up in runoff volume and its velocity and depth, which leads to erosion that would not occur on shorter slopes (Hudson 1981).

- Slope form (non-uniform slopes)

Slope form refers to the shape of the slope. Slopes can be concave, convex or a sequence of concave, convex and uniform slopes. Soil movement on a uniform slope is greater than on a concave slope, but less than on a convex slope. The physics of soil loss prediction on non-uniform slopes is discussed by Foster and Wischmeier (1974).

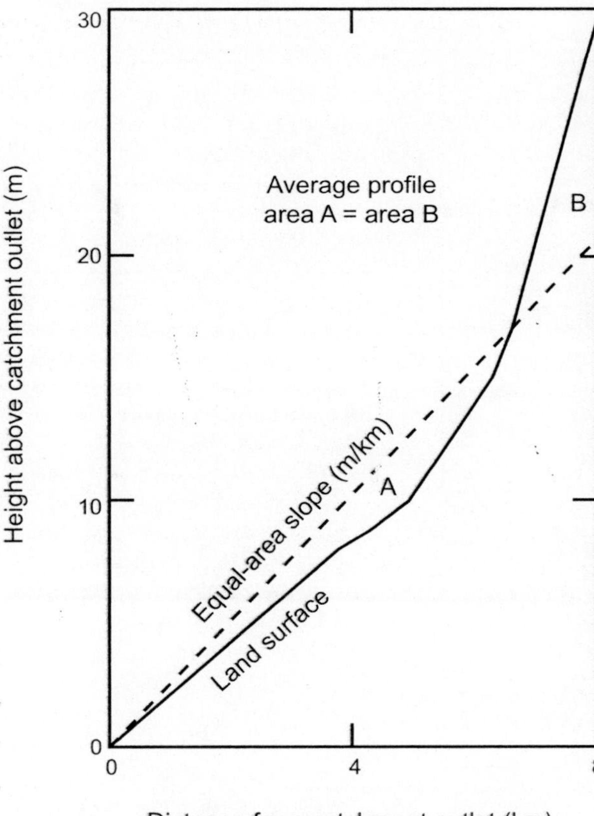

Figure 7.2.4 *Derivation of the equal-area slope (from Figure 4.1, Bligh 1989).*

The equal-area slope is a useful method for determining the equivalent uniform slope. The equal-area slope is calculated from a straight line which bisects the slope, equalising the areas above and below it (Figure 7.2.4). The equivalent uniform slope (length and angle) can then be used to determine the erosion potential of the non-uniform slope.

- Runon

Studies of erosion history using caesium-137 have demonstrated that overland flow (or runon) is a major factor affecting soil erodibility in Western Australia (McFarlane *et al.* 1992b). The highest rates of erosion were recorded from sites used for potato farming and from sites below rock outcrops. Sites below granite outcrops seem to be particularly susceptible to sheet erosion (McFarlane *et al.* 1992b). McGhie (1980) showed that a majority of gullies in the Narrogin district could be traced to runoff shed from water repellent lateritic breakaways.

Assessing the effect of topography

The effects of slope length and gradient are represented as **L** and **S** respectively in the Universal Soil Loss Equation (Wischmeier and Smith 1978). They can be evaluated as a single topographic factor **LS**, which is determined from the nomograph in Figure 7.2.5. The slope length is the distance from a defined point where overland flow starts to the point where the runoff enters a defined channel or to where the slope decreases sufficiently for deposition to occur (technically, the point furthest from the main drainage line may not be easily defined and may be defined as the centroid of a contributing area). Where there are banks on the hillslope, the slope length is the effective distance between the earthworks. The slope angle is the paddock slope, or on non-uniform slopes the equal-area slope.

> *In general, if the slope is less than 1% there is a very low risk of water erosion unless the slope is very long (>400 m), or there is significant runon.*

The nomograph can be used to compare the relative erosion risk at different sites. If all other factors are constant (e.g. rainfall, type of soil, cover level and surface condition), the LS factor represents the comparative erosion risk of slopes with different gradients and lengths. For example, a site with an LS value of 4 represents twice the erosion risk compared with a site having an LS value of 2.

Step (iii) Cover levels

- Plant cover

Vegetation has a direct effect on erosion through three mechanisms: the plant canopy absorbs the energy of falling raindrops (reducing raindrop splash); surface plant material can retard overland flow; and the plant root systems help to bind the soil. As a general rule, surface plant cover will reduce the likelihood of water erosion at a site, irrespective of rainfall and topography. There is an exponential relationship between bare ground (i.e. lack of groundcover) and soil loss (Adams 1966; Meyer *et al.* 1970).

There is a distinct difference between groundcover requirements to minimise water erosion compared with wind erosion. Plant material above the soil surface reduces raindrop impact, but will not significantly reduce overland flow which is a major mechanism of soil detachment (Wischmeier and Smith 1978). Therefore, the critical parameter is the amount of anchored plant material on the surface. The degree of cover is estimated as the percentage cover compared with photo standards viewed from above (plan view).

> *In general, if plant cover on the surface is greater than 70% and the plant material is anchored and lying on the soil surface, there is minimal soil loss. On the other hand, if there is less than 70% plant cover, the areas of bare ground tend to become interconnected, greatly increasing the erosion risk.*

- Stones or gravel

Stones or gravel on the soil surface can reduce erosion because they protect the surface (Lamb *et al.* 1950; McFarlane *et al.* 1991). This effect can be observed when stones are raised on small pedestals, a result of raindrop splash removing the surrounding soil (Hudson 1981).

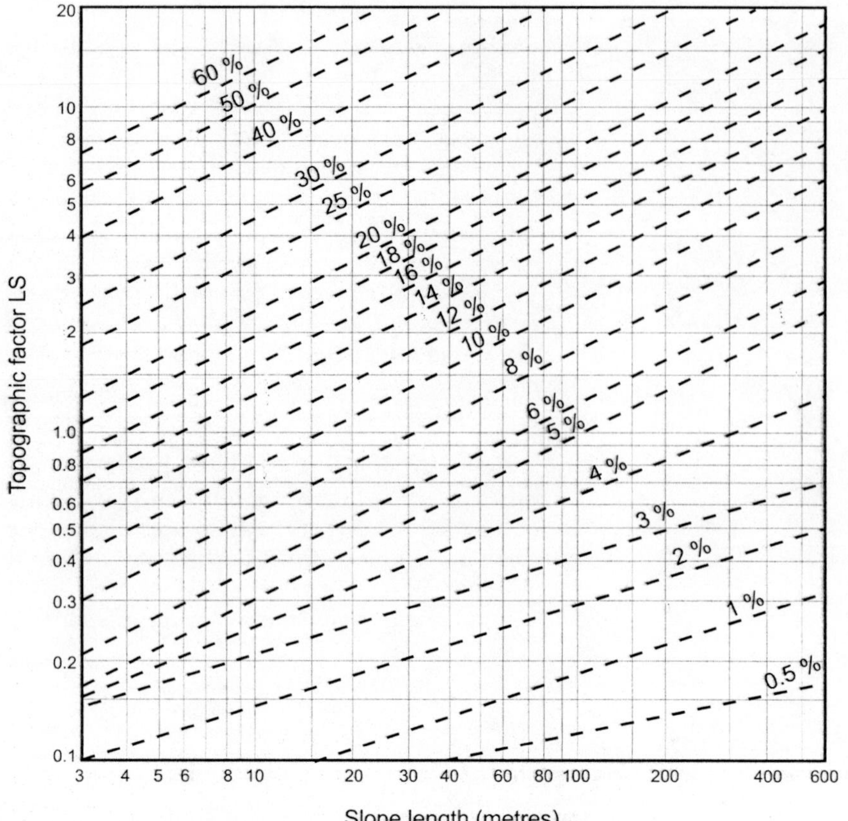

Figure 7.2.5 *Nomograph for determining the topographic factor 'LS' from the slope length and slope angle. Example: a 4% slope, 300 m long has an LS value of 1.0. (from Figure 4, Wischmeier and Smith 1978).*

Step (iv) Surface condition

One of the major influences on the volume of water that enters the soil before runoff begins is the nature of the surface (Ross 1989). Results from trials with both conventional tillage and direct-drilling indicate that surface roughness increases immediately after tillage (Steichen 1984).

Tillage increases the initial infiltration rate, loosens the topsoil, disrupts soil aggregates and compacts the subsurface soil. These conditions usually result in significant differences between the hydraulic conductivity of the surface and subsurface soil, which can promote saturation of the topsoil. The loose topsoil can then be eroded by saturation excess runoff generated during low-intensity long-duration storms.

- Loose soil

Loose soil on the surface has a low strength and is readily removed by flowing water. The detached or loose particles result from raindrop splash, cultivation, animal hooves or is an inherent condition (some sandy and self-mulching soils). Summer grazing results in soil detachment; varying from 0.1 to 2 t of soil per sheep per week, depending on the soil's surface condition and texture. The total amount of soil detached depends on the stocking rate and the number of weeks of grazing before there is sufficient rain to reconsolidate the surface.

- Soil dispersion and disruption

In some soils, the aggregates are susceptible to mechanical disruption (Powers and Skidmore 1984). This disruption causes the clays to become more dispersive, resulting in a collapse of the internal drainage. A surface crust or hardset surface will form on drying, reducing infiltration.

Cultivating moist soils increases the likelihood of mechanical disruption, especially when water contents exceed the plastic limit. This can create large clods which are devoid of coarse pores, significantly reducing the infiltration and storage capacity. Additionally, as water contents rise, the soil aggregates rapidly become plastic and flow under pressure. This significantly increases the damage to the soil structure caused by tillage, and increases runoff (McFarlane and Cox 1992). It also increases the incidence of subsurface compaction and the likelihood of waterlogging from a perched watertable (Hamblin and Tennant 1981).

Step (v) Soil erodibility

If management practices have not loosened the surface soil, then topography and the soil's inherent erodibility have the greatest influence on the amount of soil loss. Studies in the United States have demonstrated that the relative erodibility of soils can vary by a factor of 20 when measured on small plots under standard conditions (Wischmeier and Smith 1978). The properties which make a soil more or less erodible have not been defined conclusively, but there appears to be two main factors: soil texture and aggregate stability.

Soil texture. A number of studies have attempted to develop relationships between the soil particle size distribution and erodibility (Bouyoucos 1935; Barnett and Rogers 1966; Wischmeier and Mannering 1969; Evans 1980). Soil erodibility is thought to increase with silt and sand content and decrease with increasing clay (Wischmeier and Mannering 1969; McFarlane *et al.* 1991). Highly erodible soils tend to have medium textures, with clay contents between 9 and 35% (Evans 1980).

The erodibility of sands depends on both their detachment and transport. They are readily detached, but unless runoff is generated, the particles are not moved. Water repellent sands are potentially very susceptible to erosion, because they can generate large volumes of runoff and the particles are easily detached.

Stability of soil aggregates during rainfall. The stability of surface soil is important when considering the inherent erodibility of a soil (Rai *et al.* 1954; Bryan 1968; Singer *et al.* 1982) and susceptibility to soil structure decline (Section 3.2).

Very high rates of soil erosion (~78 t/ha) were measured from a structurally degraded sandy loam in NSW (Murphy and Flewin 1993). Structurally degraded soils have low infiltration rates and surface storage is likely to be reduced following rain because the unstable aggregates and clods collapse and settle rapidly, forming a crust. This greatly increases the potential for runoff.

The general erodibility of a soil (low, medium, or high rating) can be assessed from the surface texture and surface stability (Table 7.2.6).

The decision pathways for assessing runoff potential and soil erosion are shown in Figure 7.2.6 (infiltration excess runoff) and Figure 7.2.7 (saturation excess runoff).

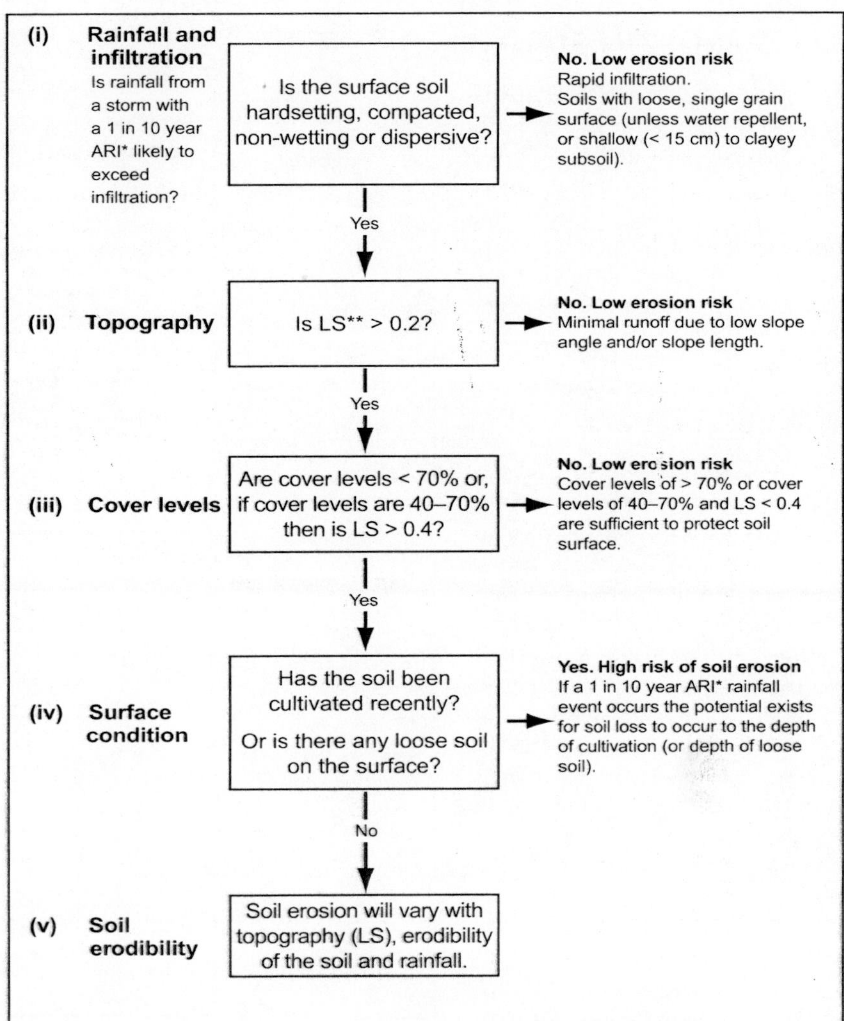

Figure 7.2.6 *Assessing susceptibility to infiltration excess runoff and soil erosion.*

* *ARI is the average recurrence interval.*

** *LS is the topographic factor which combines slope length (L) and slope angle (S), see Figure 7.2.5.*

Table 7.2.6 Estimating soil erodibility from surface texture, surface condition and soil stability*.

Surface texture	Surface condition and stability			
	Loose	**Firm-hard** (slightly or non-dispersible, slake to particles >2 mm**)	**Firm-hard** (dispersible on remoulding, and/or slake to particles <2 mm***)	**Firm-hard** (dispersible on wetting‡)
Sand	High	Moderate	-	-
Loamy sand, clayey sand	High	Low	Moderate-high	-
Sandy loam, loam, clay loam	-	Low	Moderate	High
Sandy to medium clay	-	Low	Low	High

* *Where stability is measured with the Emerson aggregate test (Emerson 1967; refer to Section 3.2.):*

** *Emerson classes 4, 6-8.*

*** *Emerson classes 3-6.*

‡ *Emerson classes 1, 2.*

WATER EROSION STATUS

The erosion status of a paddock is the degree to which it has been eroded in the past and is difficult to assess. There are two main techniques; field observations and recent erosional history which is measured by the distribution of caesium-137 within the profile. There are problems with each technique.

Field observations

Field observations of erosion are difficult in some cases. Gully erosion is probably the easiest to assess. Depth of the gullies can be measured and the rates of erosion estimated by observations with time or by interpreting aerial photographs taken periodically. However, due to the shallow soil profiles and relatively non-erosive (erosion resistant) subsoils in WA, gully erosion is less evident than in the eastern states of Australia. In WA, soil loss is predominantly the result of sheet and rill erosion. This soil loss has long-term implications for productivity, but the physical evidence is usually obliterated when the land is cultivated. Only recent erosion will be evident, which is unlikely to be a reliable indication of the erosion risk or the erosion history of the paddock.

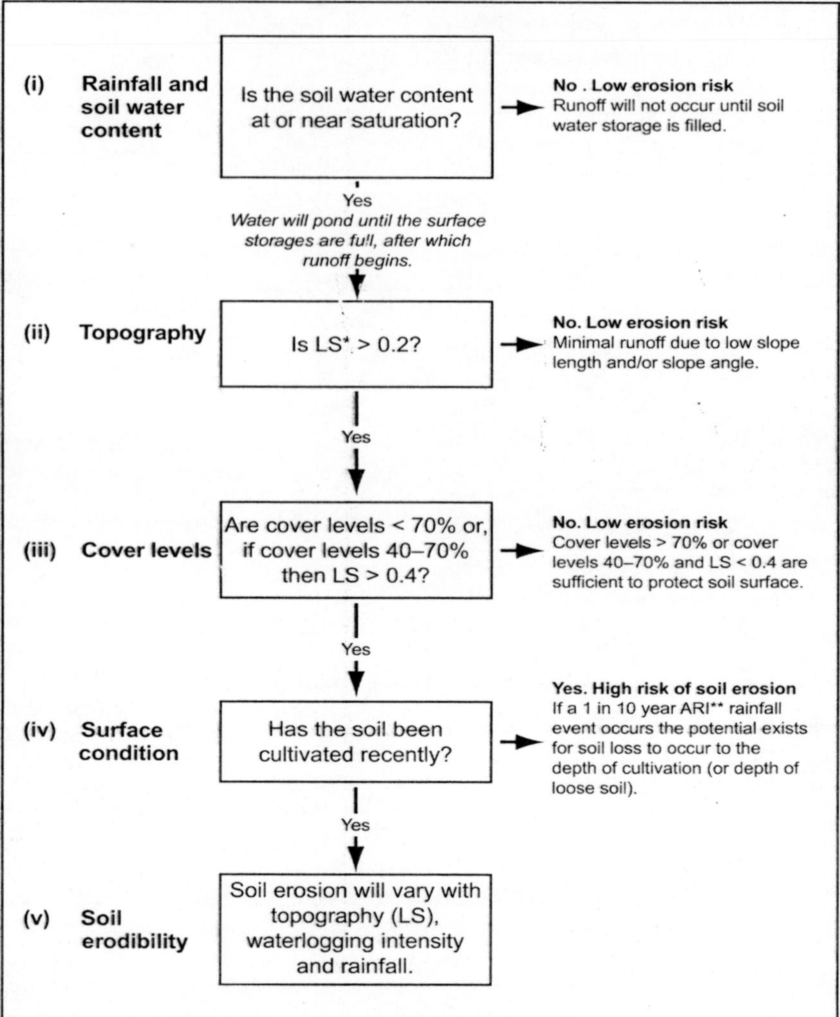

Figure 7.2.7 *Assessing susceptibility to saturation excess runoff and soil erosion.*

* *LS is the topographic factor which combines slope length (L) and slope angle (S), see Figure 7.2.5.*

** *ARI is the average recurrence interval.*

Caesium-137

Long-term soil loss from a site can be estimated using the caesium-137 (^{137}Cs) technique. The method is based on radiometric measurements of minute amounts of ^{137}Cs in the soil. The method has considerable potential for assessing erosion history at a site, providing the problems establishing datum points can be overcome.

The ^{137}Cs is a product of the atmospheric testing of thermonuclear weapons, which commenced in 1954. The ^{137}Cs is dispersed throughout the atmosphere, but when it settles it is rapidly and firmly adsorbed by the surface soil, even by the coarse-textured soils in WA (Singh and Gilkes 1990). This enables it to be used for measuring recent erosion history (Ritchie and McHenry 1990).

The ^{137}Cs technique (Figure 7.2.8) has been used widely across Australia (Loughran 1989), including Western Australia (Loughran *et al.* 1987; McFarlane *et al.* 1992b), but the results have not been universally accepted. The method requires datum points (or control areas), which are assumed to be stable and experience no erosion or deposition. Datum points are normally selected within stable areas of native vegetation. Problems arise if there is a redistribution of rain before it infiltrates into the soil, resulting in uneven labelling.

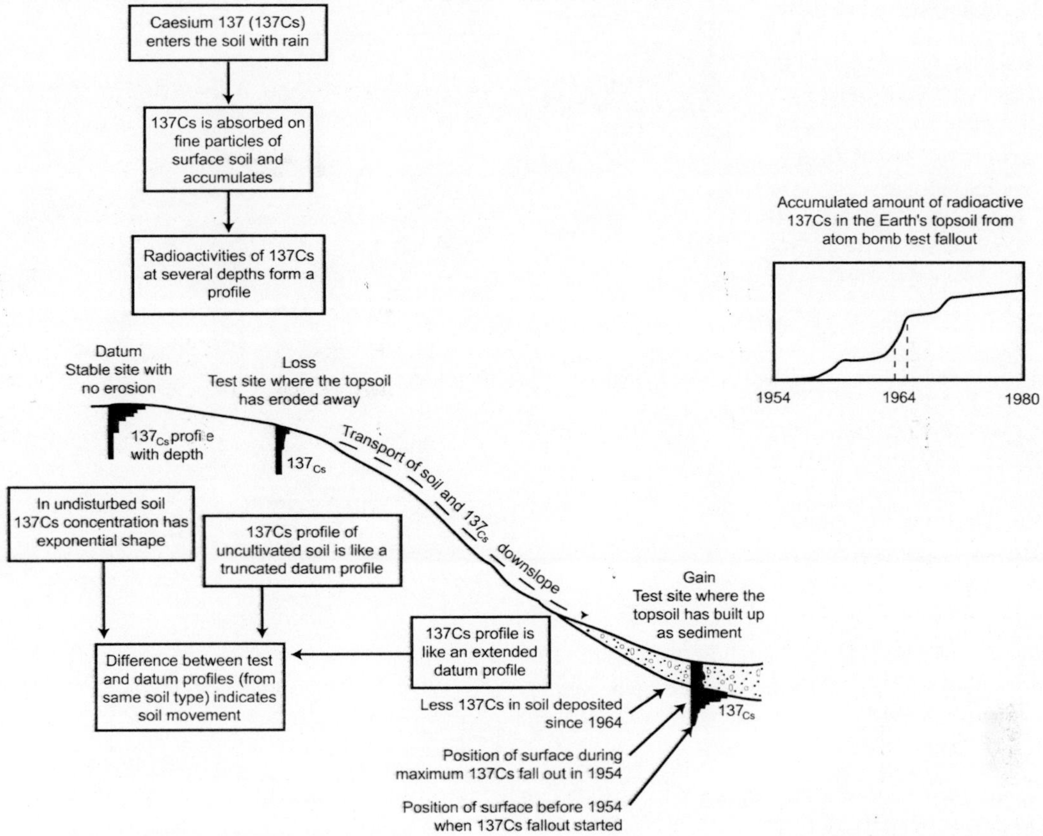

Figure 7.2.8 *Erosion and siltation measured from caesium-137 distribution within the soil profile (from Anon. 1980).*

Rainfall can be redistributed before infiltrating into the soil by:

- redistribution of rain by the tree or vegetative canopy. Field and experimental observations have established that the canopy of many native species intercept and redistribute the incident rainfall (Specht 1957; Nulsen *et al.* 1986). For example, there is considerable redistribution of rain, as stemflow, under mallee-heath vegetation (Nulsen *et al.* 1986).
- many soils under native vegetation are water repellent, resulting in an uneven wetting pattern (McGhie 1980). See Section 3.1 for more information on water repellence.
- the generation of localised runoff and redistribution of water caused by the local topography, micro-relief or surface cryptogram mats. There can be significant local runoff even when the total runoff from a catchment is minimal (Nulsen *et al.* 1986).

MANAGEMENT OPTIONS

Managing runoff and water erosion includes land management practices to reduce runoff or to increase the soil's resistance to erosion or both. Earthworks can be used to reduce water erosion, waterlogging and flooding and are useful structures to convey water to farm dams (Bligh 1989).

Management options when cropping

Mechanical disruption of soil when cultivating alters both the soil's mechanical properties (bulk density, structure) and hydrology (infiltration, soil water storage, hydraulic conductivity). This modifies the catchment's response to runoff (Ross 1989). Conventional tillage and the use of heavy farm machinery have resulted in a widespread alteration of soil properties (bulk density, hydraulic conductivity, clay dispersion) and in degradation (Nulsen 1992). Reduction in tillage with the introduction of direct-drilling and no-till sowing, combined with improved stubble management have improved the soil's structural stability and also crop yields in the agricultural area of south-western Australia (Nulsen 1992).

No-till sowing and direct-drilling usually decrease the total pore space and porosity relative to conventionally tilled soil (Ball and O'Sullivan 1982), but they also increase structural stability and the stability of the macro-pore channels, increasing the overall physical fertility of soils (Ross 1989). They are effective management strategies for many agricultural soils as the internal drainage is increased and maintained (Ehlers 1976). The amount of rainfall that infiltrates and is effectively transmitted within the soil profile increases, reducing the likelihood of runoff. Erosion problems associated with cultivation and its effect on runoff and management options are summarised in Table 7.2.7.

Water erosion and waterlogging in the Northam district.

An investigation of various tillage and cropping options for the agricultural area of south-western Australia by Bligh *et al.* (1995) concluded that:

- no-till sowing reduced infiltration excess runoff and soil loss relative to multiple or reduced tillage techniques (see box, *Contour no-till sowing minimises water erosion*).
- saturation excess runoff was unaffected by tillage treatments.
- reverse seepage interceptor banks can harvest water for farm supplies. Typically these banks are installed to reduce waterlogging but they may also perform a useful water harvesting function.
- earthworm populations increase under direct-drilling and particularly with no-till sowing.

Runoff mitigation and harvesting structures

Flood mitigation structures are used to delay or retain runoff. They help to reduce or contain flood peaks and/or volume and to divert runoff into natural drainage channels at non-erosive velocities. A number of management options are outlined in Table 7.2.7. Design options include:

- grade, level and absorption banks - to reduce peak flow and provide temporary storage (on small catchments);
- seepage interceptor banks - divert runoff while intercepting shallow seepage above a less permeable subsoil;
- water spreading structures - redistribute runoff across a slope: the thinner the overland flow, the more effective the surface roughness in preventing erosion.

CONTOUR NO-TILL SOWING MINIMISES WATER EROSION

*Kevin Bligh**

Land treatment, and in particular tillage, is a major factor affecting water erosion. Other factors being equal, erosion will usually be lower on permanent pasture than areas subjected to multiple tillage.

Applying no-till systems on the contour is an effective method for minimising water erosion on cropland susceptible to infiltration excess runoff. Even with no-till sowing, banks (i.e. contour, grade or seepage interceptor banks) are necessary on sloping sites (>3%) to safely divert large volumes of runoff from occasional intense storms, or after soil profiles become saturated during a wet winter.

No-till sowing may not be appropriate for all soils, because those with high strength (e.g. soils highly susceptible to compaction or in a degraded condition) may require some loosening, at least until structure improves, or occasional deep-ripping (0.3 m deep).

Tillage has a minimal effect on saturation excess runoff (permeability contrast soils). Many soils with high infiltration capacities experience little runoff until surface horizons become saturated (unless the soil is water repellent). Tillage then has little effect, as additional rainfall can run off the saturated areas. There is a high spatial variation in surface runoff under conditions of saturation excess, because the unsaturated areas contribute little runoff relative to saturated areas (Bligh 1984).

Effects of changing to no-till sowing on medium to fine-textured soils include:

- increased infiltration (reduced runoff). Rainfall infiltration typically increases from about the third year after changing to no-till.

  There were large differences in infiltration, water available for plant growth and runoff between treatments in a tillage trial conducted on a sandy loam at the Avondale Research Station, near Beverley, WA. For example, in 1983 the growing season rainfall was 254 mm and less than 79% infiltrated under multiple tillage (work-up, work-back and seed), compared with 86% under direct-drilling and 96% using a minimal soil disturbance sowing with a triple-disc drill. Suspended (clay and silt-sized) sediment losses decreased from 52 g/m^2 under multiple tillage, to 31 g/m^2 under direct-drilling to 6.6 g/m^2 under minimal soil disturbance sowing (Bligh 1984).

- reduced water erosion. The reduced runoff and soil disturbance combine to reduce soil loss under no-till systems. Even when infiltration rates on fine-textured soils decrease under no-till sowing, water erosion is reduced because there is less soil disturbance.

  There was significant erosion during a storm of 52 mm in six hours on a 5% slope (0.2 m loamy sand over sandy clay) at Nabawa, north-east of Geraldton, WA. This storm has an average recurrence interval of about nine years. Runoff, soil loss and nutrient loss were much higher from bays which had multiple tillage, even though the whole slope was protected by grade banks at 50 m spacing (Bligh 1994). The soil loss under multiple tillage was about 3.5 t/ha, consisting of 1.5 t/ha of clay and silt-sized sediment and 2 t/ha of coarse sediment deposited in the grade bank channels below rills. The suspended sediment contained the equivalent of 12 kg/ha of urea and the equivalent of 15 kg/ha of superphosphate.

  With no-till sowing using narrow points, the total soil loss was only 0.1 t/ha and there was no rilling or coarse sediment deposition. The nutrient loss (total nitrogen and phosphorus) was 10% of that under multiple tillage.

- increased organic matter. Organic carbon in the surface soil typically increases by the third year under direct-drilling and no-till sowing (Jarvis *et al.* 1986). Earthworm numbers and their total mass increased under continuous no-till sowing by a factor of three compared with direct-drilling and a factor of ten compared with multiple tillage at West Dale, WA (Bligh and Findlater 1996).

No-till sowing on the contour

Contour, grade or seepage interceptor banks divert runoff at a safe velocity. They also encourage sowing on or close to the contour. The seed-groove channel of each sown row therefore intercepts overland flow and sheet and splash erosion debris from the inter-row areas during bursts of intense rainfall. This runoff can then infiltrate as the rainfall rate decreases below the infiltration rate. Downslope movement of suspended and bed-load soil is therefore reduced under contour no-till sowing.

Where contour earthworks have not been constructed, no-till sowing should still be carried out on the contour. Some farmers do two or three rounds of a paddock first, then sow back and forth at least approximately on the contour, lifting the no-till seeder out of the ground for turning at the ends. Some water erosion can still occur where the initial rounds go up and downhill, but it will be less than if the whole paddock were sown around-and-around.

Even where contour earthworks have been constructed, it remains preferable to work up and back. Seeders should also be lifted out of the ground while crossing permanently-grassed waterways.

**Formerly Agriculture Western Australia, now Consultant, Busselton.*

Table 7.2.7 Management options to reduce water erosion on susceptible paddocks. The most appropriate option will vary with site details and the overall farm plan.

Erosion problem areas	Description	Recommendation
Cropped areas below rocky outcrops.	Numerous small outcrops within a paddock. Or large area of basement outcrop with adjacent shallow duplex soils.	**These areas should be classified as non-arable for cropping.** If they must be cropped, grade banks with a 1 in 10 ARI* capacity should be surveyed and constructed to divert excess runoff to a safe disposal point. If safe disposal is not available then absorption banks may be used (Farmnote 84/91), provided that salinity is not in close proximity. Direct-drilling may reduce runoff. Note: increased infiltration may also promote waterlogging and erosion in high rainfall years.
Cropped areas below lateritic breakaways.	Compact medium to gentle slopes with high runoff potential below lateritic breakaways.	Revegetate slopes to reduce water shedding potential. Contour ridging (or pasture furrows) to aid perennial establishment and reduce runoff potential. Careful species selection is required (local species preferred) and area should be fenced (particularly if remnant vegetation is left on breakaway). Absorption banks to contain runoff for a 1 in 20 ARI* could be constructed but where possible, use grade banks to divert excess runoff to a safe disposal point.
Long or steep slopes (unprotected by soil conservation structures).	Long slopes provide ideal conditions for generating runoff and they promote rilling and soil loss. Deep rills also pose problems during harvesting.	A bank system should be designed and surveyed before construction. Grade banks are preferred where stable waterways can be developed to receive the discharge. If waterways are unstable then these should be stabilised at least 1-2 years before banks are constructed. Level or absorption banks may be required where waterways cannot be developed or in higher positions in the landscape.
Cropping across or close to natural or artificial waterways.	Concentration of water in natural or artificial waterways, results in erosion and waterlogging causing loss of production.	Stable depressions, natural and artificial waterways should be left uncultivated and developed as grassed waterways which are protected with pasture cover at all times. To aid in this, waterways can be fenced or guide banks installed to delineate them, so that unwary tractor operators do not cultivate them by mistake. Waterways should be designed for a 1 in 20 ARI*.
Inadequate banking.	Banks not constructed properly or to design specification or banks not properly maintained.	Refer to earthworks design manual. Survey and advice available from Agriculture WA. Additionally, BEFORE refurbishing or re-grading banks, the existing structures should be checked to ensure that; they are on the correct grade; the correct place; they have a correct outlet and they are useful. (Farmnote 62/91).
Working corners downhill.	Traditional cultivation practices may result in soil loss as a result of concentrating runoff into corners. Double seeding reduces amount of water available for crop growth during a dry spell (i.e. first area to burn off).	**Practice is not recommended.** Do not work corners as they only constitute between 2 and 3% of a paddock and may only contribute between 1.5-2.5% of crop yield. If corners are to be worked do so before sowing the rest of the paddock (Farmnote 28/93).
Runoff from road verges.	The limited areas available to road Authorities results in water being diverted onto adjoining properties.	Liaise with Main Roads and Shires to agree on spoon drains/culvert sites so that roadside runoff can be diverted to safe disposal points. If water is diverted into a natural waterway, an uncultivated strip must be left to minimise erosion. Try to use runoff by diverting it to dams (Farmnotes 73/89; 26/93).

Table 7.2.7 (continued).

Erosion problem areas	Description	Recommendation
Erosion of waterlogged areas.	Seepage pressures and surface ponding generate runoff resulting in rills on duplex soils.	Drain these areas and use appropriate farm planning. Design and implement appropriate seepage interceptor drains or spoon/W-drains (if ponding occurs at the surface), in conjunction with high water use rotations, revegetation belts on fencelines and management of land according to its capability. A 'Notice of Intent' is required if the drainage is to prevent or alleviate salinity (Farmnotes 47/93; 106/91; 69/89; 70/89; 71/89; 78/93; 80/93).
Cultivation off the contour.	Runoff is channelled downhill by cultivation ridges. Concentration of runoff causes rilling and waterlogging.	Contour working (cultivation) is recommended, grade banks should be installed on long slopes.
Cultivation at the ends of spreader banks.	Cultivation close to grassed waterways causes rill and gully erosion.	Leave adequate areas uncultivated (Farmnote 39/83).
Hardsetting and water repellent soils.	Generates infiltration excess runoff and promotes erosion.	Refer to sections 3.2 and 3.1 for the management options. Design the grade bank system and fence to land management units.
Areas of permanent vegetation in the catchment.	Runoff from heavy or long rainfall events can be greater from vegetated areas than adjoining cultivated land resulting in erosion of the cultivated area.	Diversion banks or grade banks can be used to divert the runoff to a safe disposal point.
Cultivation practices.	Working fine-textured soils two or three times a season when very wet or too dry destroys the soil peds and breaks down aggregates resulting in the soil becoming more dispersive.	Minimise tillage, adopt a direct-drill or no-till system. Reduce cultivation speed. Do not cultivate when too wet or too dry. Use gypsum on responsive soils.
	Excessive working on coarse- and medium -textured soils significantly adds to subsurface compaction.	Minimise tillage. Deep ripping will break up traffic pans and increase subsoil infiltration reducing the likelihood of waterlogging.
Surface roughness.	In low rainfall areas, fine tilth seedbeds worked on the contour erode more than rough seedbeds on the contour.	Do not use harrows. Corrugations increase roughness, infiltration and trap rainfall reducing the likelihood of runoff and subsequent erosion.
Surface cover and compaction.	Over grazing and compaction by stock hooves result in reduced infiltration and increased runoff.	Do not overgraze pasture. Defer grazing on medium and fine-textured soils if wet and where possible, remove stock before heavy rainfall (Section 3.2).
Positioning of fences, gates and watering points.	Incorrect placement of fences, gates and watering points can result in increased erosion or runoff.	Seek advice from Agriculture WA. Attend a Farm Planning Workshop where these issues can be discussed and assessed.
Road and firebreak erosion.	If roads and firebreaks are sited across a slope and not on the contour or safe grade, they can intercept runoff from the catchment above. Water is then channelled along the road, or diverted into other areas, or the adjoining paddock at unsafe velocities resulting in gully erosion.	Locate roads or firebreaks along ridge crests or below contour banks. Do not site farm roads or firebreaks along waterways. If roads cannot be placed across slopes at safe grades it is better to place them up and down the slope. Divert runoff from roads or firebreaks at regular intervals using spur drains or spreader banks. Firebreaks can include loops at regular intervals to intercept and slow runoff.

Table 7.2.7 (continued).

Erosion problem areas	Description	Recommendation
Gateway erosion.	Headlands, roads, firebreaks and sheep pads meet at paddock corners where gates and watering points are normally sited. The soil at gateways is often bare and rutted or pulverised by stock creating an ideal spot for erosion to start.	Locate gates on ridges or localised higher ground where drainage is away from the gate. Use problem gates as little as possible and encourage pasture growth on bare areas.
Sheep pad erosion.	Sheep normally take the most direct route to water (a straight line). Once a route is established the sheep will repeatedly use the same track, resulting in structural decline of soil in this area. Wind and rainfall combine with soil loosening by stock to create an erosion problem. Erosion along sheep tracks or pads can promote dam siltation and gullying.	Provide watering points in each paddock. Where sheep pad erosion is a problem, rotate sheep between paddocks to reduce grazing pressure and the likelihood of deeper sheep pads developing (if possible). Also, site dams and troughs to the centre of the paddock to promote more even grazing or place watering points upslope. If siting dams, ensure that sufficient catchment is available to fill the proposed dam.
Natural drainage line erosion.	Runoff concentrates naturally in major and minor depressions. Many of these areas are not identified before clearing and cropping begins. As a consequence, rill erosion in the first year of cropping can progress to gullying in subsequent years if it is not addressed.	Before developing new land, survey the natural drainage patterns and incorporate designs to protect natural depressions (fencing them where possible) into the development plan. Contour/grade banks can be use to divert runoff from the natural drainage areas to grassed and fenced waterways to reduce the possibility of erosion occurring. In areas where erosion has already occurred, back-fill the protected washout or gullies and re-grass, and add banks to divert drainage and reduce erosion.
Evaporation from seepage areas.	Where overgrazing or clearing on natural seepage areas has taken place, evaporation tends to concentrate the salts and minerals that were in solution. This reduces soil productivity, vegetation cover and degrades the soil structure, leaving the area prone to erosion.	Control grazing on seepage areas and maintain vegetation cover to reduce evaporation. Fence and replant cleared areas where possible. Plant trees upslope to use/divert water away from the seepage area.

Further reading

Bligh, K.J. (1986). Water erosion in the Geraldton area during June-July 1986. Western Australian Department of Agriculture Division of Resource Management Technical Report No. 59.

Bligh, K.J. (1989). Soil Conservation Earthworks Design Manual. Internal Report. Soil Conservation Branch, Division of Resource Management, Department of Agriculture Western Australia.

Davies, J.R. and McFarlane, D. J. (1987). Estimation of runoff volumes for ungauged catchments in the agricultural areas of Western Australia. Western Australian Department of Agriculture Division of Resource Management Technical Report No. 43.

Flavell, D.J., Martin, D.K. and Belstead, B.S. (1987). Flood estimation procedures for Western Australia. Mains Roads Department of Western Australia. Main Roads Technical Report 50T.

Institution of Engineers Australia (1987). 'Australian rainfall and runoff: A guide to flood estimation'. Volume 1. (ed. D.H. Pilgrim), Institution of Engineers Australia, Barton, ACT, Australia. 373 p.

McFarlane, D.J., George, R.J. and Farrington, P. (1992). Changes in the hydrologic cycle. *In* 'Re-integrating fragmented landscapes: Towards sustainable production and nature conservation'. (eds. R.J. Hobbs and D.A. Saunders), Springer-Verlag. New York, pp. 146-186.

Nulsen, R. A. (1992). Changes in Soil Properties. *In* 'Re-integrating fragmented landscapes: Towards sustainable production and nature conservation'. (eds. R.J. Hobbs and D.A. Saunders), Springer-Verlag. New York, pp. 107-145.

7.3 SOIL FACTORS INFLUENCING EUTROPHICATION

David Weaver* and Rob Summers**

Eutrophication is essentially the nutrient enrichment of waterways leading to algal growth. It must be controlled to maintain sustainable agricultural systems. The main mechanisms of control are stabilising catchment processes and reducing nutrient output.

Eutrophication can be defined as "the nutrient enrichment of waters which results in the stimulation of an array of symptomatic changes, among which increased production of algae and macrophytes, deterioration of water quality and other symptomatic changes, are found to be undesirable and interfere with water uses" (OECD 1982). The word eutrophic is Greek and means 'well fed'. The food is plant nutrients such as nitrogen (N) and phosphorus (P) and as such the definition simply means 'an undesirable addition of plant nutrients to a water body'. The visible sign of nutrient enrichment is the proliferation of algae in waterways. Algae flourish in nutrient-rich waters, most often to the detriment of other species of aquatic plants and fauna. Algae may block light from other aquatic plants such as seagrasses resulting in reduced biomass, thus threatening parts of the ecosystem reliant on seagrass. Decomposition of algae consumes oxygen, which can lead to the death of fish, and also releases nutrients from sediments into the water column.

Eutrophication is a natural aging process. Over millennia, water bodies are slowly filled with soil and other materials entering with the inflowing waters. In this natural process water quality is usually good, there is a diverse biological community and fewer algae. Human activities accelerate these processes. Human settlement, the clearing of forests, development of cities, agriculture and industry have increased the addition of nutrients to catchments and increased water erosion and flow from catchments to downstream waterways at a rate that is too high for the systems to assimilate.

Figure 7.3.1 Map of south-western Australia showing the regions most likely to contribute to eutrophication. These are the unforested areas where the annual rainfall is more than 500 mm or more than 400 mm in the south-east. Refer to Table 7.3.4 for a more extensive list of factors influencing nutrient loss.

Eutrophication also leads to local social and economic problems including visual pollution for residents and recreational users and declining property values because of the stigma of pollution. An increase in nutrients can increase fish productivity, however algae foul nets making fishing difficult. Algal decomposition results in the release of foul smelling gases (including hydrogen sulphide) which can be a problem for some residents. At an international scale, overseas consumers demand products that are produced in a sustainable manner and so eutrophication generates barriers to trade.

The nutrient most implicated in eutrophication in Western Australia is phosphorus (P). Algal growth is usually limited in the absence of P. However, just

* Agriculture Western Australia, Albany.
** Agriculture Western Australia, Pinjarra.

as the supply of other nutrients improves pasture growth, the addition of other micro- and macro-nutrients to water bodies may also improve algal growth. The major areas of concern are the coastal zones (Figure 7.3.1), particularly in the south-west and along the south coast (Hodgkin and Hamilton 1993). These are also the most heavily populated and developed areas. Some water bodies that have received particular attention, both in the media and by research organisations, are the Peel Inlet and Harvey Estuary, Leschenault Inlet, Princess Royal Harbour and Oyster Harbour, Wilson Inlet and Swan Estuary. There are also numerous reports of the effects of nutrient enrichment on many wetlands, lakes and rivers throughout the south-west.

Soils are closely linked to eutrophication processes because their characteristics influence the delivery of soil particles and nutrients to waterways. This section describes some of the principles of eutrophication and then elaborates on some soil criteria influencing the problem.

PRINCIPLES OF EUTROPHICATION

Some features of our water bodies and environment encourage the growth of algae. These include: limited exchanges between water bodies and the ocean, which would normally flush out the nutrient-rich water; warm waters; high light intensities; shallow water bodies and ample supplies of nutrients from external and internal sources. (Nutrients are released from sediments into the water column under anaerobic conditions.)

Catchments in WA also have unique features that encourage the loss of nutrients to waterways. Some of the most commonly cited reasons for nutrient loss include a high percentage of clearing, extensive areas of sandy-surfaced soils with little capacity to retain nutrients, drainage of waterlogged soils to remove excess water, the application of highly soluble fertilisers (e.g. superphosphate) to correct nutrient deficiencies, and pastures with limited root systems which limit the uptake of applied nutrients.

Aquatic flora may respond differently to additions of nutrients than agricultural pastures and crops. A small loss of nutrients from agricultural land may not hinder crop growth in the paddock, but may greatly influence the growth of algae. Waterways generally contain low nutrient levels and flora that are adapted to efficient scavenging, so when excess nutrients are available, algal blooms can occur. Nutrients entering waterways come from a variety of sources and have their origins in two main categories, diffuse and point sources.

Diffuse sources

Diffuse sources of nutrients come from a wide area (spread throughout the catchment). Broadscale agriculture, including areas under pasture or crop is a diffuse source. These sources may contribute a large proportion of the total amount of nutrient discharged into a waterway because of the extensive nature of most catchments. The quantity of nutrient exported on a per unit basis (say kg/ha) is usually quite small (ranging from about 0.05 to 4 kg/ha/year) relative to application rates and the amount stored in the soil. Usually less than 20 kg/ha of phosphorus is applied to areas of broadscale agriculture.

It is difficult to compare rates of nutrient loss from catchments of different sizes because small catchments appear to lose more nutrient per unit area than large catchments. In addition, the retention and/or contribution from within streams, together with the variable area of sources, may lead to scale effects that confuse the interpretation of nutrient loss data.

Urban areas are major diffuse sources of nutrients because of the fertilisers applied to domestic gardens, parks and golf courses, as well as septic tanks, sewerage disposal and waste disposal sites. The proportion of the total input from these sources will depend on the extent of the urban area in comparison with other sources and the nutrient management strategies in place.

Point sources

Point sources are usually associated with intensive agriculture and discharges at a particular point. These include piggeries, sheep holding yards, dairies and horticulture, meat processing plants, vegetable processing plants, fertiliser factories and other industries.

Intensive animal industries often produce large quantities of waste water and nutrients. Nutrient concentrations in these wastes are often much higher than those leaving diffuse sources. Many of these industries currently combine ponding, irrigation and diversion to waterways to dispose of nutrient-rich waste water.

Table 7.3.1 shows approximate quantities of N and P excreted by different animals during a year. Where many animals are housed together and their waste is discharged over small areas, nutrient losses by leaching and runoff can be large because areal application rates

Table 7.3.1 Nitrogen and phosphorus output in some animal wastes (from Vanderholm 1984).

Animal	Nitrogen (kg/head/year)	Phosphorus (kg/head/year)
Broiler chickens	0.3	0.07
Laying hens	1.0-2.4	0.1-0.2
Sheep	5.5	0.9
Pigs	3.7	0.7-3.0
Feedlot beef	18-32	6.4
Dairy cows	64	13

Table 7.3.2 The fate of nitrogen and phosphorus fertiliser (in kg/ha/crop) after application to five major vegetable crops on the coastal sands (after McPharlin and Luke 1989).

Crop	Land	Applied		Crop removal		Remaining	
	status	N	P	N	P	N	P
Carrots	New	372	74	100	15	272	59
	Old	300	50	100	15	200	35
Lettuce	New	850	250	100	20	750	230
	Old	370	90	100	20	270	70
Cauliflower	New	1050	280	119	25	931	255
	Old	570	120	119	25	451	95
Onions	New	800	280	90	26	710	254
	Old	320	120	90	26	230	94
Potatoes	New	740	280	132	15	608	265
	Old	360	120	132	15	228	105

of nutrients are high and soils have limited capacity to retain applied nutrients. Nutrient losses from point sources are usually expressed as quantities (kg or tonnes) for a specific industry rather than kg/ha.

For satisfactory horticultural production, large quantities of nutrients and water are often required. They may be lost from the system before the plants can use them or be in excess of the plants' requirements. The excess nutrients are lost from the system by leaching, runoff or soil erosion (Table 7.3.2). Often two or more crops are grown in a year and recommended nutrient application rates are not always followed.

Humans discharge the equivalent of 1 kg of phosphorus annually as a result of domestic activities. Detergents make up approximately 50%. Disposal in septic tanks or improperly constructed or sited sewerage works can lead to nutrient contamination of both ground and surface waters. Table 7.3.3 shows the difference in annual nutrient input from sewered and unsewered areas of Perth.

Nutrients exported from point sources depend on the location of the source in relation to the water body, the size of the source, the soil type, groundwater proximity and direction of flow, seasonal and other environmental factors and the management strategy employed.

Table 7.3.3 Annual input of nitrogen and phosphorus from urban residential areas of Perth with 10 residences per hectare (after Gerritse *et al.* 1990).

Residential area	Nitrogen (kg/ha)	Phosphorus (kg/ha)
Sewered	80	40
Unsewered	260	70-80

FACTORS AFFECTING NUTRIENT LOSS

A commonly observed feature of nutrient transport is that it depends on rainfall and flow. Figure 7.3.2 shows P loss and flow data for a subcatchment on the south coast in 1990. Most of the nutrient was lost in one period of high flow.

Table 7.3.4 summarises factors that influence P loss and P concentrations in streams. It also illustrates whether the factor is linked to an increased or decreased risk of P loss with a brief explanation, and whether soil, hydrology or other factors (such as management or inherent natural features) are most dominant in influencing the magnitude and direction of the change in risk. Many factors are generalisations and there may be exceptions in specific situations. It is not an exhaustive list.

Nutrients usually reach water bodies by moving through soils with water. Some nutrients are discharged directly into waterways in the form of wastes but most often the nutrients are applied to soils first, in the form of fertilisers or wastes. The nutrients stored in the soil can be lost by leaching from sandy soils or by erosion from finer textured soils.

Nutrient loss depends on soil characteristics, the form of nutrient applied, rainfall, uptake by plants and water movement. Some of these factors, such as uptake by plants and form of nutrient applied, depend on management decisions which can partly control the amount of nutrients lost. Rainfall cannot be controlled, whilst soils have natural characteristics (P sorption capacity) that influence both how much is exported to water bodies and how much is available to plants. Some natural characteristics influencing a soil's sorption capacity include texture and the presence and amount of iron, aluminium and calcium compounds.

Table 7.3.4 Factors that influence the loss and concentrations of phosphorus in streams.

Factor	Risk of phosphorus loss		Soil	H_2O	Other
Years of fertiliser application	⇑	Increases with increasing years of fertilisation.	✓	✓	✓
	⇑ ⇓	Increases or decreases between adjacent water years depending on management and environmental factors.		✓	
Time since fertiliser application	⇓	Decreases as time of contact with soil increases.	✓		
Fertiliser application rate	⇑	Increases with increasing application rate.	✓		
Stream flow	⇑	Increases with increasing flow rate.		✓	
Dominant flow mechanism (runoff vs subsurface vs groundwater)	⇑	Increases with dominance of surface runoff.	✓	✓	
	⇓	Decreases with dominance of subsurface and groundwater flow, but depends on P retention of soil.	✓	✓	
Seasonal effects	⇑ ⇓	Increases in winter and spring, decreases in summer and autumn.	✓	✓	
Catchment size and stream order*	⇓	Risk of high concentrations and high unit area loss decrease as catchment size and stream order increase.		✓	✓
Catchment shape	⇓	Risk to downstream waterways decreases with elongated catchments.			✓
	⇑	Risk to downstream waterways increases with catchment shapes that favour short distances to water body.			✓
Travel time in a stream	⇓	Decreases with increased travel time in a stream.		✓	✓
Riparian (stream bank) vegetation condition	⇓	Decreases with improved condition of the riparian vegetation.		✓	✓
River/creek/stream/drain length	⇓	Decreases with increased length of the river/creek/stream/drain.		✓	✓
Spatial position - where abouts the sample is collected in a stream (laterally, vertically and longitudinally)	⇑	Concentration increases towards centre of watercourse and decreases towards edges; increases as stream bed is approached**.		✓	
	⇑	Concentrations and unit area losses increase as headwaters are approached.		✓	✓
Soil P retention	⇓	Risk of high concentrations and high unit area loss decreases as soil P retention increases, if leaching is dominant.	✓	✓	
	⇑	Risk of high concentrations and high unit area loss increases with increasing P retention if soil erodes.	✓		
	⇑ ⇓	Increases or decreases depending on dominance of subsurface and groundwater flow.	✓	✓	
		Sandy soils store little P and leach more than clayey soils which store more P and lose it mainly by erosion.			
Soil fertility	⇑	Increases with increasing soil P status.	✓		
Depth to impeding layer	⇑ ⇓	Increases with shallow soils mainly because of greater runoff; decreases with deep soils.	✓	✓	
Grass height	⇓	Decreases with increasing grass height.	✓	✓	✓
Grazing pressure	⇑	Increases with increasing stocking rate.			✓
Proximity to point source	⇓	Decreases with increasing distance from source**.	✓		
Soil moisture	⇑	Increases as soil moisture increases.	✓	✓	

Table 7.3.4 continued.

Factor	Risk of phosphorus loss		Soil	H_2O	Other
Rainfall intensity	⇑	Increases with increasing intensity if erosion is dominant.	✓	✓	
	⇓	Decreases if high rainfall intensity causes dilution.		✓	
Amount of previous rainfall	⇓	Decreases as amount of previous rainfall increases.	✓	✓	
Land use	⇑	Increases with intensity of land use, but depends on the level of management.	✓	✓	✓
Land management	⇓	Decreases as land management includes greater emphasis on soil conservation measures.	✓	✓	✓
Effective vegetation cover	⇓	Decreases as effective vegetation cover increases.			✓
Drainage	⇑	Increases with increasing surface drainage.	✓	✓	✓
	⇓	Decreases with increasing subsurface or tile drainage.	✓	✓	✓
	⇓	Decreases with increasing surface drainage if the drainage causes increased contact of water with soil, improved P retention from conditions of high oxygen and improved pasture growth.	✓	✓	
Distance to waterway	⇓	Decreases as distance from source to waterway increases.			✓

* *Low order streams equals small streams and creeks, while high order streams equal rivers.*

** *More likely to produce a measurement error or effect rather than identify a factor influencing nutrient loss*

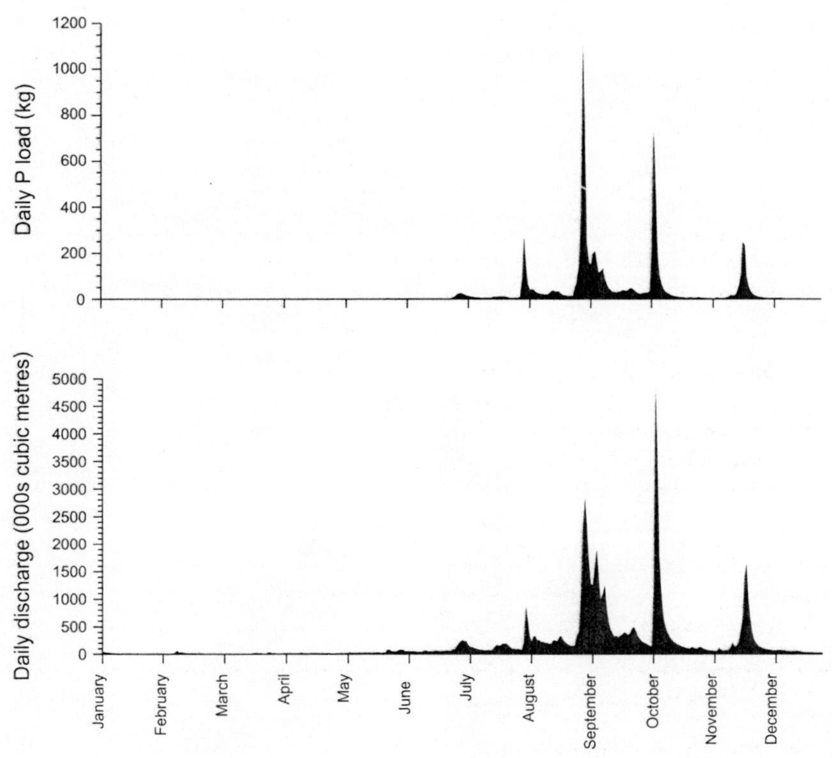

Figure 7.3.2 *Time series of phosphorus load (kg/day) and daily river flow during 1992 ('000 m³/day) for the Kalgan River near Albany.*

The amount of phosphorus leached depends on the soil type, the rate of P application and the throughput of water. For sandy soils (numbers 3 and 4 in Table 7.3.5) the time taken for P to move through the soil is short compared with lateritic soils (numbers 1 and 2 in Table 7.3.5). The capacity of each of these soils to adsorb P can be determined by measuring its Phosphorus Retention Index (PRI). The higher the PRI value the greater the capacity for sorption (see Section 6.3).

Phosphorus retention is irrelevant if management does not aim to retain soil on-site. Soils with high P retention have a tendency to store large amounts of P in the soil, but this P can then be lost if surface runoff erodes the soil particles that have been enriched with nutrients (Figure 7.3.3). Soils with high P retention reduce the amount of P lost by leaching but increase P lost by erosion.

Table 7.3.5 Calculated times (years) for phosphate to travel through 1 m of some soils of south-western Australia for different rates of accumulation of phosphorus and recharge (after Gerritse 1990).

Phosphorus accumulation (kg/ha/year)	**20**		**100**		**500**	
Recharge (cm/year)	**20**	**100**	**20**	**100**	**20**	**100**
1. Surface soil of the Darling Plateau	38,000	15,000	11,000	45,00	3,000	1,300
2. Clay subsoil of the Darling Plateau	250	170	60	45	15	11
3. Subsoil of a yellow Spearwood sand	45	30	12	7.5	4.5	2.5
4. Surface soil of a Gavin sand	6	2.5	3	1.5	1.5	<1

Smaller risk of P loss if soil is eroded — Greater risk of P loss if soil is eroded
Greater risk of P loss via leaching — Smaller risk of P loss via leaching
Soil total P (mg/kg)
Soil P retention (PRI): ≤ 2, 2-5, 5-20, 20-70, >70
increasing ability of soil to retain P

Figure 7.3.3 Box and whisker plot showing relationship between P retention and Total P indicating which soils are most likely to contribute to P loss via leaching or erosion.

(Filled circles represent the mean, bottom of box represents 25th percentile, middle of box represents 50th percentile, top of box represents 75th percentile. Top whisker represents 90th percentile and bottom whisker represents 10th percentile).

Under leaching conditions and where there is sufficient P carried over from the previous year, the best time to apply more fertiliser is after the plants have germinated. As a rule of thumb, the later the fertiliser can be applied (up to about a month after the break of the season) to established plants, the less chance there is of P being lost and the greater the efficiency of fertiliser use.

A common measure of P retention used in WA is the level of *reactive iron*. Reactive iron in a soil is constant, while soil PRI will decrease as cumulative fertiliser inputs increase over time and sorption sites in the soil become saturated. PRI indicates the soil's current ability to retain phosphorus whilst the level of reactive iron reflects the PRI before any fertiliser was applied. The PRI of some soils and red mud (bauxite mining residue) are given in Table 7.3.6.

The effect of time (surrogate for cumulative P application) on increasing the risk of P lost via surface runoff is shown in Figure 7.3.3. Phosphorus accumulates in the topsoil after many years of application of fertiliser (Figure 7.3.4) and these enriched topsoils have the greatest potential to lose P during surface runoff and erosion. In the short-term (within a season) the risk of P loss decreases with time of exposure to soil. The risk of P being lost in soluble forms from sandy soils also changes between rainstorms. The amount of P that could be lost during the next rainstorm will depend on how much P has moved into the soil solution in the time interval between rainstorms (Weaver *et al.* 1988).

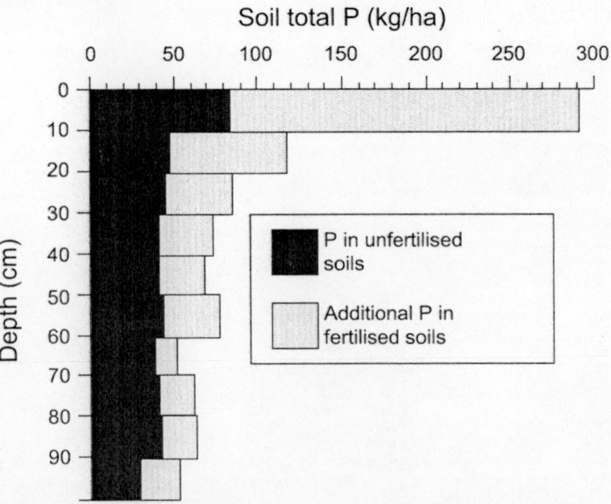

Figure 7.3.4 Average change to soil P storage with depth for soil profiles on the south coast of WA after clearing.

Table 7.3.6 Phosphorus retention indicies (PRI) of some virgin Western Australian soils and amending materials (after McPharlin *et al.* 1990).

Soil type/amendment	PRI (mL/g)
Joel sand	0
Grey Karrakatta sand	0.3-0.4
Wongan grey sand	2.7
Yellow Karrakatta sand	3.2-4.1
Spearwood sand	7.0
Wongan yellow loamy sand	13.0
Gingin red loam	70.0
Red mud	310.0

MANAGEMENT OPTIONS

There are a number of options for reducing nutrient loss, based on minimising nutrient inputs and maximising nutrient retention.

Fertilising for need

Fertilising according to the requirements of pastures rather than tradition can go far towards reducing nutrient applications in catchments. Models are available (e.g. PHOSUL-K) to predict nutrient requirements. Fertilising for need requires regular soil testing so that fertiliser applications are optimised. Reactive iron, in combination with other soil measurements such as bicarbonate-extractable P, are used to determine how much P is required by annual pastures in high rainfall areas. There are standards to determine P requirements of pastures depending on the soil's P status (Table 7.3.7). A survey by the South Coast Estuaries Project showed that more than 50% of the soil samples taken in the area had a high P status

Table 7.3.7 Soil phosphorus status in relation to levels of reactive iron and bicarbonate P in soils on the south coast.

Reactive iron (ppm)	Bicarbonate P (ppm)		
	Low status	Medium status	High status
1-100	<10	10-12	>12
101-200	<12	12-15	>15
201-400	<15	15-20	>20
401-800	<20	20-25	>25
801-1600	<25	25-30	>30
>1600	<30	30-40	>40

and could go without extra applications for at least one year. Further soil testing may indicate that some soils with a high P status may be able to do without applications for even longer.

Using alternative fertilisers

An alternative fertiliser that supplies phosphorus in a slow release form, more suitable to the needs of pasture in the high rainfall areas along the coast, may be available. It is known as coastal super and also has a high sulphur content in slow release form. In addition, particularly for soils with a high phosphorus status, other nutrients such as sulphur and potassium can be used to achieve the most economic level of production.

Treating point sources

Point sources of nutrients, particularly where effluent is discharged, can be treated to remove a large proportion of the nutrients before the effluent is discharged. Many methods of treating effluent to remove nutrients are available including passive biological treatments and more active chemical treatments. Some of these include irrigation over soils with high nutrient retention in conjunction with tree plantations. It is best to site point sources as far from watercourses as possible so that the opportunities for removal of nutrients by soil contact and biological removal are increased.

Soil amendment

Amending sandy soils with waste products from industrial processes or with soils with high capacities to adsorb phosphorus can reduce nutrient losses. Red mud (the residue produced in the process of extracting aluminium from bauxite) is one material that has been used and tested. Applications of 80 t/ha reduced the amount of P lost by 70% (Summers *et al.* 1993). In addition, both the pH of the soil and its productivity increased. Other materials, such as wastes from synthetic rutile plants are now under scrutiny as amendments for sandy soils. For very sandy soils, amendment with high PRI loams may significantly reduce P losses (Gilkes *et al.* 1992).

Water control

Water is the mechanism by which nutrients make their way to water bodies. Water carries nutrients in both dissolved and particulate forms. Much of the particulate nutrient problem can be overcome by controlling erosion. For example, in Western Europe it has been shown that a 50 m wide buffer strip of vegetation along streamlines can filter out 50 to 100% of the P (Isermann 1990). Other devices, such as detention basins or structures that slow water movement, are effective in reducing nutrient loss through sedimentation and biological processes.

Some land degradation problems such as salinity and waterlogging may be overcome by removing water from the land by drainage. This may, however, be to the detriment of nutrient enrichment problems. It would be appropriate to use methods of water control that attack these problems in unison. Subsurface drainage may reduce nutrient loss by allowing intimate contact between percolating water and soil, whilst surface drainage may contribute to increased nutrient loss if it does not improve soil water contact (Skaggs *et al.* 1994). Removal of excess soil moisture may reduce the risk of P loss from erosion.

Pasture species that use more water and have more extensive root systems can also help. Perennial pastures have root systems ready to take up nutrients applied soon after the first rains of the season. They may also assist in reducing the effects of erosive summer storms.

Planning and other control mechanisms

The mechanisms described above are mostly used for existing sources of nutrients. It is important to plan for the future. Capability of the land should be assessed, ensuring that retention of nutrients is not exceeded.

In some environmentally sensitive areas, proposals for nutrient intensive activities may need some form of environmental assessment. Proponents may need to provide detailed information on the soil's ability to retain nutrients and management measures to control nutrient loss.

7.4 HERBICIDES: MOVEMENT, PERSISTENCE AND ACTIVITY IN SOILS

*Marcus Blacklow**

Weeds cost the Australian economy about $3 billion per year. This cost is due in about equal parts to cultural and chemical methods of control, and to losses from inadequate control. In Western Australia, sufficient herbicide is used to treat the total crop area more than once per year. For example, most soils used for lupin production and about half of those used for wheat are treated with herbicides (simazine for lupins and sulfonyl-ureas for wheat) before seeding to control a broad spectrum of weeds (grass and broad-leaf) during the early weeks of crop establishment.

Chemicals will continue to be used for weed control in most agricultural systems as they have been relatively cheap, effective, and safe to the user and the environment. However, their methods of use are challenged by those who want to use them more efficiently, and by the wider community which has concerns about environmental impacts. This chapter provides a framework for thinking about the performance of herbicides in soils and specific information on some widely used herbicides. If sound methods of weed management are to be followed, all who use herbicides need to balance considerations of use with weed-response. Consequently, this chapter raises the issues of weed persistence and environmental concerns.

CHEMICAL WEED CONTROL – PERSISTENT WEEDS AND ENVIRONMENTAL CONCERNS

Modern herbicides became available in the 1940s with the synthetic auxins (2,4-D and MCPA). Since then, members of this group have been used repeatedly over large areas of many countries to selectively control a wide spectrum of broad-leaf weeds in cereals (Table 7.4.1). Despite the sustained use of herbicides, the same weeds remain problems today. Furthermore, many new groups of highly effective herbicides have been discovered (Table 7.4.1) and used, yet weed-free farms remain an illusion.

Weeds persist due to their abilities to invade disturbed habitats produced by farming systems, to complete their life cycles in competition with crops, and to adapt genetically to the selection pressures of management. For example, some weeds have developed genetic resistance to herbicides (e.g. to Groups A, B, F, G, and I in Table 7.4.1) and this is now a primary concern. If history is any indication, it is most probable that any systems of sustainable agriculture will continue to encourage their own weed problems, particularly if the available technologies are used unwisely.

The benefits from chemical weed control have received some support by those opposed to the use of pesticides. Chemical control has enabled new farming systems to be developed, e.g. crop establishment with minimal cultivation. These systems have improved soil structure, decreased erosion, increased yields from cultivars bred to exploit an extended growing season, and increased the time that green feed is available to stock before crop establishment.

There are increasing pressures to use minimal amounts of herbicides efficiently and to avoid non-target effects. Recent laws in some countries (e.g. Sweden, Denmark, and the Netherlands) require substantial reductions in the amounts of pesticides used in agriculture and some people have advocated a similar approach in Australia. There is pressure to be more informed and critical about the use of chemicals for weed control. Furthermore, the effectiveness of soil-applied herbicides is determined by processes that take place in the soil.

APPLICATION OF HERBICIDES TO SOIL

There are two main methods for applying herbicides. The first, for pre-emergence weed control, requires herbicide to be applied to the soil so that it is present in the soil solution for absorption by germinating seeds or seedling roots. The second method, for post-emergence control, requires the herbicide to be applied to the foliage of established weeds so that it is absorbed through leaf surfaces. However, the objective is to remove weeds early in the life of the crop (before a leaf-canopy covers the ground) and so a large proportion of a post-emergence herbicide can also reach the soil surface.

Most herbicides can be absorbed by the roots (Table 7.4.1) and therefore could be applied to the soil to kill weeds. Two of the most widely used groups of soil-applied herbicides are the triazines (e.g. simazine and atrazine in lupins) and the sulfonyl-ureas (e.g. triasulfuron and chlorsulfuron in wheat). Consequently, the following discussion will be illustrated with data from research on these two groups. However, data in Table 7.4.1 provide a broad perspective of all major groups of herbicides.

Herbicides in the soil can be leached, absorbed by plants and/or degraded. In plots of acid sand kept free of plants, more than half the applied simazine and triasulfuron was lost in 100 days (Figure 7.4.1). A Watheroo site (250 km north of Perth) received 105 mm of rain over the 100 days after application and the

* Faculty of Agriculture, University of Western Australia, Nedlands.

Table 7.4.1 The importance of root absorption of some major herbicides used in temperate agriculture.

Entry by roots or leaves is limited (-) or is the major path of entry (+++). Herbicides absorbed by the roots (+++) are applied to the soil where they interact with the biology and chemistry of the soil.

Persistence of phytotoxic residues in warm, moist soils may be limited (-) or continue for weeks (+) or months (+++).

The herbicides are grouped according to their modes of action in the plant. The names of the groups are standardised by the agrichemical industry. Herbicides within the same group should not be used repetitively on high densities of weeds or resistance will develop.

Absorption, selectivity and persistence differ among members within groups and 'Main uses' is only a general indication of potential uses.

Mode of action	Absorption by the plant		Name		Main uses	Persistence
	Root	Leaf	Trade	Chemical		in soil
Group A: Inhibitors of lipid synthesis					Selective control of grass weeds in cereals and legumes	
A1: Aryloxy-phenoxy-propionates ('Fops')	- +	+++	Targa	Quizalofop-ethyl		-
			Fusilade	Fluazifop-butyl		-
			Hoegrass	Diclofop-methyl		-
			Verdict	Haloxyfop-ethoxyethyl		-
A2: Cyclohexane-diones ('Dims')	- +	+++	Achieve	Tralkoxydim		-
			Select	Clethodim		-
			Sertin	Sethoxydim		-
Group B: Inhibitors of amino acid synthesis ('ALS inhibitors')					Selective control of grass and broad-leaf weeds in cereals and legumes	
B1: Imidazo-linones	++	+++	Spinnaker	Imazethapyr		+
B2: Sulfonyl-ureas	+++	+++	Ally	Metsulfuron		+
			Glean	Chlorsulfuron		++
			Logran	Triasulfuron		++
B3: Triazolo-pyrimidine sulfonamides	++	+++	Eclipse	Metosulam		+
			Broadstrike	Flumetsulam		+
Group C: Inhibitors of aromatic amino acid synthesis					Non-selective control of many weeds	
Phosphonomethyl glycine	-	+++	Glyphosate Roundup	Glyphosate		-
Group D: Inhibitors of tubulin formation (cell division)					Selective control of grass weeds in cereals and legumes.	
D1: Dinitroanilines	+++	-	Various	Trifluralin		+++
D2: Thiocarbamates	+++	-	Surflan	Oryzalin		+++
	+++	-	Avadex BW	Tri-allate		++
Group E: Inhibitors of cell division and other processes.					Selective control of grass weeds.	
E1: Amides	+++	-	Kerb	Propyzamide		++
	+++	-	Dual	Metolachlor		++
E2: Carbamates	+++	-	Carbetamex	Carbetamide		+
	+++	-	Chloro IPC	Chlorpropham		+
Group F: Inhibitors of photosynthesis (PSII) and cell membrane damage					Selective control of grass and broad-leaf weeds and non-selective, residual weed control	
F1: Ureas	+++	++	Various	Diuron		+++
F2: Triazines	+++	+	Various	Atrazine		+++
F3: Triazinones	+++	-	Various	Simazine		+++
	++	+++	Lexone, Sencor	Metribuzin		++

Table 7.4.1 continued.

Mode of action	Absorption by the plant		Name		Main uses	Persistence
	Root	Leaf	Trade	Chemical		in soil
Group G: Inhibitors of photosynthesis (PSI) and cell membrane damage					Non-selective control of many weeds	
G1: Bipyridyliums	-	+++	Reglone	Diquat		-
	-	+++	Gramoxone	Paraquat		-
	-	+++	SpraySeed	Diquat plus paraquat	Some selectivity	-
G2: Benzonitriles	-	+++	Various	Bromoxynil		-
Group H: Inhibitors of carotenoid synthesis Nicotinanilides	++	++	Brodal	Diflufenican	Selective control of *Brassica* weeds in cereals and legumes	++
Group I: Growth disruption (multiple sites of action). *I1:* Benzoic acids	++	+++	Various	Dicamba	Selective control of broad-leaf weeds in crops	++
I2: Phenoxy-alkanoic acids	+	+++	Various	2,4-D		-
	+	+++	Various	2,4-DB		-
I3: Pyridine carboxylic acids	+	+++	MCPA	MCPA		-
	+	+++	Tordon	Picloram		+
	+	+++	Lontrel	Clopyralid		+

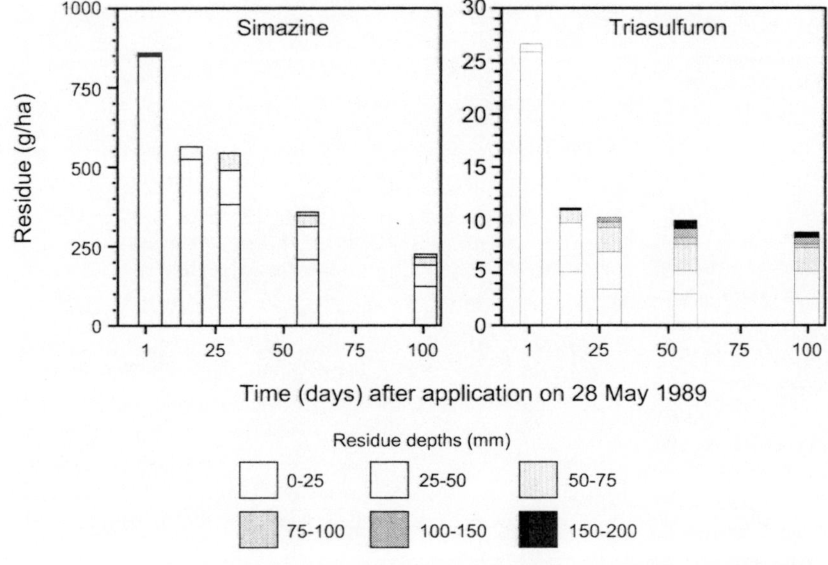

Figure 7.4.1 *Changes in the profiles of simazine and triasulfuron after application to the surface of a deep acid sand (pH_{Ca} 4.8) at Watheroo, Western Australia. Within a column the contributions of residues at each depth are added to give the total residue in the profile. Residue depths increase from the bottom to the top of the column: 0-25, 25-50, 50-75, 75-100, 100-150 and 150-200 mm. Temperatures at 13 mm below the soil surface ranged from 11 to 30°C and rainfall during the 100 days after application was 105 mm. The soil is suited to a lupin (*L. angustifolius*) - wheat rotation. After Blacklow and Pheloung (1992); Walker and Blacklow (1995).*

herbicides leached down the profile. However, 50% or more remained in the surface 50 mm. The sulfonyl-urea was more mobile than the triazine and residues were found below 150 mm. Residues, particularly at depth, are likely to carry over in the dry soil until the next rainy season when degradation* would resume. Residues equivalent to half the applied dose would be insufficient to control some weeds in the current crop but may be sufficient to damage sensitive species in the rotation (e.g. legumes that follow use of sulfonyl-urea). Different leaching profiles would be observed in other years, sites and soils, with other members of the same chemical groups and with members of other groups (Bos *et al.* 1995).

Complexity of system

Many processes in soils and plants determine whether a toxic dose of herbicide arrives and persists at sites of action in the plant (Figure 7.4.2). These processes may be sequential

* Degradation *is the chemical/physical breakdown of herbicide in soil.*

(absorption and translocation), concurrent (leaching and degradation), reversible (adsorption) or irreversible (degradation); they may improve (desorption; absorption) or diminish efficacy (adsorption; degradation; leaching). The rates are all determined by the physical, chemical and biological properties of soils and plants, the current state of the herbicide in the soil, and by past and current weather. Furthermore, the rates of these processes differ in response to the environment, both among herbicide groups and even among herbicides within the same chemical group. Despite these complexities, general statements can be made about the reactions of herbicides in soils, based on a definition of the herbicide-soil system (Figure 7.4.2). These general statements assist with the definition of more specific recommendations for weed control and crop tolerances.

FATE OF HERBICIDES IN SOILS

The model shown in Figure 7.4.2 identifies the individual processes that can affect herbicides in soils. These processes are described and their relative importance discussed below.

Additions and losses from the soil surface

Herbicides are applied in droplets of water (about 0.2 μL) to leaves, soil or both, at rates of 5 to 1,000 g/ha of active ingredient in water volumes of about 50 L/ha. These very small droplets soon evaporate and deposit the herbicide as a thin film of solid on the surface of soils and plant residues. Herbicides deposited on the soil surface can be degraded by photolysis (ultraviolet radiation) or evaporate; or be lost with eroded soil. In general, it is better to apply pre-emergence herbicides as close to the sowing time as possible and to mix them into the surface soil at seeding. This protects the herbicide from surface losses and gains maximum weed control during the early life of the crop. Trifluralin is an example of a volatile pre-emergence herbicide which needs to be incorporated into the soil surface otherwise it will be lost through evaporation. Although incorporation by cultivation makes the soil more vulnerable to erosion, some farmers need to take this risk because trifluralin may help them control annual ryegrass that has become resistant to other herbicides. The sulfonyl-urea and triazine herbicides can be degraded by photolysis but the importance of these losses under field conditions is debated.

Herbicides applied to the soil may be trapped by plant residues and need to be washed from these surfaces by rain, or be cultivated into the soil with the residues. Poor incorporation can cause losses of herbicides, which can be compensated by higher rates of application, and can cause uneven distribution which results in strips of poor weed control and crop damage.

Mixtures of herbicides are applied to broaden the spectrum of weeds controlled or to add residual weed control by adding a herbicide that can persist in the soil for several weeks. Mixtures may be formulated by the chemical industry or users may 'tank mix' their own active ingredients. A common mixture includes a broad-spectrum 'knockdown' herbicide (glyphosate, paraquat, or diquat) with a residual herbicide (triazines, or sulfonyl-ureas) to kill standing vegetation and prevent reinvasion by weeds that germinate later. This practice is the foundation of crop establishment with direct-drill and no-till systems. Pre-emergence herbicides applied to the soil are usually not good at killing established weeds and would need to be applied at higher rates to compensate for losses through degradation.

Herbicide may also be added to the surface from subsurface deposits. During a wetting cycle, herbicide is leached from the surface but these subsurface deposits may reappear at the surface during subsequent drying cycles. The driving force for upward movement of herbicides is the evaporation of water from the surface and consequent rise of subsurface water and dissolved solutes through capillarity.

Most herbicides are applied with additives that may be in greater amounts than the active ingredient. These include solvents, dispersants, wetters, emulsifiers, etc. The trend, however, is to formulate pesticides as water-dispersive granules and so minimise the use of additives. The expectation is that additives will be non-toxic and removed by degradation in the soil, plants or the atmosphere.

Movement in soil solution

Weed-kill requires absorption of herbicide from the soil solution at rates that exceed rates of detoxification by the weeds. Phytotoxicity is favoured therefore by persistently high concentrations of herbicide in the soil solution surrounding an active root system.

Herbicide applied to the soil surface moves directly or indirectly into the soil solution. The concentration in a given layer of soil will be the ratio of the mass of herbicide relative to the volume of water in the layer. If the concentration exceeds the solubility of the herbicide then some undissolved herbicide will form a suspension. Initially, a suspension may be the result of a high mass of herbicide distributed through a thin layer of surface soil. Later, a high concentration may be due to a loss of water by evapotranspiration. However, the concentration in the soil solution is in dynamic equilibrium with the mass of herbicide adsorbed to the surface of soil constituents and it is unlikely that high concentrations and suspended herbicide will persist in moist soil.

Herbicide is depleted from the soil solution by adsorption to soil colloids, degradation, absorption by plants, and upward and downward movement of water.

Adsorbed herbicide

The concentration of herbicide in the soil solution is in dynamic, reversible equilibrium with the mass of herbicide adsorbed to the surfaces of the soil constituents. The equilibrium state can be described by the Freundlich equation:

$$S <=> K_d C^n \qquad (7.4.1)$$

where S (mg/kg soil) is the mass of herbicide adsorbed, C (mg/L or µM) is the equilibrium concentration in the soil solution, K_d (L/kg) is the equilibrium distribution coefficient and n is the empirical order of the adsorption reaction. The value for n is usually near to unity and then the mass of herbicide adsorbed is proportional to the equilibrium concentration. Figure 7.4.3 shows adsorption isotherms for simazine and atrazine in a loamy sand.

After application there is an initially high concentration of herbicide in the soil solution which drives herbicide onto the adsorbing surfaces until equilibrium is reached between the mass adsorbed and the solution concentration. Subsequently, as herbicide is removed from a layer of soil, desorption re-establishes the equilibrium at a lower concentration. It is surprising that even in sandy soils with low organic matter, as much as 90% of the total herbicide can be adsorbed for herbicides with K_d values of about 1 L/kg, such as the triazines (Walker and Blacklow 1994) and the sulfonyl-ureas at low pH (Blacklow and Pheloung 1991).

Adsorption of the sulfonyl-urea herbicides decreases with increasing pH (Thirunarayanan *et al.* 1985) since they are nitrogen acids (pK_a of about 3 for chlorsulfuron and triasulfuron) and are negatively charged anions in soils of high pH (Beyer *et al.* 1988). Adsorption of simazine (and atrazine) decreases at high pH since they are cations at low pH (pK_a of about 1) and neutral molecules at high pH – although the effect is not large over the range pH 4.5 to 7.5 (Singh *et al.* 1989). Paraquat and diquat, non-selective herbicides used in minimum tillage systems, have very high K_d values (10^6 L/kg or more) and are considered to be bound irreversibly to soils (Pignatello 1989).

Adsorption of a herbicide can buffer it from leaching and degradation. The proportion of herbicide residue that is less prone to degradation has been estimated as the proportion of total residue that cannot be extracted with water (PAd). Walker and Blacklow (1994) found that the PAd increased with pH, organic matter and clay. As PAd increased from 0.2 to 0.6 the time for half the added simazine and atrazine to be degraded by chemical hydrolysis increased from 40 to 140 days (Figure 7.4.4). However, establishment of adsorption equilibria can be a slow process (Pignatello 1989; Walker and Blacklow 1994) and can have considerable influence on the dynamics of degradation and leaching in soils. Furthermore, it has been shown that rates of degradation of atrazine in suspensions of a silty clay loam increased with adsorption (Armstrong *et al.* 1967). It can be concluded, therefore, that it is possible to describe the dynamics of adsorption and some of its consequences, but it is difficult to explain the mechanisms across all groups of herbicides using one theory.

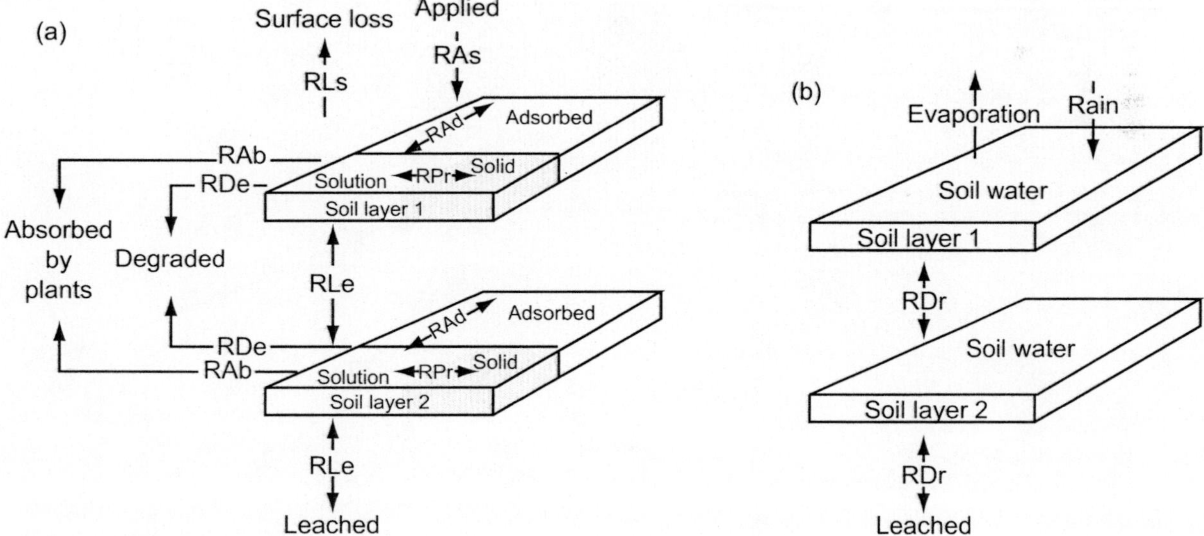

Figure 7.4.2 *Movements between states of a herbicide that occur within and between soil layers (a) and the rate of leaching between layers coupled to the rate of drainage of water between layers (b). In this model it is hypothesised that the herbicide in the soil solution phase can be precipitated, degraded, absorbed by plant roots or leached. Herbicide residues (adsorbed + solution + solid) are depleted by absorption, degradation and leaching below the sampled profile.*

Symbols: RAb = Rate of absorption by plant roots; RAd = Rate of adsorption; RAs = Rate of application; RDe = Rate of degradation; RDr = Rate of drainage of water between layers; RLe = Rate of leaching between soil layers; RLs = Rate of loss from the soil surface due to evaporation and photolysis; RPr = Rate of precipitation.

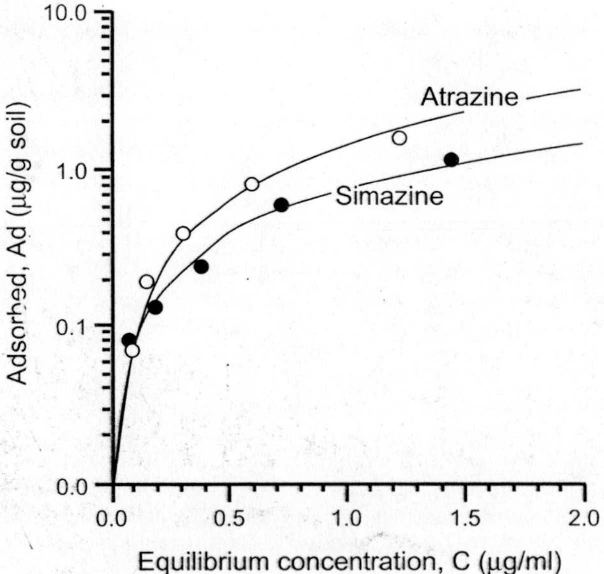

Figure 7.4.3 *Adsorption isotherms for atrazine and simazine in a loamy sand from Wongan Hills, pH$_{Ca}$ 4.6 (from Walker and Blacklow 1994).*

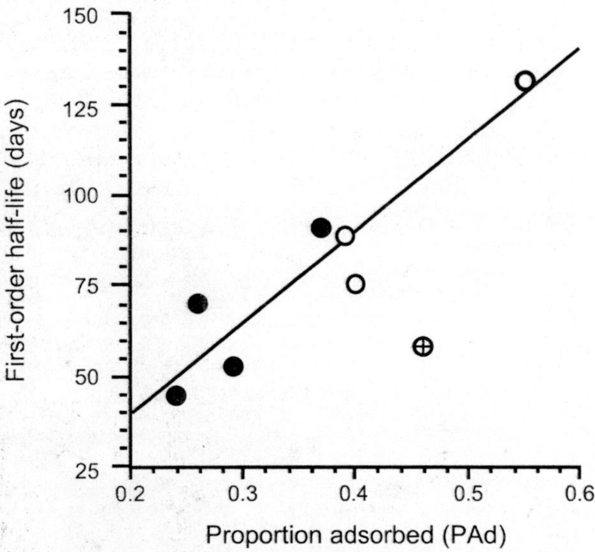

Figure 7.4.4 *Protection of simazine (closed symbols) and atrazine (open symbols) from degradation due to adsorption. The outlier is Watheroo sand (from Walker and Blacklow 1994).*

Degradation

Herbicides are degraded in soil by chemical and biological processes. The level of residues (R) declines with time (Figure 7.4.5) and may be described by an exponential function that suggests first-order reaction kinetics:

$$R = R_0 \exp(-kt)$$

where R_0 is the initial level of residue, k is the rate constant (t^{-1}) and t is time. The half-life ($t_{1/2}$), the time for 50% of the residue to disappear, is a function of the rate constant, $t_{1/2} = 0.693/k$. Departures from first-order kinetics may be due to lag-phases of microbial activity, rapid hydrolysis when solution concentrations are initially high, or if accelerations and decelerations are determined by changes in the environment such as temperature and moisture. Rates of degradation differ between layers of soil because of gradients in temperature, moisture, pH and microbial masses.

An uncritical use of the $t_{1/2}$ and associated theory can give misleading information about herbicide persistence in soils. For example, in soils incubated at constant temperatures first-order reaction kinetics was not established for a week or more, and a $t_{1/2}$ underestimated the time for half the herbicide to be lost from the soil by 3 to 21 days. This departure from first-order kinetics has also been observed by others (Armstrong *et al.* 1967; Bowmer 1991). It would be valid to refer to the time for half the residue to disappear (a half life, HL) from the soil but not to invoke first-order kinetics.

The environment has large effects on the persistence of residues in soils. For example, the $t_{1/2}$ for simazine in a loamy sand from Wongan Hills at 20°C halved from 90 to 50 days when the pH$_{Ca}$ was decreased from 6.5 to 4.5 (Figure 7.4.6). (Note, in undisturbed samples the pH may be two units lower when measured near to the exchange surfaces.)

In a loamy sand from Wongan Hills with a pH$_{Ca}$ of 4.6, the $t_{1/2}$ decreased from 150 to 20 days when temperature increased from 10 to 28°C (Figure 7.4.7), a change likely in the soil surface during a single day.

Chlorsulfuron and triasulfuron are degraded in soils by chemical hydrolysis (Blacklow and Pheloung 1991). Like the triazines, the rates of degradation of these sulfonyl-ureas are much slower at near neutral pH and low temperatures (Beyer *et al.* 1988; Blacklow and

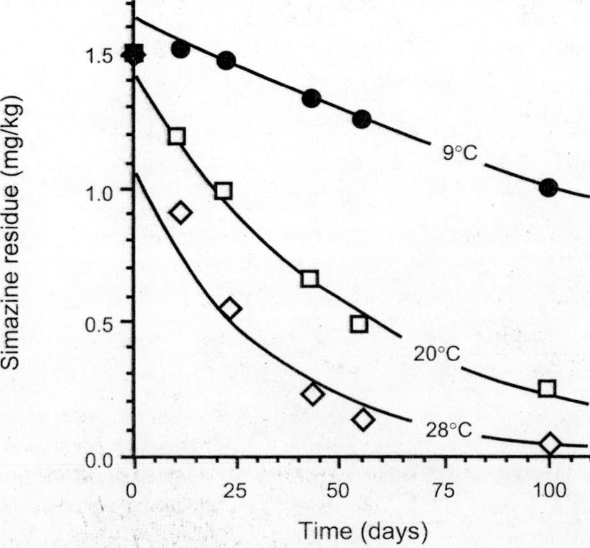

Figure 7.4.5 *Residue decay curves for simazine at constant temperatures in a loamy sand from Wongan Hills, pH$_{Ca}$ 4.6 (from Walker and Blacklow 1994).*

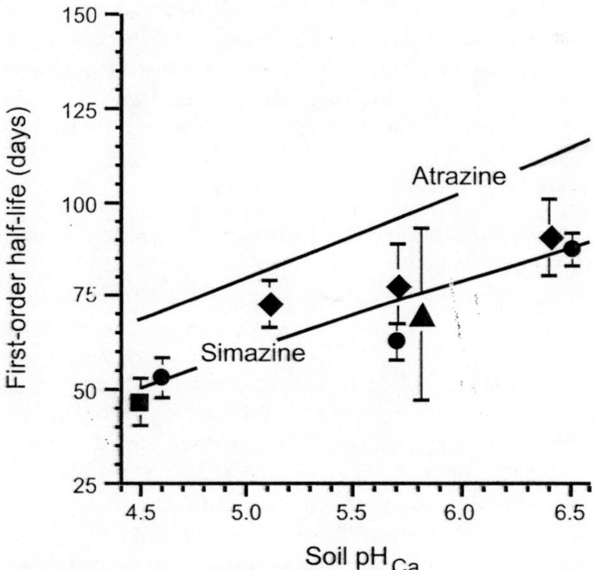

Figure 7.4.6 *Increase in persistence of triazine residues at 20°C in soils with pH_{Ca} differences (from Walker and Blacklow 1994).*

Pheloung 1991). Farmers are able to use lower rates of these herbicides in near neutral soils if they are concerned about residues that can persist into the next growing season and damage sensitive crops. In highly acid soils, however, the residues may not persist long enough to prevent late cohorts of weeds invading the crop. Other members of the sulfonyl-urea group degrade faster or slower than chlorsulfuron and microbial degradation may be more important, particularly at near neutral pH (Beyer *et al.* 1988). The half-life of trifluralin was 14 to 30 weeks at Victorian sites (Bos *et al.* 1995); microbial degradation is relatively unimportant, and pH, clay and organic matter have no consistent effect on the half-lives. Low

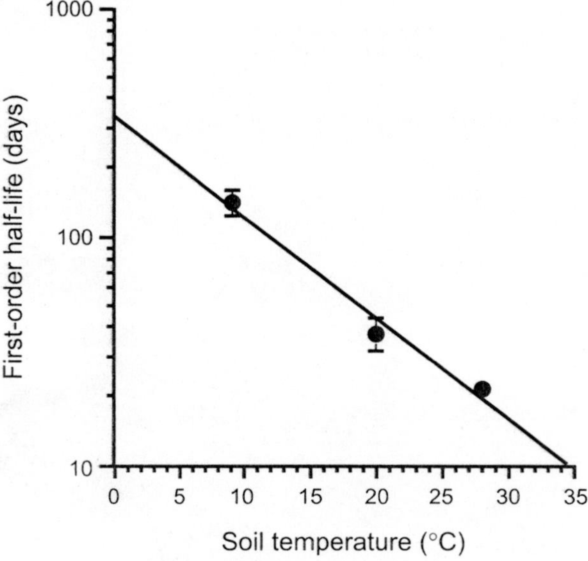

Figure 7.4.7 *Decrease in persistence of triazine residues with temperature in a loamy sand from Wongan Hills, pH_{Ca} 4.6 (from Walker and Blacklow 1994).*

temperatures and soil moisture are important determinants of degradation but care is needed to exclude their effects on losses due to volatilisation.

Soil moisture can also influence the persistence of herbicide residues. When the mode of degradation is microbial then moisture is important because of its indirect effect on microbial activity; hence degradation will be faster in warm, moist soils. However, if degradation is due to chemical hydrolysis then, very little water is needed to hydrolyse a small mass of herbicide and degradation can continue in a relatively dry soil (Blacklow and Pheloung 1992; Walker and Blacklow 1995). In an air dry soil, all the herbicide will be adsorbed (driven by the increasing herbicide concentrations in a drying soil) and so protected from degradation. It can be imagined, however, that as soil dries, herbicide will be concentrated at the surfaces of soil particles and there the reaction-kinetics may differ from those in the macro-pores (Pignatello 1989).

Leaching

Herbicide in soil solution can be leached from the profile when water in excess of the upper storage limit moves through the soil pores. However, the potential for a herbicide to be leached is unlikely to be predicted by the product of its solubility in water and the volume of water that moves through the profile, because a high proportion of the herbicide may be immobilised by adsorption. Furthermore, the soil macro-pores are readily flushed by water draining through the soil, and herbicide dissolved in the water of the micro-pores may be relatively immobile. Most of the herbicide therefore remains in the upper levels of the soil profile (Figure 7.4.1) and does not move with the wetting front established by drainage water (Walker and Blacklow 1995).

Redistribution of herbicide in the soil solution after a drainage event will lower the concentration of herbicide and desorption will establish a new, lower equilibrium. This redistribution and desorption may take several days to be completed. Consequently, water draining through the soil after desorption is complete will leach the greatest amounts of herbicide. The re-deposition of herbicide in the soil surface by upward movement of water and solute has been discussed in the section on *Additions and losses from the soil surface.*

Herbicides that are adsorbed weakly (i.e. they have a low value for K_d in the Freundlich equation; equation 7.4.1) will be more mobile in soil. The sulfonyl-urea herbicides, chlorsulfuron and triasulfuron, are nitrogen acids (pK_a near 3) and in soils of high pH they are found mostly as the anion and, therefore, are much more mobile than in acid soils.

Herbicide that leaches below the root zone of weeds may be wasted. There is concern also that leached herbicide may contaminate groundwater. In some European countries and parts of the USA, atrazine has been banned because it has been found in groundwater. In Australia recommendations for the use of atrazine in non-agricultural situations have been changed to avoid groundwater contamination.

Integration of complexity – computer simulation

If we wish to make quantitative predictions about changes in residues during the growing season, then we need to know the relative importance of the components shown in Figure 7.4.2 and how they change with the states of the herbicide. An option for reaching this objective is to develop computer simulation models that describe the dynamics of herbicides in soil-plant-environment systems.

Simulation models require specific information about the soil, the herbicide and the environment if they are to be applied to local situations. Models that are adapted with specific information then need to be calibrated against local measurements of the environment and levels of herbicide in the profile over a selected period of time. Adapted and calibrated (validated) models may then be used as management aids to make more accurate decisions on:

- rates and frequencies of herbicide applications
- persistence of potentially damaging levels of residues in crop rotations
- the extent of leaching under simulated rainfall intensities.

A simulation model based on the system shown in Figure 7.4.2 gave good agreement with the observed changes in the levels and states of herbicide in soil (Figure 7.4.8).

There was rapid degradation of herbicide during the first five days after application, while adsorption equilibria were being established. Subsequent degradation was dominated by soil temperature at the particular pH of the soil. The simulation model also gave good agreement with changes in the levels of residues in the field where temperatures fluctuated (Figure 7.4.8b). The expected exponential decay of residue with time was not observed in the field because increasing temperatures accelerated decay (Figure 7.4.8b).

Losses of herbicide from the surface 25 mm of soil during the first weeks following application can be high (25%) and were underestimated by simulation (Walker and Blacklow 1995). In acid sands, most of the sulfonyl-urea and triazine herbicides remained in the surface layers (Figure 7.4.1) where residues were subject to the extremes of a fluctuating environment (water, temperature, sunlight, wind erosion) and were modified by farming practices (cultivation, animal movements). Processes of herbicide degradation and movement in and on the soil surface are difficult to describe quantitatively yet can account for considerable losses. However, imprecise rates of application can cause unexpected results and studies of herbicides in soils should include an accurate measurement of applied herbicide.

Soil biology

Herbicides interact in two ways with the living constituents of soils (micro-organisms, earthworms and arthropods). First, they can be substrates for metabolism and, as a consequence, are broken down to less complex molecules and become herbicidally inactive. For example, glyphosate is broken down into CO_2 and H_2O, and the nitrogen and phosphorus components are incorporated into the microbial biomass (Grossbard and Atkinson 1984). Second, herbicides may affect the living constituents of the soil.

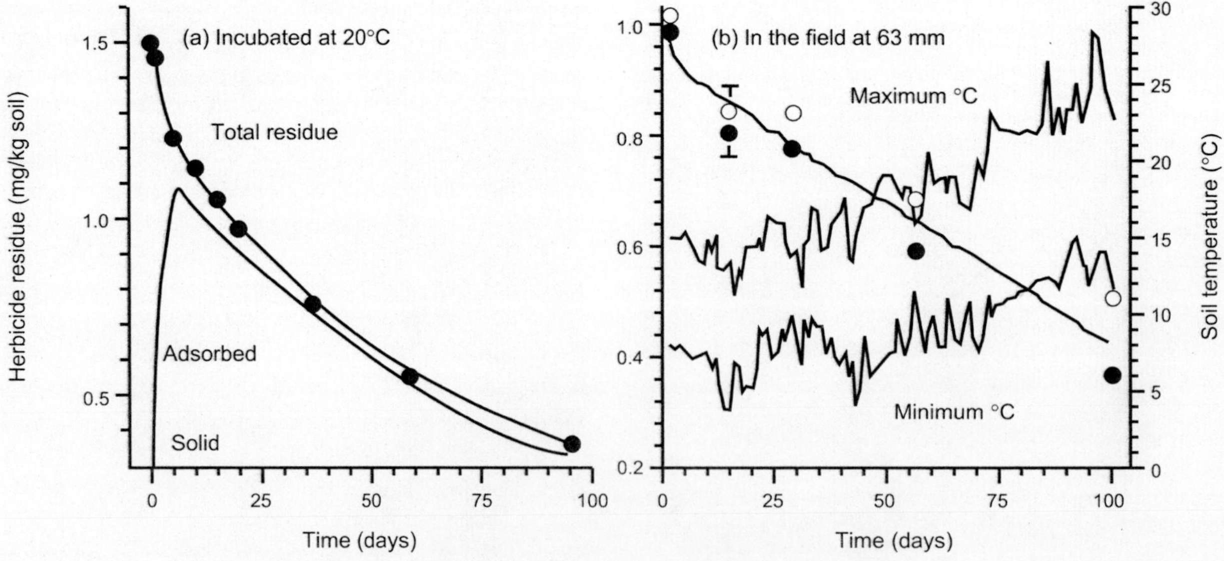

Figure 7.4.8 Simulated (continuous lines) and observed (●, ○) changes in the states of herbicides after application on 24th June 1989. (a) shows simazine in a loamy sand from Wongan Hills (pH_{Ca} 4.6) incubated at 20°C and (b) shows simazine (●) and atrazine (○) at a depth of 63 mm, in an unleached sandy soil at Watheroo (pH_{Ca} 4.7). The maximum and minimum temperatures are also plotted (from Walker and Blacklow 1994, 1995).

For example, soil bacteria can adapt to exposures of 2,4-D and carbamate herbicides and after initial lag phases they can accelerate degradation of subsequent applications (Cain and Head 1991). Concerns about deleterious effects of herbicides on the living constituents of soils are generally discounted. This is because herbicides are relatively non-persistent and many either target biochemical processes of green plants or are substrates for growth of soil organisms. Furthermore, soil organisms are resilient to changes in their environment. However, specific tests are required to show that herbicides intended for registration are not damaging to soil organisms and the vital processes they perform, such as the mineralisation of organic matter and nitrogen conversions (Riley 1991).

SUMMARY

Herbicides are applied to most soils used in agriculture and horticulture, and in many other situations where weeds are problems. Some herbicides are applied directly to the soil (for pre-emergence weed control) while others reach the soil because they are not intercepted by foliage. In both cases, chemical and biological processes in the soil are responsible for removing them from the environment.

There are two concerns about herbicides in soils. First, whether they actually kill the weeds and, second, whether their persistence in the soil will damage valued species that follow in the cropping systems. Soil is not an inert receptacle for herbicides. Rather, there are chemical and biological reactions between soil and herbicide that determine the extent herbicides move, persist and are phytotoxic.

Although there is a wide diversity of soil types and herbicides it is possible to make general statements about the interactions of herbicides with soils. At one extreme there are herbicides which are inactivated through chemical bonds with soil colloids (e.g. paraquat and diquat). At the other extreme there are herbicides that can exist as negatively charged species, are repelled by soil colloids, and remain in the soil solution for absorption by plant roots or are leached to depth (e.g. some sulfonyl-ureas at high pH). The reaction of most herbicides is between these two extremes – they are bound reversibly to the soil colloids and are desorbed as herbicide is removed (e.g. the triazines and sulfonyl-ureas at low and neutral soil pH). Herbicide that remains on the surface of dry soil or plant residues may be lost by volatilisation and photolysis before interacting with the soil.

The rate of herbicide degradation is also determined by interactions between the soil, the herbicide and the environment. The processes of degradation may be chemical or biological or both. In general, degradation of reactive species is fast in warm moist soils. However, some herbicides are inherently persistent. Chemical hydrolysis can be catalysed by low or high soil pH and can continue in relatively dry soils. Degradation by both chemical and biological reactions can cease at low temperatures and increase dramatically (exponentially) as temperatures increase.

Herbicides move down and up the soil profile with water. However, they are not as mobile as water because they can be adsorbed to soil colloids and because water drains preferentially through the macro-pores of the soil; herbicide dissolved in the water of the micro-pores is relatively immobile. The re-establishment of equilibria (desorption and diffusion into the macro-pores) after a drainage event may take several days and so intense rainfall may not leach as much herbicide as intermittent rainfall.

In general, pre-emergence herbicides are applied as preventive treatments, against the anticipated weeds within tolerant crops. It is helpful to remember that the objective is to ensure that the weeds absorb a toxic dose of the herbicide. This is achieved if the weed has a prolonged exposure to the herbicide in the soil solution surrounding the root system. For reasons of minimising costs and crop damage, however, the rates of application need to be minimised. If soil conditions favour sustained uptake by the weeds then lower rates can be used. Conversely, if adequate concentrations of a herbicide do not persist (because the herbicide is removed by degradation, leaching or adsorption) then higher rates of application may be needed. For example, triasulfuron can be used at lower rates in alkaline soils because it degrades more slowly and less is adsorbed than in acid soils. For similar reasons, simazine should be applied at lower rates in lupins grown on coarse-textured sands otherwise the crops will be damaged, but higher rates are needed on finer textured soils with a low pH. Pre-emergence herbicides should be applied as near to the time of crop establishment as possible and be incorporated into the surface layers of soils by cultivation or following rainfall.

The response of weeds and crops to a given level and state of herbicide in the soil may depend ultimately on the responsiveness of the plants and not the herbicide dose.

Further reading

Dodd, J., Martin, R.J. and Howes, K.M. (eds.) (1993). Management of agricultural weeds in western Australia. Department of Agriculture Western Australia Bulletin No. 4243.

Summers, L.A. (ed.) (1980). 'The bipyridinum herbicides'. Academic Press, London, 449 p.

Worthing, C.R. and Hance, R. (eds.) (1991). 'The pesticide manual: A world compendium'. The British Crop Protection Council, 1081 p.

HERBICIDE RESIDUE PROBLEMS IN WESTERN AUSTRALIAN FARMING SYSTEMS

*Terry Piper**

Current rotational cropping systems practised in WA always run some risk of herbicide carryover. Generally, wheat is alternated with narrow-leafed lupins (on acid coarse-textured soils) or another pulse crop (on neutral to alkaline, finer textured soils). Simazine (sometimes with atrazine) at 1 to 2 kg/ha is a universal basal herbicide for weed control in lupins, but 15 to 30% of this rate can damage wheat. Sulfonyl-urea herbicides (SUs) are widely used as a pre-plant herbicide in wheat; chlorsulfuron at 7 to 15 g/ha or triasulfuron at 20 to 25 g/ha are the most common. At 5 to 10% of these rates, severe damage can be caused to pulses.

These herbicides are all root-absorbed, relatively persistent in the soil and known to cause residue problems elsewhere. Given our relatively short growing season and thus limited time for herbicides to break down, there is a high potential in WA for carryover damage. However, this only occurs rarely, in seasons of low rainfall.

Simazine in pulses

Simazine is usually applied just before seeding and incorporated into the topsoil by the action of the seeding machine. It is degraded mainly by the action of micro-organisms, with some chemical hydrolysis catalysed by clays. It is not leached appreciably, so degradation tends to cease as the topsoil dries out. If applied immediately after seeding or emergence, or if no-till seeders are used, the chemical will remain on the soil surface and be even more prone to soil drying.

Cases of residue carryover seem quite soil type dependent.

Lupin crops on acid sandplain soils

Simazine carryover is rare in these soils, although it has been reported. Important factors in degradation are temperature and moisture. With early sowing, temperatures are relatively high and degradation is rapid. Trials were conducted at Newdegate to establish a safe interval for re-sowing cereals following a failed lupin crop. Simazine was applied and incorporated as if lupins were to be seeded, then cereals were sown into the site at weekly intervals. In 1985, less than 10% of the cereal seedlings survived even when sown 10 weeks after the simazine had been applied. In contrast, in 1986, 80% of the cereal seedlings sown after four weeks survived, while those sown after five weeks showed no growth or yield reduction. Soil type and rainfall for the two trials were almost identical. The major difference was in simazine application time: July in 1985 and May in 1986 (Piper 1985, 1986).

Other trials have shown a half-life of between three and five weeks for simazine applied in May (Piper 1986). Thus half the chemical has gone before soil temperatures drop over winter, leaving some five months for the next half to fall below the danger level.

* *Agriculture Western Australia, South Perth.*

Moisture is not a big factor in degradation in these soils. The clay content is very low, so the soil wets quickly to the point of having free water available and adsorption of simazine is low. Degradation can then commence. Current research is indicating that once moisture contents reach about 3 to 4%, the degradation rate is near its maximum and increasing the moisture content to the upper storage limit has little additional effect (T. Piper unpublished data).

In an average season, an initial application of 1000 g simazine/ha will degrade to around 50 to 100 g/ha before the end of the season, well within safety limits. Growing season rainfall would have to be as low as 200 mm before any problems would be expected.

Alternative pulse crops on neutral to alkaline, fine-textured soils

These are the soils on which we see most instances of carryover damage, which can be attributed to their higher clay content.

First, these soils need more rainfall before a crop can be sown. Rather than seeding on an opening rain of 10 to 15 mm as for the sandplain country, a farmer may have to wait an extra few weeks until a cumulative total of 25 to 30 mm has fallen, by which time the soils are considerably cooler, slowing degradation. Furthermore, the degradation rate only reaches its maximum at a water content of ~15%, and as the surface dries out between rainfall events degradation will slow dramatically. Simazine does not leach appreciably, so it remains in the topsoil where it is subject to periodic drying.

Second, these soils adsorb simazine more strongly and thus reduce the base rate of breakdown. They also have a higher pH which reduces degradation (see *Herbicides: Movement, persistence and activity in soils*).

Thus in an average season, 1000 g/ha might degrade to around 150 to 300 g/ha. A mitigating factor is that cereals growing in these soils can tolerate a higher level of residue before being damaged. Cereal variety trials over a number of years to assess simazine tolerance indicate that rates of 300 to 500 g/ha are the maximum safe level for yellow deep sand at East Chapman, but on the alkaline red shallow loamy duplex soils at Merredin this level is 500 to 1000 g/ha. The explanation is probably the same as for reduced degradation: higher clay content means less free water and greater simazine adsorption. Thus for any given level of residue, there is less available for uptake by the cereal crop. For residue damage to occur, there must be low rainfall in the season of application and the opening to the next season must be relatively wet.

Root disease interactions

In most cases of simazine carryover causing damage, the damage has presented like a root disease problem, with circular patches of dead crop surrounded by healthy plants. The transition from dead to healthy plants is very sharp, characteristic of rhizoctonia, Eradu patch and other fungal diseases. Nevertheless, the plant symptoms are those of simazine damage and moderate levels of simazine residue have been detected in the soil. We surmise that some level of fungus survives after seeding, possibly beneath the cultivated layer. This fungus then retards root growth, killing those roots that try to extend into the subsurface soil. The cereal seedlings are then forced to survive entirely on surface roots which are in a band of soil containing residual simazine. The level of simazine is effectively magnified and kills the plants.

This effect was illustrated at a variety tolerance trial at East Chapman in 1995. The trial site had an area of Eradu patch across it. Within this area simazine at 125 to 250 g/ha was killing >90% of wheat seedlings, while outside the area seedlings were surviving residues of 375 to 500 g/ha.

Sulfonyl-ureas in wheat

Chlorsulfuron (Glean, Seige, Tackle etc.) and triasulfuron (Logran, Nugran etc.) are the two sulfonyl-ureas most at risk of causing carryover damage. Metsulfuron (Ally, Associate etc.) has a much shorter residual life and would not normally be expected to cause problems.

Chlorsulfuron and triasulfuron are usually applied just before seeding and are incorporated by seeding, although they may be applied immediately after seeding and some is used post-emergence. They are degraded mainly by chemical hydrolysis, with some microbial degradation. They are leached readily, so surface-applied chemical is quickly moved into the root zone of weeds and into moist soil where breakdown can commence.

In northern USA, chlorsulfuron residues have damaged crops even six years after application (Moyer *et al.* 1990). Admittedly, their soils are close to freezing for several months each year and the damaged crop was sugar beet which is particularly sensitive.

In WA by contrast, reports of damage are very rare. The rate of breakdown is much faster in acid soils (see *Herbicides: Movement, persistence and activity in soils*) and most WA soils are acidic. Also, lupins and field peas are at least an order of magnitude less sensitive than sugar beet.

The ready leaching of SUs means that any residues are mostly in the subsoil by the end of the season which remains moist far longer than the surface, so the time available for degradation is greatly extended. Leaching can present problems in duplex soils, especially when the subsoil is alkaline. The SUs are leached quickly through the sandy top layers, but retarded by the clay layers of lower permeability. Degradation is much slower in an alkaline environment and enough can persist to affect legumes the next season. The legumes, either crop or pasture, will not be killed, nor will they show any symptoms. Rather their roots will be pruned as they attempt to enter the soil containing SU residues. Growth will be normal until later in the season when subsoil moisture is necessary for pod filling or seed set. Without access to subsoil moisture the plants will hay off earlier than expected. Such behaviour has been observed when SU trial sites on duplex soils have been resown the next season.

Leaching also presents problems if SUs are sprayed onto coarse-textured soils some time in advance of the break of season in preparation for seeding. If opening rains are intense, the SU can be leached beyond the reach of germinating weeds. By the time the weed roots reach the chemical, the weeds will be too large to be controlled by the rate applied. It is never good practice to apply SUs ahead of seeding, in contrast to simazine which is used this way quite successfully.

New crops

It is possible that the current low incidence of herbicide carryover damage will change with developments in the farming rotation. Some of these are considered below.

Triazine-tolerant (TT) canola

Triazine-tolerant canola is grown to overcome wild radish, which cannot be selectively controlled in standard crops of canola. Atrazine is applied soon after emergence (application is two to four weeks after seeding) and provides good control of most weeds. Used this way however, the atrazine has less time to degrade and is left on the soil surface in an environment less conducive to breakdown (dry for a larger proportion of the season). Atrazine is at least as toxic to cereals as simazine, so the potential for crop damage is high.

Several possible solutions are currently being evaluated, but none are ideal. Sowing a pulse rather than a cereal is an option, as the pulse will tolerate the atrazine residue. Controlling volunteer canola in the pulse may prove difficult. No-till seeding of the cereal should be successful, as the atrazine will be left on the surface rather than being incorporated with the roots of the cereal seedlings. As a last resort, paddocks with excess levels of residue can be seeded at the end of the cropping program, thus giving the residue an extra few weeks to degrade.

Alternative pulses

Until recently, the only pulse widely available for alkaline soils where SU carryover is a risk, has been field peas. These are not widely grown, therefore large areas of crop have not been exposed to SU carryover in any one year. The newer species of pulses are likely to expand the area cropped, so that statistically there will be more chance of damage. In addition, these alternative pulses are significantly more susceptible to SUs than field peas so we may find that cases of damage increase.

The only solution is to avoid the problem. If the average rainfall or soil properties indicate a high risk of carryover, do not apply SUs before planting, but leave them until after emergence. Then, if the season looks like being drier than normal, alternative herbicides can be used. Be prepared to change the planned rotation if SU residues are present and sow a second cereal crop.

CHAPTER 8

CROPS: SOIL AND CLIMATIC REQUIREMENTS

CHAPTER 8.
CROPS: SOIL AND CLIMATIC REQUIREMENTS

This chapter summarises soil and climatic requirements for the major rain-fed crops grown in south-western Australia. Crop requirements are described in terms of species tolerance, although differences between varieties in their tolerance to soil conditions are mentioned. For recommendations on varieties refer to the *Crop Variety Sowing Guide* published annually by Agriculture Western Australia.

A reliable assessment of the potential for a crop to grow at a site can be made providing:

- the climate and soil properties at the site have been assessed
- the climatic and soil requirements for the crop are known.

Chapters 1 to 7 describe how to assess the soil properties at a site. This chapter describes the soil and climatic requirements for: wheat, barley, oats, narrow-leafed lupins, field peas, canola, faba beans, chickpeas, grape vines and blue gums.

The information for each crop is presented under the following headings:

Introductory paragraph

A brief description of the history and present role of the crop in Western Australian agriculture plus key references describing the agronomy and management of the crop.

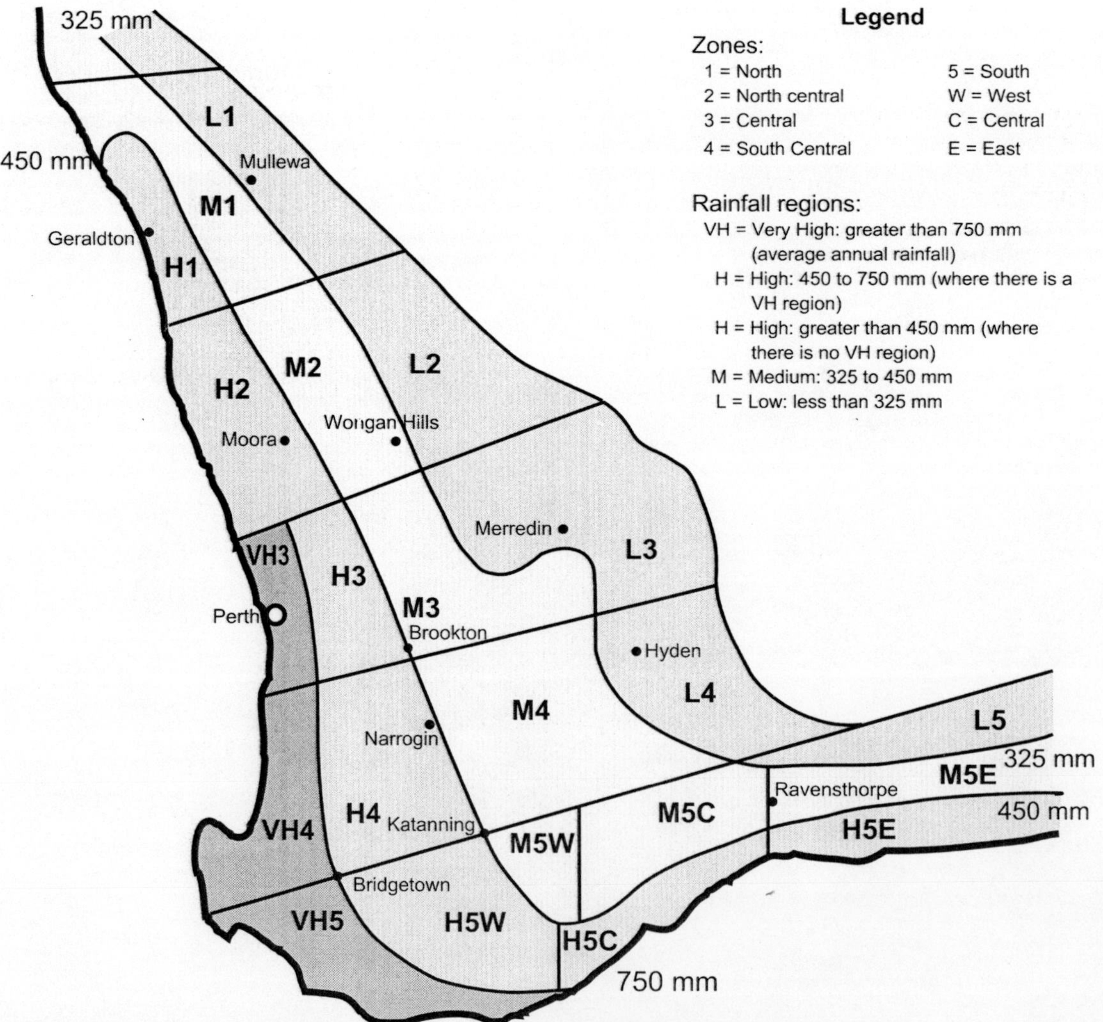

Figure 8.0.1 *The agroclimatic zones for south-western Australia (from Brown 1994).*

RELIABILITY OF INFORMATION

- The amount of information relating performance to soil and climate varies enormously between crops. For a major crop such as wheat there is a large body of information on almost every facet of growth and development, while the same level of information is not available for a relatively new crop like blue gums.
- We have attempted to indicate the reliability of the information by referencing its source and the confidence of the relationship. Some relationships are well established and widely accepted (e.g. compacted soils slow the penetration of roots), while other relationships may involve conjecture (e.g. effect of high concentrations of soil boron on barley yields).

Root growth

An understanding of the characteristics of the root system is essential to relate crop performance to the soil properties at a site. For instance, a taprooted plant is likely to grow more easily through a compacted soil layer than plants with a fibrous root system. The maximum rooting depth of a crop is important when the soil has a low water-holding capacity. Deep-rooted crops have a distinct advantage on uniform coarse-textured soils providing there are no other limitations to root growth.

Climate requirement

The agroclimatic zones used are from the *Crop Variety Sowing Guide* (Brown 1994) except for grape vines and blue gums. There are four rainfall divisions based on the average annual rainfall: *very high* (>750 mm), *high* (450-750 mm), *medium* (325-450 mm) and *low* (<325 mm). There are also five divisions which reflect changes in solar radiation and temperature in a north-south direction through the agricultural area of Western Australia, from Kalbarri to Albany (Figure 8.0.1). Specific relationships between temperature and growth rates are noted for each crop where the information is available.

Summary of soil requirements

A summary of the soils to which the crop is adapted and those to avoid.

Soil factors affecting productivity

The tolerance of the crop to a range of soil conditions is described. These soil conditions either directly or indirectly affect plant or root growth and/or the ability of the root system to supply the plant with water and nutrients.

8.1 WHEAT: SOIL AND CLIMATIC REQUIREMENTS

Wal Anderson and Geoff Moore*

Wheat (*Triticum aestivum*) is the major crop grown in Western Australia. Its management is summarised in *The Wheat Book* (Perry and Hillman 1991).

Root growth

Wheat has a similar system of root development to other cereals. The temperate cereals possess two distinct root systems: the seminal roots which develop from primordia within the seed; and the nodal or adventitious roots which develop from the nodes within the crown (Figure 8.1.1; Troughton 1962). Under ideal conditions when there is no restriction to root growth, wheat roots can penetrate to a depth of about 3 m. However root density is low below 2 m (Hamblin and Tennant 1987). Other studies on uniform coarse-textured soils report the effective maximum depth of root growth to be 1.6 to 1.7 m (Tennant 1976; Hamblin 1982; Hamblin and Hamblin 1985).

Climatic requirements

A general discussion of climates for wheat growing in Australia is given by Nix (1975). In WA, the suitable area has an average annual rainfall from less than 325 to 750 mm, incorporating all of the low, medium and high rainfall agroclimatic zones (Figure 8.0.1).

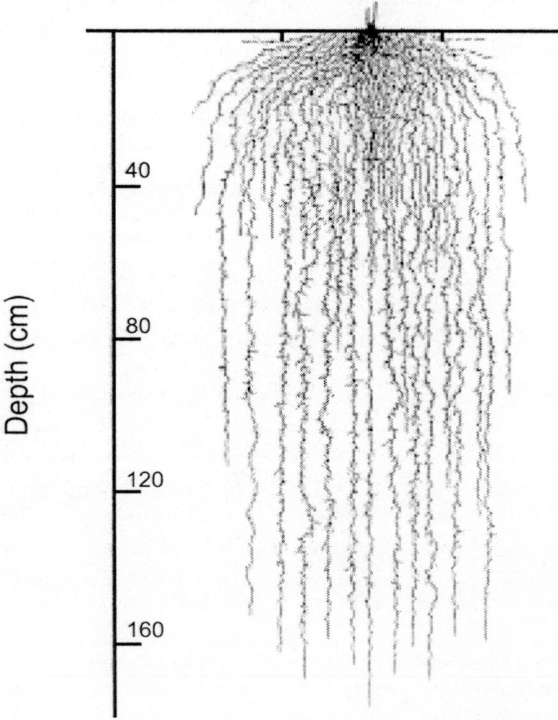

Figure 8.1.1 *Diagram of wheat root growth (from Plate 4.2, Wild 1988).*

* *Agriculture Western Australia, Northam.*

The average levels of solar radiation in the agricultural area during winter are unlikely to limit crop growth, although growth will be retarded on certain overcast days.

Temperature is the fundamental property which controls the development of a crop, while photoperiod and vernalisation also have a significant role (Perry and Belford 1991). The temperature regime for growing wheat in south-western Australia was categorised as ranging from 'hot' in the northern wheatbelt, to 'warm' and then 'cool' in the south (Nix 1975).

Winter temperatures restrict growth, because in general the optimum temperature for wheat is about 20°C (Milthorpe and Moorby 1975). The temperature requirements vary depending on the growth stage and the particular plant organ. For instance, the optimum temperature for maximum *length* of the leaves is about 20°C, which is higher than the optimum temperature for leaf *width* (15°C). The optimum temperature for *leaf area* is about 20°C (Friend *et al.* 1962). The rate of leaf emergence and the number of tillers depends on the *thermal time* rather than the ambient temperature, providing the plant is not stressed. Thermal time (or accumulated temperature) is the sum of the mean daily air temperatures above some base temperature, usually 0°C (Perry and Belford 1991).

The grain protein is influenced mainly by the environment and management and to a lesser extent by genotype. High temperatures plus water stress induce higher temperatures in the plant than the ambient temperature. These conditions promote rapid grain development which reduces carbohydrate content more than protein accumulation, resulting in higher

SUMMARY OF SOIL REQUIREMENTS

Wheat is grown successfully on a wide range of soils because it is comparatively tolerant of different conditions. Acid soils (pH_{Ca} <4.5) are unsuitable for some cultivars. Be aware of varietal differences in sensitivity to acid soils and boron toxicity (*Crop Variety Sowing Guide)*.

If yields are to approach their climate potential, then soil conditions need to be considered carefully. To grow high yielding crops the soil should be well to moderately well drained; with good physical characteristics and no barriers to root penetration; have no extremes of pH; be non-saline and have an adequate nutrient supply.

grain protein. If conditions are cooler and more humid, grain filling is prolonged producing plump grain of low protein content (Nix 1975).

Wheat is sensitive to frost after ear emergence. Cultivars are sown to avoid flowering during times of high frost risk (Anderson and Smith 1990).

Soil factors affecting the productivity of wheat

Soil conditions	Tolerance
Soil water deficit	Soil water storage will usually only prevent production if very low (<30-40 mm/m). Otherwise healthy crops are very tolerant of water stress during establishment and can produce some grain even if water in the root zone is extremely limited after flowering. In the eastern wheatbelt, grain yield was reduced by at least 25% (due to moisture stress) when ferricrete (duricrust) was within 30 cm of the surface (Sullivan 1991).
Waterlogging	Mild waterlogging affects yield, while severe waterlogging can kill plants (i.e. bare patches in crop). Wheat is more sensitive than oats, but less sensitive than barley. In general, sensitivity decreases as plants mature. The most sensitive stages are between germination and emergence, the seedling stage before nodal roots have developed and at stem elongation (Setter and Belford 1990). Roots can adapt to waterlogging by forming aerenchyma. Wheat may only be able to form aerenchyma in short roots less than 20 cm long (Thomson *et al.* 1990), which would have implications for waterlogging deeper in the soil. Root diseases, especially take-all, can be more severe in waterlogged crops. Crops grown at high levels of nutrition, especially nitrogen, are able to recover rapidly from the stress imposed by waterlogging (Barrett-Lennard *et al.* 1988).
Soil salinity	Moderate tolerance. Under controlled conditions yield is affected when ECe is >600 mS/m, with a yield decrease of 7.1% per 100 mS/m increase above 600 mS/m. There is less tolerance during emergence and seedling stages, when ECe should not exceed 400 mS/m (Maas and Hoffman 1977).
Salinity and waterlogging	More sensitive to salinity if waterlogged, when ECe of 200 mS/m can kill the crop (Barrett-Lennard 1986; Barrett-Lennard *et al.* 1988).
Acidity: minimum pH_{Ca}	Tolerance to acidity or aluminium varies. Sensitive varieties (e.g. Aroona, Cranbrook, Schomburg, Wilgoyne) are usually affected by pH_{Ca} 4.2-4.5, while tolerant varieties are affected by pH_{Ca} 4.1-4.4 (Brown 1994).
Alkalinity: maximum pH_w	Roots can tolerate moderately alkaline conditions. Very high pH_w (>9.0) is likely to decrease growth, because sodium carbonate may be present (Section 5.2). Presence of calcium carbonate does not normally affect growth unless a cemented pan (calcrete) physically impedes roots. Highly alkaline soils may be highly sodic, which can be toxic to plants. Growth will be reduced by 50% if the ESP is >30-50 (Hira *et al.* 1980; Gupta and Abrol 1990).
Key nutrient requirements	N and P are the most important nutrients. The most profitable fertiliser level for a specific crop can be determined using the NP-Decide model (Burgess 1988; Bowden 1989). *Potassium.* Most cereals show either no yield response or an unprofitable response. *Copper.* Highly responsive (Gartrell 1981). Will be affected by low Cu concentrations before other crops. Deficiency can cut yield by 20% before leaf symptoms are evident. *Molybdenum.* Lower requirement than pasture legumes. Deficiency only reported on the acid yellow sandy earths and gravelly sands in the Zone of Ancient Drainage (Riley 1984). *Manganese.* Deficiency can occur on severely deficient soils (Teakle and Wild 1940; Smith and Toms 1958). *Boron.* Effect of high soil B is a point of conjecture in WA, although in South Australia toxicity is recognised as reducing grain yield. Considerable variation in tolerance between varieties.
Compacted soils	Subsurface compaction decreases the rate of root elongation through the compacted zone (Tennant 1986). It also greatly restricts seedling growth (Materechera *et al.* 1991). Effect on yields varies with season and degree of compaction (Jarvis and Porritt 1985; Jarvis 1986; Henderson 1991).
Root growth into clayey subsoils	Varies enormously depending on the subsoil structure. In duplex soils the depth of penetration into clayey subsoil varied from 0.1 to 1 m (Dracup *et al.* 1992).
Soil properties affecting germination	A surface crust or hardset surface can reduce germination, particularly on degraded soils (McIntyre 1955; Chambers 1962; Stoneman 1962).
Erosion risk	Sand blasting affects growth. Compensatory growth may offset this if the seedling is cut once from sand blasting, but if cut a second time, yield can be reduced by 33% (McFarlane and Carter 1990).

8.2 BARLEY: SOIL AND CLIMATIC REQUIREMENTS

*Kevin Young**

Barley (*Hordeum vulgare*) follows wheat as the second major crop grown in WA. In many ways its management is similar to wheat and is summarised in *The Barley Book* (Young 1995).

A mature barley crop.

Root growth

Barley has a root structure similar to wheat, consisting of two distinct systems, the primary and nodal roots. The primary or seminal roots arise from the germinating seed. Usually five to seven emerge and branch freely as they extend downwards. The seminal roots are generally the deeper system and under favourable conditions have reached 2.1 m (Briggs 1978). Maximum growth is reached about the time of ear emergence.

The upper soil layers tend to be packed with secondary roots known as nodal or adventitious roots. These are derived in a series of irregular whorls at the base of the crown from closely packed basal nodes. Development of nodal roots is closely linked to tillering and if a tiller becomes detached from the plant it can be supported by these roots only.

Under conditions of extreme drought or nutrient deficiency the nodal roots may not develop, as tiller buds remain dormant when exposed to a high level of stress. Barley roots will not grow into dry soil (i.e. soil below the lower storage limit) and will not grow appreciably into a stationary watertable, or survive long periods of anaerobiosis (Greenwood 1968). Under conditions of poor aeration, such as waterlogging, nodal roots are more numerous, shorter and thicker, and the cortex develops large air passages separated by thin strands of parenchyma which enable roots to receive more oxygen from above. This is more of a survival mechanism than a means of coping with waterlogging. Consequently, barley is poorly adapted to soils that waterlog regularly or for prolonged periods. As with wheat, root growth is restricted by traffic pans and poor soil structure.

SUMMARY OF SOIL REQUIREMENTS

Barley can grow successfully on a wide range of soil types. It is often grown on coarser textured soils with a lower nitrogen status, because a low to moderate protein content is favoured for malting.

Barley tolerates higher salinity than other cereals, but is more sensitive to waterlogging and soil acidity. The pH_{Ca} of the surface soil (0 to 10 cm) should be higher than 4.5.

Climatic requirements

Barley is a crop of temperate climates. It evolved in marginal winter rainfall areas in which the life cycle is completed rapidly. Traditional Australian cultivars were early maturing and adapted to an annual average rainfall of 500 mm or less. In lower rainfall areas where wheat is the most important crop, farmers use the early maturity of barley varieties such as Stirling by sowing it late in the cropping program. Recently, later maturing varieties have been released which are well adapted to the 500 to 750 mm zone.

In higher rainfall areas, the cool moist climate and long growing season favour development of a wide range of foliar diseases and insect pests. In general, barley diseases are easier to control by plant resistance and limited use of fungicides than are wheat diseases. The cereal root disease, take-all, one of the major constraints to yield in the higher rainfall zone, is also far more severe in wheat. Greater resistance to both foliar and root diseases makes barley a more reliable, higher yielding crop than wheat in the high rainfall zone.

* Agriculture Western Australia, Esperance.

Soil factors affecting the productivity of barley

Soil conditions	Tolerance
Soil water deficit	Barley is often regarded as having a high level of drought resistance. There is little evidence to suggest that it has a higher water use efficiency than wheat. It is more a case of drought avoidance through early maturity, than drought resistance.
Waterlogging	Susceptible, although it can withstand transient waterlogging. When waterlogging is only for a short duration, crops can still yield more than 3 t/ha. The development stage at which waterlogging has its largest effect on yield has not been defined clearly. The most sensitive stages appear to be between germination and emergence, the early seedling stage and when the plants are elongating rapidly, which is the most rapid growth stage (Belford and Thomson 1981; Stepniewski and Labuda 1989a, b).
Soil salinity	Has the highest tolerance of the cereals. Under controlled conditions yield is decreased when electrical conductivity (ECe) >800 mS/m, with a yield decrease of 5.0% per 100 mS/m increase above 800 mS/m. Less tolerance during emergence and the seedling stage (Maas and Hoffman 1977).
Salinity and waterlogging	Saline areas are often waterlogged and dominated by barley grass which is a host of the root disease, take-all. If barley is grown on saline areas there must be an attempt to drain the area and reduce root disease. The most practical method may be applying a high rate (approx. 200 kg/ha) of a compound fertiliser containing ammonium sulphate (e.g. Agras No. 1) at seeding.
Acidity: minimum pH_{Ca}	More sensitive to aluminium toxicity than other cereals. Acid sandplain soils (e.g. wodjil) that increase in acidity with depth and have high Al levels are unsuitable. In the medium/high rainfall areas, high yielding crops are commonly grown on coarse-textured soils that are mildly acidic at the surface. These soils generally become neutral with depth and have naturally low levels of Al. If the pH_{Ca} is <4.5 and crops have a history of poor yields, barley is likely to respond to liming (Dolling *et al.* 1991).
Alkalinity: maximum pH_w	More tolerant than other cereals (Bower and Fireman 1957), but most cultivars are very susceptible to boron toxicity. A feature of some alkaline soils in WA is an increasing concentration of boron with depth. On these soils there is a significant yield advantage for boron-tolerant varieties such as Skiff from SA. This advantage tends to be greatest in drier years confirming the SA observation that root growth of boron susceptible varieties into the subsoil is impeded, reducing their capacity to extract moisture from depth in a season with a dry finish (Section 6.10).
Key nutrient requirements	The amounts of major and trace elements for optimum growth are very similar to those required by wheat. An important consideration is the optimum grain protein content for malting barley. High protein is desirable in wheat, but not in malting barley as it leads to low malt extract and a protein haze which must be precipitated out of the resulting beer. Therefore, soils and rotations which result in low protein are more suited to malting barley than to bread wheat.
Compacted soils	Roots are restricted by traffic pans, so barley can respond to deep ripping (Crabtree 1989).
Root growth into clay subsoils	There is little information on the extent of root growth into clayey subsoils in WA. A study on the Esperance sandplain in 1993 (D. Tennant and K. Young unpublished data) showed that barley extracted moisture to a depth of 1.3 m on a gravelly sand that had a clayey subsoil at 50 cm. On the alkaline mallee soils, tolerance to high boron is likely to influence root penetration in the subsoil.
Soil properties affecting germination	Crop establishment principles are essentially the same as for wheat. Barley experiences the same problems with crusting surface soils and water repellent sands.
Erosion risk	Appears more susceptible to sand blasting than wheat.

8.3 OATS: SOIL AND CLIMATIC REQUIREMENTS

*Wal Anderson** *and Geoff Moore*

Oats (*Avena sativa*) is the third most important of the temperate cereal crops grown in Western Australia. In 1992 the area sown to oats was about 450,000 hectares with about 72% used for grain production, 19% for hay and 9% for green feed and silage.

Root growth

The root system is similar to other cereals, as described for wheat and barley.

Climatic requirements

Suitable varieties are available for all the agroclimatic zones of south-western Australia (Figure 8.0.1). Oats tolerate low temperatures and will grow if the mean temperature is higher than about 0°C. Oats are susceptible to high temperatures (>35°C); at anthesis high temperatures increase the number of empty spikelets, while during grain development they result in premature ripening and lower grain weights (Brown 1975).

Oats is susceptible to frosts during the early growth stages and at anthesis, but is less susceptible than other cereals. The tall oats grown for milling are susceptible to lodging under windy conditions on highly fertile sites.

SUMMARY OF SOIL REQUIREMENTS

Requirements are similar to wheat, but there are some observed differences.

Oats are more tolerant of acid soils, waterlogging, low temperatures and poor seedbed preparation than the other cereals. These observations probably give rise to the belief that oats can withstand more stressful conditions and rougher management than wheat or barley. However, they will respond to good soil conditions and management (Anderson and McLean 1989).

Oats.

* *Agriculture Western Australia, Northam.*

Soil factors affecting the productivity of oats

Soil conditions	Tolerance
Soil water deficit	Oats are usually grown in the high rainfall districts, but are not more susceptible to the effects of soil moisture stress than other cereals.
Waterlogging	More tolerant than wheat or barley and less affected by root rots such as take-all. In a trial at Mount Barker, the yield was not reduced by waterlogging until the SEW_{30} index was greater than 500 cm.days (Cox 1988), but in another trial at Narrogin, yields were reduced at a lower intensity of waterlogging (McFarlane *et al.* 1992).
Soil salinity	Not as tolerant as barley, but probably comparable to wheat.
Salinity and waterlogging	Mildly saline and waterlogged areas often have high levels of the take-all fungus. Oats can be very useful on these areas, because they are more tolerant of take-all than other cereals.
Acidity: minimum pH_{Ca}	Less sensitive than barley or wheat. The pH_{Ca} should be >4.0-4.3 in the surface (0-10 cm).
Alkalinity: maximum pH_w	Will grow in calcareous soils. More sensitive to relatively high levels of exchangeable sodium than wheat or barley. When ESP is 15-30, yields will be reduced significantly (Abrol 1982).
Key nutrient requirements	*Nitrogen.* Similar response to N fertiliser as per wheat and barley (Mason and Glencross 1980). *Phosphorus.* Requirement is generally less than for wheat on a range of lateritic soils, although it is not known whether oats is more efficient in taking up P from deficient soils or in using it to produce grain (Bolland 1992). *Copper, zinc.* Sensitive to both low Cu and Zn (Mengel and Kirkby 1987). *Manganese.* Symptoms have been observed in highly deficient soils, and susceptibility differs between cultivars. *Iron.* Deficiency has been reported on acid organic sands in WA, and can be corrected with foliar applications of iron sulphate (Fitzpatrick and Sprivules 1963).
Compacted soils	Roots are restricted by traffic pans. The grain yield response of cereals to deep ripping is likely to be more affected by the vigour of early growth, time of flowering and absolute yield potential than differences between species in sensitivity to compaction (Henderson 1991).
Root growth into clayey subsoils	Probably similar to wheat, although it has not been studied.
Soil properties affecting germination	Soil crusting and water repellence affect germination as for wheat. Oats can withstand damage from wetting and drying of the seedbed better than wheat.
Erosion risk	Sand blasting affects growth, in the same way as wheat.

8.4 NARROW-LEAFED LUPINS: SOIL AND CLIMATIC REQUIREMENTS

Peter Nelson, Geoff Moore and Miles Dracup***

The recent development of sweet narrow-leafed lupins (*Lupinus angustifolius*) as a major crop in Western Australia is the result of the efforts and vision of Dr John Gladstones, with support of other plant breeders, agronomists, marketeers, farmers and extension personnel (Nelson 1993). The area sown to this crop in 1994 was 900,000 ha.

The production and management of lupins has been summarised in *Producing lupins in Western Australia* (Nelson and Delane 1991); their nutrient deficiency symptoms are described in *Symptoms of Nutrient Deficiencies: Lupins* (Snowball and Robson 1986); and their growth and development has been described in the *Lupin Development Guide* (Dracup and Kirby 1996).

SUMMARY OF SOIL REQUIREMENTS

Lupins are adapted to a wide range of soils provided they are well drained and have no free lime in the profile. The deep taproot system gives them a comparative advantage over other grain legumes on deep, uniform coarse-textured soils.

In the Mediterranean region, where plants were collected for the lupin breeding program, narrow-leafed lupins (*L. angustifolius*) are adapted to a wide range of textures, from coarse sands to clays; pH ranges from strongly acidic (pH_w 4.2) to alkaline (pH_w 8.0). However, they are predominantly found on freely drained, sandy loams or loamy sands with pH_w between 6.0 and 7.5 (Gladstones 1974; Gladstones and Crosbie 1978).

Root growth

Lupins have a single main taproot with lateral branches. Under favourable conditions the taproot can penetrate to 2.5 m or more (Perry *et al.* 1986). In general, significant water extraction is limited to about 1.9 m (Hamblin and Hamblin 1985).

The root system is illustrated in Figures 8.4.1 and 8.4.2.

First order lateral roots appear two weeks after sowing and grow horizontally for a few centimetres before turning downwards. The laterals are much thinner than the taproot and give rise to even thinner second and third

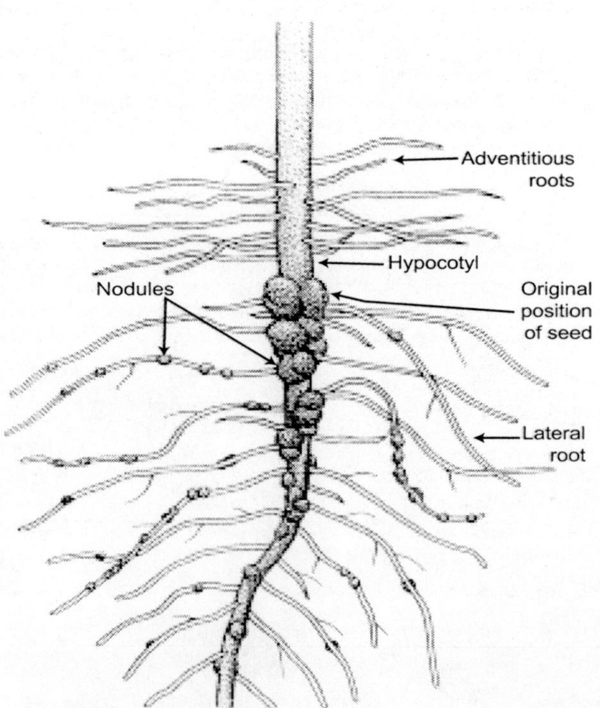

Figure 8.4.1 *The root system of 10 week old* Lupinus angustifolius *showing development of nodules on the taproot and adventitious roots arising from the hypocotyl (from Figure 9.2, Dracup and Kirby 1996).*

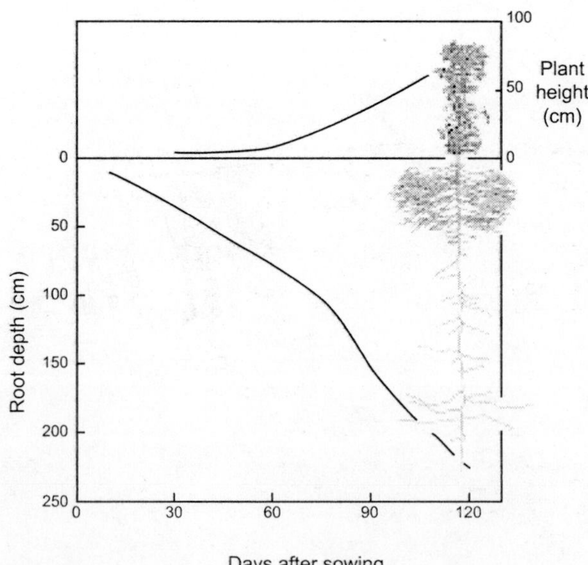

Figure 8.4.2 *Plant height and root growth of narrow-leafed lupins from sowing to maturity (from Figure 9.4, Dracup and Kirby 1996).*

* Formerly Agriculture Western Australia, now Grain Pool of Western Australia, Perth.
** Agriculture Western Australia, South Perth.

Soil factors affecting the productivity of lupins

Differences between narrow-leafed lupin (*L. angustifolius*), yellow lupin (*L. luteus*) and albus lupin (*L. albus*) are noted where applicable.

Soil conditions	Tolerance
Soil water deficit	The deep root system enables growth on soils with a low water-holding capacity (e.g. AWC 40-50 mm/m). Slight moisture stress shortly after flowering can improve seed yield compared with adequate moisture supplies. Ideal growing conditions often result in highly vegetative crops which do not always translate into high seed yields. Rapid droughting on shallow soils, e.g. duplex soils, leads to rapid senescence and poor yields.
Waterlogging	Narrow-leafed lupins are sensitive to waterlogging, albus lupins more sensitive and yellow lupins less sensitive. Well established nodulated lupins can tolerate transient waterlogging for two to three weeks, although yields may be reduced.
Soil salinity	Very sensitive. Growth is adversely affected if ECe is >250 mS/m, with nil growth >400 mS/m (Jeschke *et al.* 1986).
Salinity and waterlogging	Sites which are both saline and waterlogged should be avoided.
Acidity: minimum pH_{Ca}	Narrow-leafed lupins are relatively tolerant and yellow lupins are more tolerant. In general, root growth is not restricted if the pH_{Ca} >4.0, restricted at pH_{Ca} 3.5-4.0 and greatly restricted at <3.5.
Alkalinity: maximum pH_w	Growth is generally poor and they suffer from chlorosis on calcareous alkaline soils (Jessop and Mahoney 1982; Walton 1982; Nelson *et al.* 1986; White and Robson 1989a). Can be grown successfully where pH is relatively high but lime is low (e.g. alkaline soils in the Victorian Mallee, White 1990), because they are adversely affected by free $CaCO_3$ (lime), rather than high pH *per se* (Jessop *et al.* 1990; White and Robson 1990). Several interacting mechanisms cause poor growth on alkaline and calcareous soils (White 1990). Iron deficiency can cause chlorosis and root growth is severely impeded which restricts water and nutrient uptake. Nodulation is reduced in alkaline soils (Tang *et al.* 1995).
Key nutrient requirements	*Phosphorus.* Lupins require a good P status for satisfactory seed yields. On high fixing soils and those low in P, research has shown that banding below the seed is efficient (Jarvis and Bolland 1991). *Potassium.* Deficiency is common on pale deep sands. Lupins do not give seed yield responses when levels in the top 10 cm are greater than 40 ppm. On some duplex soils K levels in the surface may be low but can be extracted from the clayey subsoil. *Manganese.* Narrow-leafed lupins are sensitive to low Mn and deficient lupins exhibit split seed disorder (Gartrell and Walton 1989). Deficiency is likely in every lupin crop on highly deficient soils, and in years with a dry spring on marginally deficient soils. *Copper, zinc, molybdenum.* Similar requirement to wheat.
Compacted soils	Root growth is restricted by high soil strength (subsurface compaction), although not to the same extent as many other crops. The use of lupin roots as a *biological plough* has been proposed, with variable results depending on the plant density (Henderson 1989). Materechera *et al.* (1991) found that high soil strength affected the root growth of lupin seedlings least when compared with a wide range of other field crops and pasture species.
Root growth into clayey subsoils	The taproot is advantageous for penetrating hard soil layers, but not for growing into weakly structured clayey subsoils. The presence of a single taproot (1–2 mm diameter) limits the chances of finding cracks or macro-pores in comparison to wheat, which has numerous thinner roots (Dracup *et al.* 1992). Therefore, the depth of lupin roots in duplex soils is highly variable (45–160 cm) depending on subsoil conditions (Dracup *et al.* 1992). In fine to medium-textured soils, root growth is frequently restricted more by the chemical properties than the physical properties of the soil (Atwell 1991a, b).
Soil properties affecting germination and emergence	Seedling emergence is significantly reduced by a surface crust, because with its epigeal emergence pattern, the comparatively large cotyledons need to be forced through the soil surface. If seedlings emerge, they are frequently (10–20%) damaged (White 1988; White and Robson 1989b). Optimum conditions for germination and emergence are when the soil is close to 20°C, well drained and the water content is near the upper storage limit (Dracup *et al.* 1993).
Erosion risk	During the first four weeks of growth, lupins are very susceptible to sand blasting. Unlike wheat, the growing point of the plant is above the ground so it is easily damaged. On erosion-prone sandy soils, lupins should always be established by planting into surface trash of at least 1 t/ha, which is equivalent to about 50% surface cover (Leonard 1993).

order laterals which follow no defined path. Adventitious roots can appear on the hypocotyl, below the soil surface five weeks after sowing. These roots can be numerous, have little branching and are possibly a response to high soil moisture (Dracup and Kirby 1996). Cluster (or proteoid) roots can also occur on the sandplain blue lupin, yellow and albus lupins but not on narrow-leafed lupins. These are a dense proliferation of rootlets from a lateral root covered in root hairs.

Lupins form nitrogen-fixing nodules on the roots. These are found mainly on the top 5 to 8 cm of the taproot where, in a healthy crop, they form a dense collar around the root. Nodules are sparse on the lateral roots unless the taproot is lost.

Climatic requirements

Narrow-leafed lupins are suitable for areas with a growing season of four to eight months and growing season rainfall from 200 to 750 mm. Their development and rate of growth respond mainly to temperature. The optimum temperature for growth is about 20°C, but in the vegetative stage, lupins can tolerate temperatures down to -6°C. High temperatures (>26°C) increase flower abortion on the main shoot. Higher temperatures (>30°C) and rapid soil drying during pod filling hasten maturation and leaf loss, reducing yield. Lupins are sensitive to frost at flowering, but it has very little effect during pod filling (Nelson and Delane 1991).

8.5 FIELD PEAS: SOIL AND CLIMATIC REQUIREMENTS

Ian Pritchard[*]

Field peas belong to the same species as the garden pea (*Pisum sativum*). They are one of the oldest cultivated crops and originate from the mountainous areas of Afghanistan, India and Ethiopia. In WA, field peas were first grown in the early 1900s. The crops were mainly grown in the high rainfall zones for seed production, green manure and stock feed. Seed production was limited by the arrival of the pea weevil in 1932, following which the main growing area moved from the high rainfall south-west to the Northam region. Production declined in the 1970s, but gradually increased during the 1980s, with an estimated 30,000 ha yielding 23,500 t in 1990.

The general agronomy and management of field peas is summarised in *Growing field peas* (Pritchard 1993) and the *Grain legume handbook* (Lamb and Poddar 1992).

Root growth

Field peas have a relatively shallow fibrous root system and a fine taproot which extends to depth. The root system has well developed lateral roots.

Varieties differ in their maximum rooting depth. On a uniform loamy sand with no chemical or physical restrictions to root growth, the maximum rooting depth of field peas ranged from 1.2 to 2 m (Armstrong *et al.* 1994). The roots of the fully-leaved genotype, Wirrega, extended deepest and extracted water from greater depth than the other genotypes (Armstrong *et al.* 1994).

Climatic requirements

Suitable varieties are available for all the agroclimatic zones (Figure 8.0.1). The minimum requirement for growing season rainfall is about 150 mm, depending on the amount of stored moisture in the soil.

Development is largely controlled by temperature and the optimum temperature depends on the stage of development. During early growth (emergence to six nodes), the optimum temperature is 20/13°C (daily maximum/minimum), while from the six node stage to flowering it is 16/10°C and during flowering 20-21/10-12°C (Pritchard 1993). These temperature optima are relatively low when compared with other legumes and probably reflect the crop's origins. Consequently, their growth in winter is more vigorous than lupins, chickpeas or faba beans.

Field peas are sensitive to frosts, especially late frosts during flowering and podding. High temperatures (>26°C) during flowering can cause flower abortion. Strong winds at crop senescence result in the crop lying in one direction, which makes it easy to harvest.

SUMMARY OF SOIL REQUIREMENTS

Field peas are an alternative legume for moderately well to well drained soils with an alkaline reaction. The four main soil requirements for a successful crop are good drainage, a relatively stone-free paddock, pH_{Ca} >4.5 and low susceptibility to wind erosion (e.g. categories (i)-(iii); Section 7.1). Paddocks need to be relatively stone-free because the cutter bar on the harvester has to be set low to handle the prostrate growth habit.

[*] *Agriculture Western Australia, Northam.*

Soil factors affecting the productivity of field peas

Soil conditions	Tolerance
Soil water deficit	Field peas can tolerate moisture stress. On fine-textured soils they appear more tolerant than faba beans or lupins (French and Ewing 1989). Ability to tolerate moisture stress appears to be related to early maturity rather than an intrinsic physiological adaptation (E. Armstrong personal communication). Their shallower root system limits moisture extraction on uniform coarse-textured soils compared with narrow-leafed lupins.
Waterlogging	Comparable to lupins and depends on growth stage. Particularly sensitive from emergence to the six node stage. At later stages of development, waterlogging will not kill the plants, but fungal diseases like blackspot and downy mildew are likely to be more severe. Field peas have no adaptation (e.g. aerenchyma, adventitious roots) to waterlogging, but if affected during early vegetative growth can produce new roots (Cannell *et al.* 1979). The emerging radicle is particularly susceptible to low concentrations of oxygen (Eavis *et al.* 1971). Short periods of surface waterlogging (e.g. two days) can result in root death, leaf chlorosis and significant yield losses (Cannell *et al.* 1979; Jackson 1979; Belford *et al.* 1980).
Soil salinity	Sensitive to salinity at all stages of development.
Salinity and waterlogging	Avoid sites which are both saline and waterlogged.
Acidity: minimum pH_{Ca}	The rhizobia, not the plant, limit tolerance to soil acidity. Rhizobia do not survive if the pH_{Ca} is <4.5 (Evans *et al.* 1993). Low pH especially inhibits nodule formation (Lie 1969).
Alkalinity: maximum pH_w	Grow well on calcareous soils as they are less sensitive to iron deficiency than lupins (White and Robson 1989b; Atwell 1991a, b). On highly alkaline soils (pH_w >9.5), field peas can develop iron chlorosis (Lamb and Poddar 1992). Nodulation is still active at pH_w 9.2 (Jessop and Mahoney 1982).
Key nutrient requirements	Symptoms of deficiency are illustrated in Snowball and Robson (1991). *Phosphorus.* Providing plants have been inoculated with the correct rhizobia so that nodulation and nitrogen fixation occur, the main nutrient requirement is soil P >15 ppm (Colwell bicarbonate test). *Iron.* A suspected deficiency has been reported on an alkaline grey clay (pH_w >8.0). *Manganese.* Deficiency has not been reported in WA. *Copper, zinc, molybdenum.* Few reported trace element deficiencies. Not usually grown on soils with coarse-textured A horizons because of their high risk of wind erosion. Zinc deficiency has been reported overseas. *Boron.* Toxicity has been reported on alkaline soils in SA (Lamb and Poddar 1992).
Compacted soils	Sensitive to poor soil physical conditions. High soil strength will impede elongation of seedling roots (Materechera *et al.* 1991). Root elongation is restricted in compacted soils and they are also very sensitive to reduced aeration (Hebblethwaite and McGowan 1980; Castillo *et al.* 1982; Dawkins *et al.* 1984). Low oxygen concentrations can occur in highly compacted soils resulting in plant stress (Dawkins *et al.* 1984).
Root growth into clayey subsoils	No information.
Soil properties affecting germination	Peas have a hypogeal emergence pattern, with cotyledons remaining below the soil surface. The emergence pattern and small diameter of the emerging plumule allow it to push through surface crusts (White and Robson 1989a), but a compacted surface, strong surface crust or hardset surface can restrict emergence (Hebblethwaite and McGowan 1980).
Erosion risk	Stubble is fragile and easily broken by stock or strong winds, leaving paddocks bare (Russell 1993). Field peas should only be grown on soils which form a surface crust or hardset surface over summer, i.e. wind erosion susceptibility classes (i)-(iii). Soils with susceptibility classes of (iv) or (v) are not suitable for growing field peas because these soils have a high to extreme risk of eroding (Refer to Section 7.1). Soils with a susceptibility to wind erosion of class (iii) are suitable for growing field peas provided the stubble is not grazed. Can regenerate if sand blasting cuts the seedling because the cotyledons remain below the soil, although delayed emergence will result in a yield loss even if sufficient seedlings re-emerge.

8.6 CANOLA: SOIL AND CLIMATIC REQUIREMENTS

Paul Carmody and Graham Walton***

Canola is the name given to the low erucic acid (<2%), low glucosinolate varieties of rapeseed, following the nomenclature from Canada. The recommended varieties are all from the species *Brassica napus,* although some older varieties were from *Brassica campestris* (summer turnip rape).

A flowering canola crop (G. Moore).

The canola industry was first established in Western Australia in the early 1970s, but a severe outbreak of the fungal disease, blackleg, in 1972 effectively wiped it out as a commercial crop (Poole 1970; Bokor *et al.* 1975). Release of varieties with a shorter growing season, higher yields and higher resistance to blackleg (Roy 1984) in the late 1980s resulted in renewed interest. It is now one of the fastest expanding broadscale crops in WA, with production in 1994 of 116,000 tonnes (Carmody 1995a) and the area sown may increase substantially with the release of triazine-tolerant varieties.

Management is summarised in *Canola cache: the farmer's handbook for growing canola* (Casey and Cooke 1992).

Root growth

Canola has a taproot which penetrates to depth and a dense mass of roots in the top 35 to 50 cm. Taproots penetrated to more than 1.3 m in a uniform coarse-textured soil at Merredin (Suanda 1992).

Climatic requirements

Canola has traditionally been grown in areas with an average annual rainfall of more than 450 mm (high and very high rainfall zones, except for H1; Figure 8.0.1). The release of new short season varieties allows successful cultivation in the medium rainfall zones (M1-5). Canola can be grown as an opportunist crop in the low rainfall zone when the break of the season allows sowing before the 15th of May.

Canola prefers mild temperatures (20-25°C), and higher temperatures (>32°C) during flowering can cause flower abortion. The low temperatures normally experienced in early winter can delay seedling emergence (McKay 1987). They also limit growth rate and leaf appearance (Thurling and Vijendra Das 1977; Acharya *et al.* 1983).

Severe frosts at flowering can cause flowers to abort (Scott *et al.* 1973). In general, the indeterminate flowering pattern of canola reduces the frost risk when compared with wheat. Frosts can also kill seedlings (rare) and prevent seed development in pods during spring. In general, only severe frosts on dry, coarse-textured soils cause significant frost damage. Areas with a frost risk at flowering of one in four years or higher are marginal.

SUMMARY OF SOIL REQUIREMENTS

A high yielding canola crop can be grown on soils with a range of textures providing the soil:

- is well to moderately well drained;
- has no extremes of pH, with the pH_{Ca} of the surface soil >4.5;
- has a soil water storage of at least 70 mm within the root zone;
- is non-saline;
- has a good supply of nitrogen and sulphur;

If it has a loose, single grain surface, then canola should be sown into cereal stubble.

* *Agriculture Western Australia, Northam.*

** *Agriculture Western Australia, South Perth.*

Soil factors affecting the productivity of canola

Soil conditions	Tolerance
Soil water deficit	More sensitive to moisture stress than wheat. A critical time appears to be from flowering to when pod growth ceases. Moisture stress at this growth stage reduces both pod numbers per plant and seed numbers per pod (Richards and Thurling 1978a, b). Soil water deficits immediately after anthesis have a large effect on yield, because this is when seeds are most likely to abort (Tayo and Morgan 1978). Moisture stress after anthesis is a major factor restricting yields in areas with an average annual rainfall of <500 mm (Lewis 1993; Lewis and Thurling 1994). Even when soil is waterlogged before flowering, yield is primarily restricted by moisture stress after anthesis. Waterlogging restricts root development, which can exacerbate moisture stress in the maturing crop (Lewis 1993).
Waterlogging	An important factor that can restrict yield, e.g. a waterlogged site may achieve only 50% of the yield of a well drained site (Perry and Hawley 1983). Highly sensitive during the seedling stage and at flowering. Mature plants can tolerate short periods of transient waterlogging (Cannell and Belford 1980) and are thought to have a similar tolerance to barley. The roots can adapt by turning upwards into aerated soil.
Soil salinity	Some tolerance, hence its use as a pioneer crop on the polders in the Netherlands (Weiss 1983). Thought to be similar to barley. Yield is affected when ECe is >400 mS/m, although it can tolerate 500-600 mS/m (Thomas 1984).
Salinity and waterlogging	Sites which are both marginally saline and waterlogged for 2-3 weeks during the first 10 weeks of growth can kill crops.
Acidity: minimum pH_{Ca}	The minimum pH_{Ca} is thought to be 4.5. Below 4.5 the crop is likely to be stunted and not reach its yield potential.
Alkalinity: maximum pH_w	Canola is grown on calcareous soils and can tolerate pH_w 8.5 with no harmful effects. Soil pH_w >8.5 is likely to restrict growth or induce nutrient deficiencies (Thomas 1984).
Key nutrient requirements	High nitrogen and sulphur are the main requirements. In general, 50-90 kg N/ha, 10-20 kg P/ha, 20-30 kg K/ha and 10-25 kg S/ha are required. See Holmes (1980) and Carmody (1995b). *Nitrogen.* Canola was always considered to have a higher N requirement than cereals. However, research in WA has that shown canola requires no more N, although it may have a greater vegetative response (Mason 1995). *Sulphur.* A high sulphur requirement; the amount of S removed in 1 tonne of canola grain is 7-10 kg, compared with 2.5-4.0 kg in narrow-leafed lupins and 2-4 kg in wheat. The likelihood of S deficiency is mainly related to previous fertiliser history. Deficiency is unlikely if fertilisers such as single superphosphate (10.5% S) are used, but is more likely after using low S containing fertilisers (Mason and Walton 1993). *Calcium.* Deficiency has been identified as a collapsing of flower stalks just below the inflorescence, causing the flowerhead to wither, known as 'tippletop' or 'withertop' (Mason 1992). Usually only 1-2% of plants are affected. *Copper, molybdenum, manganese.* Cu deficiency is rare in brassicas (they are less sensitive than wheat). Mo deficiency can occur on acid soils. Observations suggest canola is less sensitive to Mn deficiency than narrow-leafed lupins.
Compacted soils	Traffic pans or plough pans slow root elongation in canola.
Root growth into clayey subsoils	Limited information available, but it is likely that growth will be good in well structured soils or if there are cracks to follow. Crops establish in shallow ironstone gravelly soils where cereals have difficulty.
Soil properties affecting germination	Soil should be moist at 2-3 cm, with good soil/seed contact for a high germination rate (obtained by using press wheels or rollers). Surface crusts or hardset surfaces reduce germination. Canola has small seeds and so is sensitive to the strength of the soil crust. Emergence is improved with shallow sowing and a moist soil (near the upper storage limit; Nuttall 1982).
Erosion risk	Fairly tolerant of high winds, but susceptible to sand blasting during the first 4-6 weeks. Sites with a high to very high susceptibility to wind erosion (categories (iv) and (v); refer to Section 7.1) are at risk unless sown into cereal stubble. Wind erosion caused by pre-frontal winds in the Jerramungup district during May 1995 appeared to cause less damage on paddocks of canola stubble compared with paddocks of lupin stubble.

8.7 FABA BEANS: SOIL AND CLIMATIC REQUIREMENTS

Stephen Loss and Kadambot Siddique**

The faba bean (*Vicia faba*) is an alternative grain legume. There are two groups: *Vicia faba major* (broad bean) with large seeds, and *Vicia faba minor* (tick or horse bean) with small seeds.

Faba beans are a relatively new crop in Western Australia with approximately 7,000 ha sown in 1993. The beans are used for human consumption and stockfeed. Management is summarised in the *Grain legume handbook* (Lamb and Poddar 1992) and Farmnotes (34/93, 35/93).

Root growth

Faba beans have a mainly fibrous root system, although they do have a taproot. The maximum rooting depth, 0.8 m, is fairly shallow (S.P. Loss and K.H.M. Siddique unpublished data).

Climatic requirements

Faba beans are suitable for all of the medium and high rainfall agroclimatic zones (M1-5 and H1-5; Figure 8.0.1) and for early sowing in the low rainfall zone. Growing season rainfall needs to be in the range of 200 to 800 mm.

The optimum temperature range for growth is 15 to 25°C. They can tolerate frosts but are sensitive to high temperatures (>28°C), especially near flowering which can cause flower abortion.

SUMMARY OF SOIL REQUIREMENTS

Faba beans are best suited to neutral to alkaline soils with medium to fine textures. The advantage of faba beans over other grain legumes is their greater tolerance of waterlogging. They appear to have a niche on soils which are susceptible to transient waterlogging which restricts the yield of field peas, chickpeas, lentils and narrow-leafed lupins.

Faba beans.

* *Agriculture Western Australia, South Perth.*

Soil factors affecting the productivity of faba beans

Soil conditions	Tolerance
Soil water deficit	Sensitive to moisture stress. The comparatively shallow root system means that uniform coarse-textured soils and other soils with low plant available water (<60 mm within the root zone) should be avoided.
Waterlogging	Can tolerate mild waterlogging with no major adverse effects, but stressed plants may be susceptible to disease. Similar tolerance to oats. In a series of comparative trials, faba beans were compared with other grain legumes: narrow-leafed lupins, field peas and chickpeas. Faba beans yielded 1.7 to 4.4 t/ha in trials where waterlogging either caused crop failure or severely reduced the yields of the other grain legumes (Siddique *et al.* 1993).
Soil salinity	Faba beans are classified as moderately sensitive to salinity. Yield is affected when the ECe exceeds 160 mS/m. There is a 10% reduction in yield for each 100 mS/m increase in ECe above 160 mS/m (Maas and Hoffman 1977).
Salinity and waterlogging	No information.
Acidity: minimum pH_{Ca}	Moderate tolerance of surface acidity and prefer pH_{Ca} >5.0 in the 0-10 cm layer. Sensitive to subsurface acidity and pH should increase down the profile. The subsurface pH_{Ca} (15-25 cm) should be >6.0 (K.H.M. Siddique and S.P. Loss, unpublished data).
Alkalinity: maximum pH_w	Faba beans are grown successfully on calcareous soils, consequently pH_w up to 8.5 has no adverse effects (Hebblethwaite 1983; Siddique *et al.* 1993). If the pH_w >9 there are likely to be nutrient deficiencies induced or a direct adverse effect on growth.
Key nutrient requirements	For every tonne of grain produced, faba beans require 33% more P and twice as much K and Ca as wheat. Cu and Zn requirements are also higher than wheat.
Compacted soils	Roots of seedlings are severely restricted by strongly compacted soil, but less so than cereal crops. High soil strength causes root thickening (Materechera *et al.* 1991).
Root growth into clayey subsoils	The root growth in clayey subsoils is thought to be moderate, although there is little information available.
Soil properties affecting germination	With its large seeds, the emerging seedlings are able to push through most surface crusts. Soil conditions required for germination are similar to those required for wheat.
Erosion risk	Soils under faba bean stubbles are less prone to wind erosion than those under field peas. Faba beans produce more biomass, and when grazed the stubbles are less prone to being detached and then removed by the wind.

8.8 CHICKPEAS: SOIL AND CLIMATIC REQUIREMENTS

*Kadambot Siddique**

The chickpea (*Cicer arietinum)* is an annual legume with the potential for greater production in WA. There are two groups, the *desi* and *kabuli* types, which are differentiated by seed size, shape and colour. The small-seeded, desi type accounts for 85 to 90% of the world's production. Management is described in Farmnotes 31/93 and 33/93.

Root growth

Chickpeas have a strong taproot system with well developed laterals. In a Merredin sandy loam the roots extended to a depth of 1.0 m (Siddique and Sedgley 1987). The depth of penetration is likely to be greater on a uniform coarse-textured soil with a low strength. Chickpeas have an indeterminate root development pattern, so there is considerable root growth after anthesis.

Climatic requirements

Chickpeas can be grown in many agroclimatic zones (L1-5 and M1-5; Figure 8.0.1) provided the growing season rainfall is 150 to 500 mm. Early sowing is recommended in low rainfall zones. Suitable varieties of kabuli chickpeas are available for areas with an average annual rainfall greater than 400 mm, while the desi chickpeas can be grown in drier areas.

The plants tolerate high temperatures during flowering better than other grain legumes. The optimum temperature is about 25 to 30°C, but they can withstand temperatures between 2 and 40°C.

A fungal disease, *Ascochyta blight*, can devastate sensitive varieties in many countries. It develops under humid, overcast conditions when the canopy is wet. Fortunately in the eastern wheatbelt, periods of rain are frequently followed by clear, sunny skies (Siddique 1985). So far *Ascochyta blight* has not been reported in Australia.

Frost at flowering and podding affects flower and pod development and can result in crop failure (Siddique *et al.* 1993).

SUMMARY OF SOIL REQUIREMENTS

Chickpeas are grown on a wide range of soils including: dune sands, cracking clays, sandy loams and calcareous soils (Saxena 1987). In WA, their potential niche is as an alternative grain legume on alkaline soils unsuitable for narrow-leafed lupins.

For good production the soil should be well drained, non-saline, have a water storage of >60 mm in the 0 to 1 m depth interval, and have a pH_{Ca} 5.0 to 9.5.

Harvesting chickpeas.

* *Agriculture Western Australia, South Perth.*

Soil factors affecting the productivity of chickpeas

Most research on the tolerance of chickpeas to a range of soil conditions is from the Mediterranean region and the Indian sub-continent. The general principles should apply in Western Australia because the varieties used are from the same gene pool.

Soil conditions	Tolerance
Soil water deficit	Able to withstand moisture stress (Siddique 1985; Siddique and Sedgley 1985, 1986).
Waterlogging	Sensitive to poor soil aeration. They have a low tolerance of transient waterlogging and are especially sensitive during the seedling stage and at flowering. Tolerance is similar to narrow-leafed lupins (Krishnamurthy *et al.* 1983; Saxena 1987; Siddique *et al.* 1993).
Soil salinity	Highly sensitive compared with other grain legumes (Chandra 1980). Genotypic improvement for tolerance is not possible at this stage as no useful variability has yet been identified in the germplasm, including wild species.
Salinity and waterlogging	Soils which are both waterlogged and saline should be avoided.
Acidity: minimum pH_{Ca}	Prefer neutral to alkaline conditions. They can tolerate a surface pH_{Ca} 4.5 (0-10 cm) providing the subsurface soil is not acidic (Siddique *et al.* 1993). Acid conditions may reduce nodulation and increase problems with *Fusarium* wilt (Kay 1979).
Alkalinity: maximum pH_w	Grow well in alkaline conditions and are commonly grown on calcareous soils (Saxena 1987). Iron deficiency may occur at pH_w>10. However, this can be easily overcome by a foliar application of 0.5% (w/w) aqueous solution of ferrous sulphate, because of the presence of highly acidic exudates on the foliage. Current varieties have been selected for their more efficient use of Fe, as there is large genotypic variation available for this trait. Sodium carbonate reduces growth. Sodicity can have an adverse effect on growth if ESP is >30, with negligible growth at ESP 58 (Chandra 1980).
Key nutrient requirements	*Phosphorus.* Very efficient in extracting P by acidifying the rhizosphere in calcareous soils. Response to P applications is variable and generally low. *Potassium.* Response to fertiliser is usually negligible. *Manganese.* Very sensitive to Al and Mn toxicity. *Zinc.* Field deficiency has been observed in certain parts of northern India. Can be corrected with a soil application of 10-25 kg/ha of zinc sulphate or a foliar spray at 0.5% mixed with 0.25% lime. *Boron.* Toxicity has been reported.
Compacted soil	Compacted subsoil does not greatly affect root penetration.
Root growth into clayey subsoils	Normally good, provided it is not poorly structured (Siddique and Sedgley 1987).
Soil properties affecting germination	Prefers soil temperatures >15°C, but will tolerate to 5°C. Soil should have a good tilth, avoiding large clods. A surface crust or hardset surface reduces emergence.
Erosion risk	Plants are erect at maturity (and thus easier to harvest than field peas) which should reduce harvest losses. They have strong stems that remain anchored in the soil, offering more protection against wind erosion after harvest than field peas.

8.9 GRAPE VINES: SOIL AND CLIMATIC REQUIREMENTS

Officers of the Viticulture Branch

Grape vines (*Vitis vinifera*) can be grown for table grapes, wine or dried vine fruit. Both table grape production and the wine industry are expanding in Western Australia, but the dried vine fruit industry has been in steady decline since the 1950s. To redress this decline, a group was recently formed to increase industry size. This initiative has already resulted in new plantings.

The viticulture industry was initially centred around the Swan Valley, but the work of Olmo (1956) and Gladstones (1965) described the similarity between the Margaret River region and the Bordeaux area in France and this led to the establishment of a thriving wine industry in the south-west. Table grape production has expanded to new regions including Carnarvon, Harvey, Vasse and Margaret River. The wine industry has expanded to new areas including the Warren Valley.

The general agronomy of grape vines is summarised in Winkler (1970), Weaver (1976), Coombe and Day (1988, 1992) and *Viticulture principles and practice* (Anon 1992).

Grape quality is of paramount importance and many environmental and cultural (management) factors contribute to wine quality. Their relative importance is illustrated in Figure 8.9.1.

SUMMARY OF SOIL REQUIREMENTS

Grape vines are fairly adaptable to different soil conditions, because a range of rootstocks are available. In general, the best yields are obtained on deep, well drained soils, providing the available water capacity is more than 60 mm/m.

The most important requirement is good drainage. Soils that are waterlogged in spring should be avoided or drainage installed.

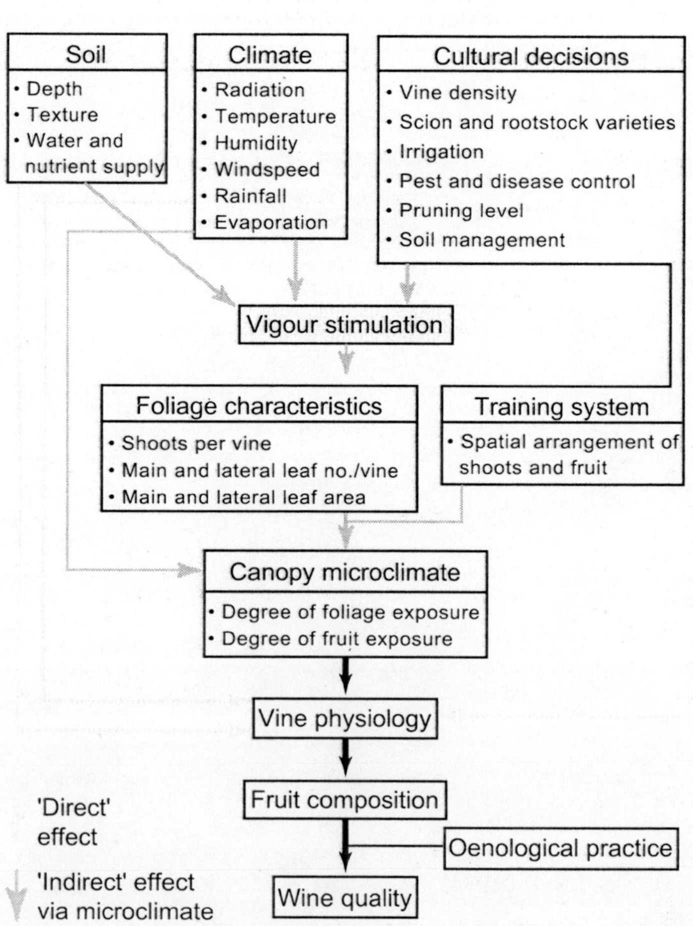

Figure 8.9.1 A schematic diagram illustrating factors affecting wine quality (after Smart et al. *1985).*

Root growth

Root growth varies according to the variety or rootstock, but in general vines have a large, spreading root system. The fine lateral roots are mainly within the top 0.6 m. Under favourable conditions where there are no impeding layers, roots can grow to a depth of 10 to 20 m.

Climatic requirements

Irrigation is essential for commercial grape production in all areas of south-western Australia, because it increases the quality and quantity of production.

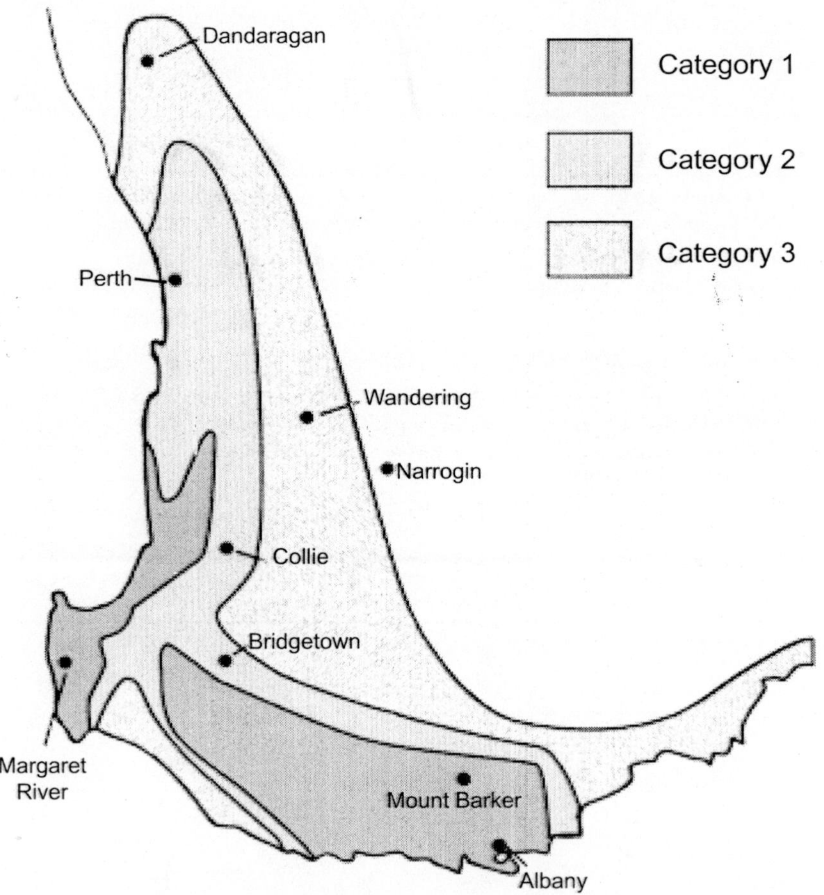

Figure 8.9.2 *A regional classification of the environmental suitability for growing commercial wine grapes in Western Australia (after Figure 2, Gladstones 1992).*

Gladstones (1992) has distinguished three categories of suitability for growing wine grapes in south-western Australia. The classification is based mainly on the climate but also considers soils and topography (Figure 8.9.2).

The coastal and near coastal areas (<60 km from the coast) in WA are superior for viticulture and have a number of significant advantages including higher relative afternoon humidity in summer, moderating effect of sea breeze on temperature, low frost risk and higher rainfall. Solar radiation is generally considered non-limiting, except on the south coast (e.g. at Denmark) for slopes with a southerly aspect (Gladstones 1992).

In general, a suitable area should be frost free after bud burst and have a low probability of rain during harvest. Protection from wind is important especially in exposed sites. Strong winds over 6 m/s reduce vine performance, while winds over 10 m/s damage the vines, particularly during spring (Campbell-Clause 1994; Morris and Campbell-Clause 1994). Light winds can be advantageous because they maintain air circulation in the canopy thus reducing humidity and helping to prevent frosts (Gladstones 1992).

There are no exact standards for the EC of irrigation water, because it depends on the soil and management, but in general, the EC of water to irrigate grape vines can be broadly categorised as:

<100 mS/m	no adverse effect on growth
100-200 mS/m	slight to moderate effects on productivity
200-700 mS/m	severe effects on productivity

Vines can be grown without supplementary irrigation in areas receiving >550 to 600 mm rainfall providing the soil has a high water-holding capacity, but the vineyard may not be commercially viable.

For wine grapes, climate has a large influence on the quality of the vintage, as well as yield. There is a detailed description of the suitability of the climate in WA for growing wine grapes in *Viticulture and Environment* (Gladstones 1992).

The agroclimatic zones used for annual crops (Figure 8.0.1) are not applicable. The temperature when the grapes are ripening, especially during the final month, has a large influence on the style of wine that can be produced (Amerine and Winkler 1944; Gladstones 1965).

KEY FOR FIGURE 8.9.2

Category 1. Regions with the best and most reliable climates, with good water supplies and/or rainfall, and enough suitable soils and topography to support substantial vine industries producing wines of consistently very high quality at minimum cost.

Category 2. Regions capable of producing high to very high quality wines, but over smaller areas and perhaps less regularly or with greater difficulty and risk. Extra care is needed in site selection and vine management.

Category 3. Viticulture is possible, but the environment is clearly marginal, or risky, or both. A few carefully selected sites may give good wines on a small to medium scale. Major commercial development is not recommended (Gladstones 1992).

Soil factors affecting the productivity of grape vines

The requirements for table and wine grape production are similar, but any differences have been noted.

Soil conditions	Tolerance
Soil water deficit	Tolerant of moisture stress and can survive in soils close to the lower storage limit (i.e. permanent wilting point), but survival does not equate with economic production. The optimum moisture supply for wine grape production can include slight moisture stress up to veraison and an adequate moisture supply through to crop maturation. Moisture stress up to veraison reduces vegetative growth and promotes grape production and juice quality. Moisture stress during grape ripening is undesirable, reducing berry fresh weight and dry weight (yield). Grape quality is also reduced with less sugar accumulation and elevated pH. A moderately deep soil with a moderate to high water-holding capacity enables development of an extensive root system. Coarse-textured soils with a low available water capacity (<60 mm/m) are generally less suitable unless irrigated. The soil-water relations are less critical when there is partial or full irrigation. *Commercial table grapes in WA require irrigation. Soil moisture should not fall below 50% available water and preferably not below 75%.*
Waterlogging	Grape vines tolerate transient waterlogging during winter when dormant, but waterlogging in spring affects growth. Good drainage in spring is essential for high yields.
Soil salinity	A soil is categorised as saline for grape vines if the ECe is >400 mS/m. Production is reduced by 10% at ECe of 400 mS/m, and by 50% at 800 mS/m (Bernstein 1965). *No commercial vineyard should be planted on saline soils or soils which may become saline.*
Salinity and waterlogging	Saline and waterlogged sites should be avoided.
Acidity: minimum pH_{Ca}	Root growth is reduced if the pH_{Ca} <5.0 (Robinson 1993). The rootstocks 34EM and 140 Ruggeri are more tolerant of acidity than others.
Alkalinity: maximum pH_w	Suitable rootstocks are available for alkaline soils (pH_w <8.5). Calcareous soils are used for grapes in many regions (e.g. Champagne). However, alkaline soils may induce iron chlorosis.
Key nutrient requirements	The key requirements for neutral and acidic soils are N, P, K, Mn and Cu, while in alkaline soils (pH_w >7.5) Fe may also be deficient (Cripps *et al.* 1990). For wine grapes, soils with high N promote excessive vine growth, which may reduce both grape yield and quality. Tissue analyses of the leaf petiole, sampled in spring, are used to diagnose the nutrient status of vines (Cripps *et al.* 1990). This is an important monitoring method to determine nutrient requirements to optimise grape quality and quantity.
Compacted soils	Vineyards are highly susceptible to the development of traffic pans, because all vehicle traffic is confined to the inter-row wheel tracks. The effect of traffic compaction was demonstrated in the Swan Valley, where high bulk densities (1.85 g/cm^3) were reported in the inter-row spacing (Smith *et al.* 1969). A plough pan can also develop if the soil is regularly cultivated to the same depth. Both of these compaction processes can greatly reduce the effective soil volume available for root growth, especially traffic pans adjacent to the vine rows.
Root growth into clayey subsoils	The growth of roots into clayey subsoils depends on the soil structure, but where conditions allow, vines will develop extensive root systems. Deep ripping (0.75-1 m) before establishing the vineyard can allow deeper root penetration in soils with a restrictive subsoil.
Soil temperature and stony soils	The optimum temperature for root growth is 15 to 30°C and the soil temperature can affect the distribution of growing roots (Woodham and Alexander 1966). With wine grapes, soil properties affect not only the vine's vigour and yield, but also wine quality. The widespread growing of wine grapes on stony or gravelly soils is thought to be related to their greater capacity to absorb solar radiation to depth and then re-radiate heat at night thus moderating the soil temperature. This reduces the risk of frost and allows fuller ripening. Another facet of stony or gravelly soils is their low fertility which reduces vegetative growth. In addition, stony or gravelly soils are usually moderately well to well drained (Gladstones 1992).
Erosion risk	The risk of wind erosion is minimal once the vines are established. Water erosion can be a problem on sloping sites, especially if the rows are aligned up and down the slope.

8.10 BLUE GUMS: SOIL AND CLIMATIC REQUIREMENTS

*Richard Harper**, *Keith Smettem*** and *John Bartle**

Tasmanian blue gums (*Eucalyptus globulus*) are native to the east coast of Tasmania, Flinders Island in Bass Strait, the southern and highland areas of Victoria and parts of New South Wales. They are a relatively new commercial crop in WA with the first plantations and shelterbelts established on private land in 1988 (Inions 1991).

Integrating a commercial tree crop into farming systems is attractive from the view of both enterprise diversity and development of sustainable farming systems. The trees have a comparable economic return to traditional agricultural pursuits (Eckersley 1990), but also have secondary benefits such as reducing groundwater recharge, providing stock shelter, control of erosion, waterlogging and salinity, plus off-site benefits such as reduced stream salinity and eutrophication (Bartle 1989).

Root growth

There is a dense surface root development within a radius of up to 12 m. Well structured clayey subsoils encourage deep roots to penetrate more than 8 m. The roots of seven year-old blue gums have been shown to extract moisture from the phreatic zone at 6 m (Greenwood *et al.* 1985).

Climatic requirements

In general, *E. globulus* requires a minimum rainfall of 600 mm per annum in regions with a Mediterranean climate, like south-western Australia. They have been planted in areas with less than 600 mm average rainfall, however production is likely to be poor and the trees prone to drought death.

Blue gums will grow if the air temperature is above 5°C so they are able to continue growing throughout WA winters. Growth is very active in summer, usually only limited by the amount of stored soil water. Blue gums are mildly frost tolerant and frosts in WA are unlikely to be a problem.

SUMMARY OF SOIL REQUIREMENTS

Water availability is the critical requirement for growing blue gums successfully through to maturity. They are sensitive to water stress, especially during the peak growth period from two to six years (leaf area index 4-5). Widespread tree deaths on shallow soils following the summer drought of 1993-94 emphasised water availability as the critical factor for growth and survival in the Mediterranean climate of WA. Blue gums are also sensitive to excess water or waterlogging, especially when combined with salinity.

The solution is to equate the long-term water demand of the trees with the water supply in dry years, remembering tree survival is of paramount importance.

The following site selection guidelines are recommended:

- All soils should be deeper than 2 m. Soil depth is defined as the depth of the sand and clay horizons overlying basement rock (i.e. granite, gneiss, dolerite, spongeolite, limestone) or saprolite (partially weathered rock), whichever is shallower. As saprolite may be penetrated by rotary drills, inspection via coring or backhoe pits is recommended to assess likely root penetration. Laterite (ironstone, ferricrete) boulders and gravel are treated as soil and not basement rock. Laterite can occur as a continuous sheet (duricrust, laterite cap) and backhoe inspections are necessary to determine whether this can be penetrated by roots.
- Poor growth is expected on soils with a sandy horizon (available soil water storage <70 mm/m) that is deeper than 2 m, so avoid these sites (Edwards and Harper 1996). The exception is deep sand in water-gaining areas.
- The effect of waterlogging has not been resolved. General indications are that sites that are waterlogged for several months each year (but dry in summer), which cannot be surface drained and which have duplex profiles with gleyed subsoil clays should not be planted.
- Avoid marginally saline sites with EM38 readings of >50 mS/m, or >25 mS/m on deep sands (Bennett and George 1995).
- The possibility of future salinity should be assessed by drilling to 6 to 10 m. Sites where the watertable is within 4 to 5 m and saline (>1000 mS/m) should be avoided.

* *Conservation and Land Management, Como.*

** *Faculty of Agriculture, University of Western Australia, Nedlands.*

Soil factors affecting the productivity of blue gums

Blue gums are a new crop and research is continuing on their tolerance to many soil conditions.

Soil conditions	Tolerance
Soil water deficit	Blue gums are sensitive to moisture stress and are poor at reducing transpiration when the moisture supply becomes limiting. Moisture demand is controlled by planting configuration (strips vs. blocks, stand density), nutrition (more fertile sites have larger canopies) and potential evaporation (latitude, aspect, distance from coast). Trees grown on sites with poor fertility (especially nitrogen) have smaller canopies, lower growth rates, lower transpiration and therefore greater ability to survive drought. However, their productivity is also reduced. In general, the larger the soil volume available for root growth the better. Soils formed on deeply weathered laterite profiles promote tree survival more than soils formed on fresh rock e.g. granite (R. Harper, unpublished data). When groundwater of suitable quality is within the root zone, the moisture supply is greater than simply rainfall plus soil water storage. If groundwater is to be a major source of water for the trees, then consider the following: • Determine whether groundwater is confined to a localised area. Blue gums will lower the watertable and have been known to dry up perched watertables. In one example, the trees subsequently died from moisture stress, because their roots could not penetrate an organic hardpan, which was initially below the watertable. • Ensure the watertable will not rise to within 0.5 m of the surface during winter and kill tree roots. • Test the salinity of the groundwater (EC suitable if <1000 mS/m).
Waterlogging	Sensitive to waterlogging in the top 0.5 m. If waterlogging occurs for more than two to three months a year, trees may die. Mounding along the tree lines can reduce the effect of waterlogging.
Soil salinity	Low tolerance. Growth is suppressed above 50 mS/m (as measured with an EM38) on finer textured soils and above 25 mS/m on coarse-textured soils (Bennett and George 1995). Blue gums can be used for lowering watertables, because they are 'phreatophytes' (plants whose roots reach the watertable). For example, the evapotranspiration from seven year-old blue gums with access to fresh groundwater was about four times the annual rainfall (Greenwood *et al.* 1985).
Acidity: minimum pH_{Ca}	Good, will grow into acid kaolinitic clayey subsoils.
Alkalinity: maximum pH_w	Prefer neutral to acid soils, but exact tolerance to alkaline conditions is unknown. Appear to grow well on Spearwood dunes with alkaline subsoils, where soil depth is not constrained by limestone.
Key nutrient requirements	*Nitrogen.* Highly responsive. In plantations on infertile sites, an initial application of DAP (50 kg/ha) is followed by 200 kg/ha every second year. Fertiliser responses do not occur on fertile ex-pasture sites. *Phosphorus.* May be required on sites with <10 ppm bicarbonate P. If applying N and P, ensure trace elements are adequate, otherwise higher growth rates from the additional N and P can induce trace element deficiencies. *Potassium.* Only likely to be required on leached sands (<20 ppm bicarbonate K), however, these soils are generally unproductive due to poor water relations. Symptoms of major nutrient deficiencies and critical nutrient tissue concentrations are summarised in Dell *et al.* (1995).
Compacted soils	Will restrict root growth, which is critical when roots need to exploit sufficient soil to extract subsoil moisture during the first summer. It is essential to plant the seedlings along rip lines.
Root growth into clayey subsoils	Good, providing there is no shallow perched watertable. In a well structured clayey subsoil, roots have been observed at more than 8 m (Greenwood *et al.* 1985). Where the clayey subsoil is massive or weakly structured roots have difficulty penetrating.
Soil properties affecting germination	Not applicable (always propagated by seedlings).
Erosion risk	Generally wind stable, although can be prone to windthrow in soils that have restricted the rooting depth. Strategically placed tree-belts can act as windbreaks. Water erosion can occur along tree lines at establishment, if trees are not placed on the contour.

CHAPTER 9

PASTURES: SOIL AND CLIMATIC REQUIREMENTS

CHAPTER 9. PASTURES: SOIL AND CLIMATIC REQUIREMENTS

Officers of the Pastures and Revegetation Branch

This chapter describes the soil and climatic requirements for the main annual and perennial pasture species grown commercially in south-western Australia. For each species the climatic requirements are described in terms of rainfall, temperature and frost tolerance. The soils section summarises the main soils on which they grow and their tolerance to a range of conditions affecting productivity.

The amount of information available on the requirements of individual species varies considerably from species such as subterranean clover which have been studied extensively, to minor species such as rose clover for which there is minimal information.

The pasture species have been subdivided into four groups: annual pasture legumes, herbaceous perennial legumes, perennial grasses and pasture shrubs.

9.1 Annual pasture legumes

Introduction to annual pastures

Annual clovers (*Trifolium* spp.)

- Balansa clover
- Persian clover
- Subterranean clover
- Other clovers: arrowleaf clover, cupped clover, rose clover

Serradellas (*Ornithopus* spp.)

- Yellow serradella

Annual medics (*Medicago* spp.)

- Barrel medic
- Burr medic
- Murex medic
- Other medics: disc medic, strand medic

9.2 Herbaceous perennial legumes

- Lucerne
- White clover

9.3 Perennial grasses

Temperate grasses

- Phalaris, cocksfoot, tall fescue, perennial ryegrass

Subtropical grasses

- Kikuyu, Rhodes grass, couch, paspalum

Salt-tolerant grasses

- Puccinellia, tall wheat grass, saltwater couch

9.4 Pasture shrubs

- Tagasaste
- Halophytes or plants for saltland

9.1 ANNUAL PASTURE LEGUMES

INTRODUCTION TO ANNUAL PASTURES

*Mike Ewing**

Annual legumes are a pivotal component of farming systems in the Mediterranean climatic zone of Western Australia. The diverse environmental conditions for which they are required has led to the development of many commercial cultivars, mostly selected or bred from annual species of *Trifolium* and *Medicago*.

Climatic and soil factors directly influence the selection of annual legumes most suited to a particular location. They also influence the choice of farming system into which the pastures fit. Successful legumes are those which are adapted to the physical environment and have characteristics that enhance their performance in the chosen farming system. In practice, farming systems vary widely; near-permanent pastures are common in high rainfall zones, while systems involving intensive rotation with crops are more common in lower rainfall zones.

Despite the undoubted importance of the current annual pasture legumes and the record of success in selecting cultivars, new species and cultivars are still required. The continuing search is based on the need to identify annual legumes capable of:

- *Growing and persisting under conditions of severe soil stress (acidity, waterlogging, salinity, nutrient deficiency and toxicity).*

  Successful annual legumes have been identified for large areas where soil conditions are least stressful. However, these plants are often poorly adapted to conditions associated with extreme stress. Despite this, their natural diversity provides opportunity to identify particular species or cultivars that tolerate most conditions.

- *Contributing to improved sustainability of systems.*

  Hydrological instability is a serious problem and increased water use associated with greater capacity to extract water from deep in the profile is desirable. Variability in rooting depth can be exploited as part of the solution to this problem.

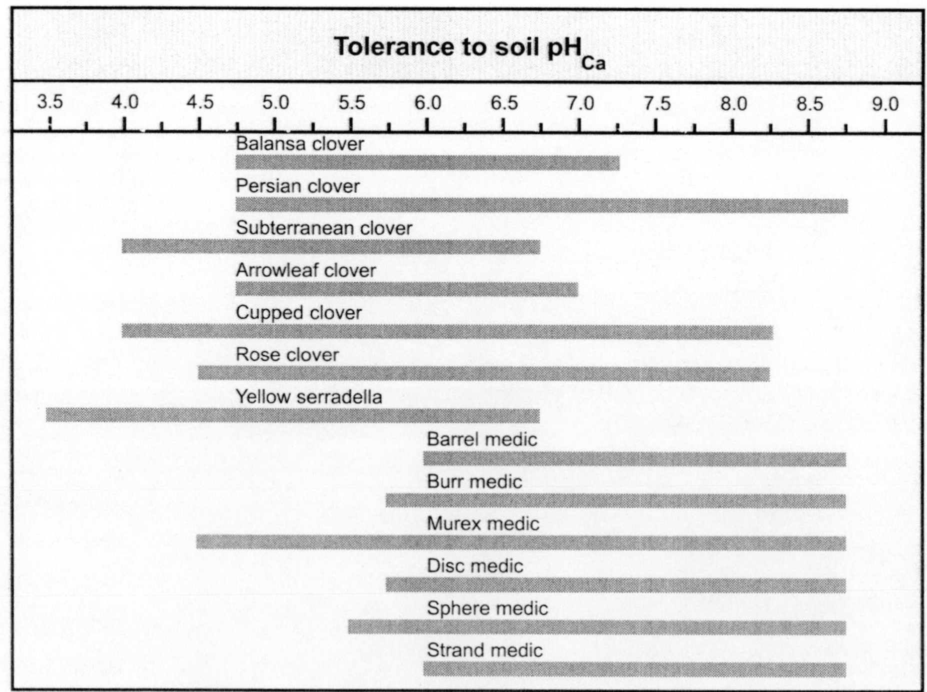

Figure 9.1.1 *Tolerance of annual pasture legumes to soil pH.*

* *Agriculture Western Australia, South Perth.*

- *Productivity within a widening array of farming systems.*

  Farming systems change because of changing technology and economic circumstances. Changes in other elements of the system may give rise to the need for pasture legumes with different characteristics. Increased emphasis on cropping has increased the need to select legumes able to regenerate in an intensive cropping system, or that can be introduced as a phase between crop sequences.

- *Incorporating resistance to biotic factors (insects and disease).*

The need to identify pasture legumes capable of overcoming specific constraints, combined with the desirability of increased genetic diversity, will lead to greater numbers of commercial species and cultivars. Making the best use of these resources requires knowledge of individual characteristics.

Relative tolerance to soil pH, waterlogging and salinity

Figures 9.1.1, 9.1.2 and 9.1.3 summarise the relative tolerance of annual pasture legumes to the three key soil properties of pH, waterlogging and salinity. For more information refer to descriptions of the individual species.

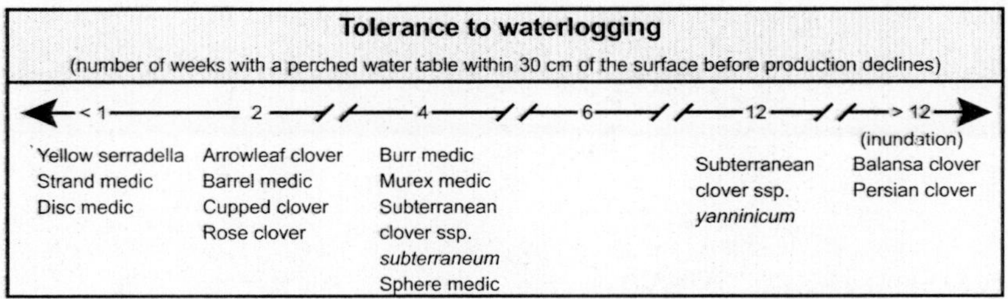

Figure 9.1.2 *Tolerance of annual pasture legumes to waterlogging.*

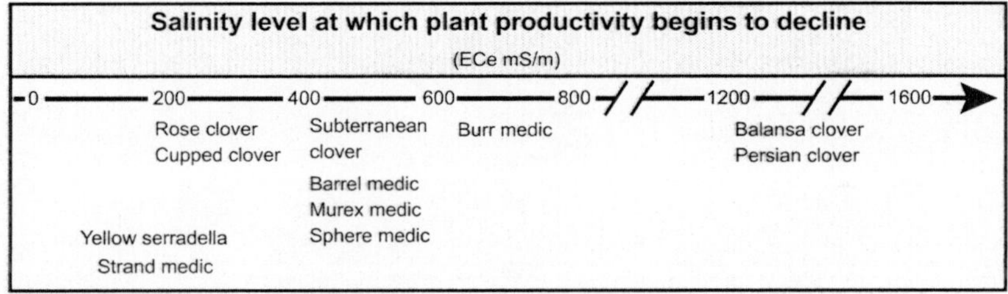

Figure 9.1.3 *Tolerance of annual pasture legumes to salinity.*

ANNUAL CLOVERS (*TRIFOLIUM* SPP.)

Balansa clover

Pedro Evans[*]

Trifolium balansae (syn *T. michelianum)* or balansa clover is a small-seeded annual. It can grow prolifically, but is susceptible to red-legged earth mite especially at the seedling stage. The general agronomy is described in Snowball (1994), Evans and Snowball (1993) and Cransberg (1990).

Root growth

The estimated maximum depth of root growth is about 0.5 m.

Climatic requirements

Balansa clover can be grown in the medium (M3-5) and high (H3-5) rainfall zones (Figure 8.0.1). The minimum annual rainfall is generally considered to be 350 mm, although it does depend on the position in the landscape.

The optimum temperature range for growth is 20 to 25°C, so balansa clover grows vigorously when there is an early break to the season. No growth is expected when the temperature falls below 4°C or exceeds 30°C. Measurements from Katanning in WA have shown that good production (60 kg/ha/day) can be obtained from balansa pastures at very high densities when the daily air temperature fluctuates between 5° and 13°C (P. Evans unpublished data).

[*] *Formerly Agriculture Western Australia, now Agriculture Victoria, Hamilton.*

SUMMARY OF SOIL REQUIREMENTS

Balansa clover is particularly suited to sites which are both waterlogged and saline It is highly tolerant of waterlogging and will tolerate moderate salinity (e.g. barley grass flats). Balansa suits acidic to neutral sites where the pH_{Ca} is above 4.7.

Balansa clover.

Soil factors affecting the productivity of balansa clover

Soil conditions	Tolerance
Soil water deficit	Sensitive, but has capacity for rapid seed development under conditions of moisture stress (Cransberg 1990).
Waterlogging	Very high tolerance (no effect after three months of inundation). Plants adapt to waterlogged conditions by producing aerenchyma.
Soil salinity	Moderately tolerant with a threshold ECe of ~1000 mS/m (Rogers and Noble 1991). In trials at Katanning the biomass and seed yield were not reduced by an electrical conductivity (EC 1:5) of 80-180 mS/m. The seedling stage is more sensitive and an EC of 100 mS/m will kill the plants.
Salinity and waterlogging	Tolerant providing the site is only moderately saline.
Acidity: minimum pH_{Ca}	pH_{Ca} >4.7. Growth reduced to 70% of maximum at pH_{Ca} 4.2 and to 90% of maximum at 4.7 (Evans *et al.* 1990).
Alkalinity: maximum pH_w	Can tolerate slightly alkaline conditions, pH_w 7.5-8.1 (Gillespie 1989; Evans and Snowball 1993).
Key nutrient requirements	High requirement for P and K (250 kg/ha of superphosphate, 150 kg/ha of potash; Bolland 1993).
Compacted soils	Probably slow root growth.
Effect of surface crust	Reduced seedling emergence after sowing, but not in regenerating stands because it is an aerial seeder.
Soil conditions required for germination	Moisture, oxygen and a soil temperature of 15-20°C (Jansen and Ison 1994). High temperatures (>30°C) prevent germination. Deep cultivation is likely to reduce emergence in fine-textured soils.

Persian clover

*Pedro Evans**

Persian clover (*Trifolium resupinatum*) is a tall, erect annual suitable for hay production especially where irrigation is available. It is a minor pasture species in Western Australia, but is being used increasingly to replace subterranean clover on irrigated dairy farms in eastern Australia (Stockdale 1994). A potential problem is its high susceptibility to attack by red-legged earth mites. The general agronomy is described in Snowball (1994), Lee and Reed (1994) and Evans and Snowball (1993).

Root growth

Persian clover is shallow-rooted and the estimated maximum depth is about 0.5 m.

Climatic requirements

The available cultivars (e.g. Kyambro) are suitable for areas receiving more than 500 mm annual rainfall (H3-5 and VH4-5). An early maturing variety suitable for areas receiving more than 350 mm annual rainfall (M3-5 zones) will soon be released commercially.

The temperature requirements are similar to those for balansa clover. The optimum temperature range for growth is 20 to 25°C, with no growth when temperatures fall below 5°C or exceed 30°C.

SUMMARY OF SOIL REQUIREMENTS

Persian clover is well adapted to acid to highly alkaline (pH_{Ca} >4.7), waterlogged sites which can be marginally saline. It responds well to irrigation.

Soil factors affecting the productivity of Persian clover

Soil conditions	Tolerance
Soil water deficit	Unknown.
Waterlogging	Very high tolerance; can withstand waterlogging for three months.
Soil salinity	Moderately tolerant. Threshold ECe is 1000 mS/m, below which growth not affected, but much more sensitive during seedling stage.
Salinity and waterlogging	Very tolerant.
Acidity: minimum pH_{Ca}	In general, pH_{Ca} >4.7 can be used as a guide. Growth is reduced when the pH_{Ca} is <5.0 and root nodule formation and weight is reduced at pH_{Ca} <6.0 (Evans *et al.* 1990). Production has been reported as 4.7 t/ha at pH_{Ca} 4.9, and 7.2 t/ha at pH_{Ca} 5.4 (Hertzch *et al.* 1974). Good production has been reported from a grey clayey soil with pH_{Ca} 5.1 (Cunningham and Feely 1987).
Alkalinity: maximum pH_w	Very tolerant; will grow at pH_w <10.5 (Kumar *et al.* 1981).
Key nutrient requirements	Phosphorus and potassium.
Compacted soils	Probably slow root growth.
Effect of surface crust	Reduced seedling emergence after sowing, but not in regenerating stands because it is an aerial seeder.
Soil conditions required for germination	Soil temperature 15-20°C (Jansen and Ison 1994).

* *Formerly Agriculture Western Australia, now Agriculture Victoria, Hamilton.*

Subterranean clover

Phil Nichols[*]

Subterranean clover (*Trifolium subterraneum*) is the most widely sown pasture species in south-western Australia. The subspecies most widely grown in WA are *T. subterraneum* ssp. *subterraneum* and *T. subterraneum* ssp. *yanninicum*. The general agronomy is described in Collins *et al.* (1984), Francis *et al.* (1976) and Morley (1961).

Root growth

Subspecies *subterraneum* has a lateral branching and sub-branching taproot, but is relatively shallow-rooted. In deep sand the roots can extend 0.8 to 1.2 m (Humphries and Bailey 1962; Ozanne *et al.* 1965), although in most soils the effective rooting depth would be much shallower.

Subspecies *yanninicum* has a shallower root system with a maximum rooting depth of 0.4 to 1.2 m, depending on soil texture (Ozanne *et al.* 1965; Hamblin and Hamblin 1985). Root concentration is high in the top 10 cm layer because of large numbers of superficial roots (Humphries and Bailey 1962). Their proliferation probably helps growth under waterlogged conditions.

Climatic requirements

Suitable cultivars are available for all of the agroclimatic zones (Figure 8.0.1), although the growing season rainfall should be at least 200 mm (or annual rainfall 275 mm) for stands to persist. In the agricultural area there are very few good stands where the growing season rainfall is less than 250 mm (i.e. annual rainfall less than 325 mm) and it is almost never found in its native Mediterranean environment under these conditions. Subspecies *yanninicum* is only suitable for the VH3-5 zones, and the wetter parts of the H3-5 zones where the growing season rainfall is more than 450 mm.

Soil factors affecting the productivity of subterranean clover (*T. subterraneum* ssp. *subterraneum*). Assume the same for ssp. *yanninicum* unless stated.

Soil conditions	Tolerance
Soil water deficit	Low to moderate tolerance; only marginally suitable for uniform coarse-textured soils with AWC <80 mm/m. Moisture stress during flowering reduces seed yield (Andrews *et al.* 1977). Ssp. *yanninicum* does not tolerate low AWC because of its very shallow root system.
Waterlogging	Ssp. *subterraneum* has moderate tolerance. Francis and Devitt (1969) reported that dry matter production was reduced after three weeks of inundation. Ssp. *yanninicum* has high tolerance. Inundation for three weeks did not reduce dry matter production (Marshall and Millington 1967; Francis and Devitt 1969).
Soil salinity	Moderately sensitive. Maas and Hoffman (1977) reported that under controlled conditions, growth was reduced when ECe exceeded 150 mS/m, with a 12% decrease in yield per 100 mS/m increase above this level. Under controlled conditions West and Taylor (1981) found differences between cultivars and ssp. *brachycalycinum* was the most tolerant.
Salinity and waterlogging	Low tolerance.
Acidity: minimum pH_{Ca}	In general, will tolerate acidity to pH_{Ca} 4.2. Evans *et al.* (1990) reported that growth was 90% of maximum at pH_{Ca} 4.2. Yeates (1988) found poor nodulation at pH_{Ca} 4.3, and only low numbers of rhizobia at pH <4.3.
Alkalinity: maximum pH_w	Low tolerance, does not like pH_w >7.5.
Key nutrient requirements	P, K, Mg, Mo, Zn, S, Fe.
Compacted soils	Root elongation is slowed (Nutt 1989).
Effect of surface crust	Seedling emergence is reduced, but the greatest effect is in spring when seed burial is hampered, thereby reducing seed production.
Soil conditions required for germination	About 10 mm rain over two days.

[*] *Agriculture Western Australia, South Perth.*